SPRINGER TRACTS IN MODERN PHYSICS

Ergebnisse
der exakten Natur-
wissenschaften

Volume **45**

Editor: G. Höhler

Editorial Board: P. Falk-Variant S. Flügge J. Hamilton
F. Hund H. Lehmann E. A. Niekisch W. Paul

Springer-Verlag Berlin Heidelberg GmbH 1968

Manuscripts for publication should be addressed to:

G. HÖHLER, Institut für Theoretische Kernphysik der Universität, 75 Karlsruhe, Kaiserstraße 12

Proofs and all correspondence concerning papers in the process of publication should be addressed to:

E. A. NIEKISCH, Kernforschungsanlage Jülich, Institut für Technische Physik, 517 Jülich, Postfach 365

ISBN 978-3-662-16106-7   ISBN 978-3-540-35937-1 (eBook)
DOI 10.1007/978-3-540-35937-1

Library of Congress Catalog Card Number 25-9130

Regge Poles in Particle Physics

P. D. B. Collins and E. J. Squires

Contents

Introduction

Although the generalization of angular momentum, l, to a complex variable, and the subsequent transformation of the partial-wave series to an integral in the l-plane, was introduced by *Watson* in 1918 and revived by *Sommerfeld* in 1949, it was not until the work of *Regge* in 1959 that the possible usefulness of the idea in particle physics was realized. Regge showed that, for a wide class of potentials, the only singularities of the non-relativistic scattering amplitude in the l-plane were poles which move with energy, i.e. poles at positions $l = \alpha(s)$. These are now called Regge poles and $\alpha(s)$ is called a Regge trajectory. The method was used by Regge to close an important (if not physically very interesting) gap in previous proofs of the Mandelstam representation for potential scattering. However the fact that the Regge trajectories correspond to physical particles (or resonances) when $\alpha(s)$ equals an integer for positive s, and also determine the high energy behaviour of the cross-channel amplitudes, attracted the attention of elementary particle theorists, and since 1961 the subject has had a vigorous, controversial, and exciting history.

The early discussions were very promising. The belief that total cross-sections tend to constants at high energy led to the suggestion of a new Regge trajectory (the Pomeranchon) not associated with any known particles or resonances, with the quantum numbers of the vacuum and having $\alpha(0) = 1$. Very soon a particle of spin 2, the f at 1250 Mev and well placed to lie on this trajectory, was discovered.

An interesting feature of the hypothesis of Regge pole dominance of cross sections at high energy was that the differential cross sections would have simple power behaviours, the power depending on the momentum transfer. It predicted a "shrinking" of the diffraction peak with energy, which in potential language would correspond to a radius of interaction which was an increasing function of energy. This rather odd behaviour was found to hold for p-p scattering, and attempts to correlate the data on elastic total cross sections with the known particles appeared to be successful. The trajectories found were approximately straight lines on the so called Chew-Frautschi plot of $\text{Re}\{\alpha\}$ versus real s.

Meanwhile there was a great deal of work done on the nature of Regge trajectories in potential scattering; and attempts were made, with partial success, to show that the theory could be generalized to relativistic amplitudes satisfying the Mandelstam representation. It was also successfully extended to embrace particles with spin, where the total angular momentum, J, becomes the appropriate variable for continuation.

The Regge pole concept also permitted clarification of the distinction between a "bound state", which lies on a Regge trajectory, and an "elementary" particle, which does not lie on a trajectory but corresponds to a Kronecker δ-function at the value of J equal to the spin of the particle. This gave a great stimulus to the notion of "nuclear democracy", that all strongly interacting particles are bound states, and to the bootstrap hypothesis.

This first blossoming of theory and phenomenology was short lived. Further experiments in late 1962 and early 1963 showed that shrinking was not apparent in differential cross sections other than pp. It was still possible to fit the available data by including several Regge poles, but this meant that there were too many parameters for the fit to be convincing. Also, on the theoretical side, Mandelstam demonstrated that there were cuts in the J-plane in relativistic scattering, arising from the presence of the third double spectral function. These do not exist in potential scattering where only two-particle unitarity is involved. In trying to come to terms with these cuts a lot of work was done on the three-body problem, but many difficulties were encountered. It was also found that, when particles with spin are considered, it is possible for elementary particles to be on Regge trajectories, because δ-functions in the J-plane appear, which can cancel those mentioned above. Enthusiasm for the whole subject languished except among a handful of devoted disciples for about two years.

More recently, since 1965, the increasing amount of experimental information, particularly on inelastic processes at high energies, has brought about a revival of interest in Regge pole phenomenology, and many successful fits to the data have been achieved. There has also been more work on the theoretical aspects of the complex J-plane, and problems involving particles with spin, and unequal mass scattering, have been studied in detail, leading to the introduction of "conspirators", and "daughters". Cuts and fixed poles in the J-plane, as well as the Reggeization of elementary particles, have also been clarified in recent work.

In this book we attempt both to review the theory as it stands at present, and to show how well it agrees with experimental information.

In the first chapter we give a brief introduction to analytic S-matrix theory, and maximal analyticity of the first kind. Its application to the two particle scattering amplitude, leading to the Mandelstam representation, is covered in some detail. We hope that this chapter will be comprehensible even to the reader who has not met these things before (though we also hope that he will be encouraged to follow up the references).

The second chapter goes on to discuss partial-wave amplitudes, partial-wave dispersion relations, and the concept of complex angular momentum, and Regge poles and cuts. The significance of the postulate of maximal analyticity of the second kind is brought out.

In chapter III we combine the Mandelstam representation with the Regge pole analysis, and are able to deduce some of the properties of the trajectory and residue functions. We also discuss various Regge representations which satisfy the Mandelstam analyticity requirements, and look in detail into the difficulties of unequal mass scattering. We show why an attempt to reconcile Mandelstam analyticity with Regge asymptotic behaviour has led to the suggestion that, associated with each trajectory, there must be an infinite sequence of "daughter" trajectories, which are separated from each other by one unit of angular momentum at $s = 0$.

Though we do not believe that the arguments for daughters are completely convincing such trajectories certainly ought to be looked for, and we discuss such evidence as is available.

In chapter IV we introduce spin for the first time, and develop in detail the formalism for general spin. We use the helicity crossing matrix to locate the various kinematical singularities of amplitudes, which have to be taken into account when making fits to experimental data. We also look at the kinematical constraints on the Regge representation, and the "conspiracies" which may be invoked to get reound them. Brief consideration is also given to the use of group theoretical methods in Regge theory, and the $SO(4)$ symmetry at vanishing 4-momentum is noted.

Some further properties of the complex J-plane are discussed in chapter V. We endeavour to determine how far the results of potential scattering are applicable to S-matrix theory, and find that though we are able to obtain some restrictions on the possible singularities the situation is much more complicated, because of the third double spectral function. For example cuts and fixed poles must be present in some amplitudes, though the latter do not affect the asymptotic behaviour. We also prove the factorization theorem, and derive "superconvergence relations".

In chapter VI we try to relate maximal analyticity of the first and second kinds, and introduce the bootstrap hypothesis. Bootstrap calculations involving Regge trajectories are discussed in some detail.

Chapter VII stands somewhat apart from the rest in that we examine the use of old-fashioned perturbation theory techniques for understanding the complex J-plane. In particular we try to clarify the arguments for Regge cuts, and the distinction between "elementary" particles and those lying on Regge trajectories.

We believe that all strongly interacting particles will be found to lie on trajectories, and are encouraged in this by the evidence presented in chapter VIII, where the whole experimental situation is reviewed. It is found that essentially all the known particles can be convincingly fitted on trajectories, and that these trajectories provide a satisfactory explanation of the data on high energy cross-sections. One surprising feature is the straightness of the trajectories over the whole available energy range, and the degeneracy between trajectories of opposite signature. Some possible explanations for this behaviour are discussed. The attempts to fit superconvergence relations with single particle states are also reviewed, and we close with a brief section on amplitudes for electromagnetic and weak processes, where some curious features appear.

Perhaps one or two of the conventions which we use require explanation. We have numbered the chapters with Roman numerals, and sections and equations with Arabic ones, so that (II. 3.4) refers to equation 4, of section 3, of chapter II. However, many of the references are to equations in the same chapter and we have then abbreviated them by omitting the chapter number, so that in Chapter II only, this equation will be referred to as (3.4). We have tried to give as many cross-references as possible, so that it will not be necessary to read the book consecutively.

References to other works quote the name(s) of the author(s) and the year only, and the details are collected together at the end of the book. In such a large and rapidly expanding subject one can not hope to be comprehensive, but we have been as thorough as seemed reasonable, particularly referring to papers which have been published since the earlier reviews were written. A much more complete list of references up to about the end of (1963), which of course includes the whole of the formative period of the subject, and especially on potential scattering, which is outside our purview, is to be found in *Newton* (1964).

One exception to our scheme is that very frequent recourse is had to various volumes of the "Bateman manuscript project" [*Erdelyi* et al. (1953)], the bible of all Regge theorists. We have abbreviated reference to these, so that "Higher Trancendental Functions" volume 1, chapter 2, paragraph 3, equation 4 (which appears on p. 75 of that book) is referred to as [B 1, 2.3 (4)], while a similar equation in the "Tables of Integral Transform" would be given as [TIT 1, 2.3 (4)].

Two mathematical points may also by worth spelling out in detail. We are very often concerned with the asymptotic behaviour of functions, and use the following three expressions:

$A(s, t) \sim t^{\alpha(s)}$, meaning $\lim_{t \to \infty} \dfrac{A(s, t)}{t^{\alpha(s)}}$ = non-zero constant (independent of t);

$A(s, t) \to \Gamma(s) \, t^{\alpha(s)}$, meaning $\lim_{t \to \infty} \dfrac{A(s, t)}{t^{\alpha(s)}} = \Gamma(s)$;

and occasionally,

$A(s, t) = O(t^{\alpha(s)})$, meaning $\lim_{t \to \infty} \dfrac{A(s, t)}{t^{\alpha(s)}} \leq$ constant (independent of t).

The other point concerns the (very frequent) use of the word "analytic". When mathematicians say that a function is "analytic" in a given region, they mean that it is free of singularities (holomorphic) in that region. Physicists tend to use the word much more loosely, to mean that the function contains only isolated singularities, such as poles and branch points, but not essential singularities, step functions, δ-functions, or any other such pathological "non-smooth" behaviour. Most of the functions with which we shall be concerned are (or are postulated to be) analytic apart from isolated poles or branch points, and we shall use the word "analytic" to mean just this, preserving "holomorphic" for the complete absence of singularities (and of course "meromorphic", for functions containing only poles in a given region).

Some apologies are also probably called for. In trying to keep the book up to date we have made use of work, which, at the time of writing, existed only in preprint form. We have tried to obtain the author's permission in each case, but occasionally may have neglected to do so. For these omissions, and for cases where we have misrepresented, over-looked, or failed to appreciate the true importance of an author's work, we can only apologize. We may not for example have done justice to $SO(4)$, and have largely ignored the three-body problem, and many of the applications are treated rather briefly. Our excuse is simply that the book is already far longer than was originally intended.

We have discussed the Regge model to the virtual exclusion of all else, but we may as well state frankly (either in mitigation or provocation) our belief that it has no serious rivals in high-energy physics. The absorptive peripheral model rightly enjoyed a period of popularity, and it certainly contains elements of truth, but no remedy has been found for its failure to cope with the exchange of particles having high spin. Putting absorptive corrections into the Regge theory can not be justified at present, though it is possible that some reformulation of the problem, probably involving cuts, might be successful. The quark model is of course in no sense a rival, and studies the problem in a completely different way. It seems clear to us that the successful correlation of high-energy scattering with the known particles, which is given by the Regge theory, is not accidental, and that the Regge hypothesis has a much better theoretical foundation than any other suggested scheme. We anticipate that it will have a permanent value, and hope that this book will succed in explaining why we believe this.

It is regretted that we have been inconsistent in what we have required of the reader, in that we have tried to develop S-matrix theory (at least in outline) from the beginning, but have assumed familiarity with SU_3 etc., and even, in one chapter, with the analysis of Feynman diagrams. There seems to be no way round this difficulty short of writing a complete encyclopaedia, but we have tried where necessary to give useful references to introductory works, as well as to the original papers.

Finally our thanks. These are due to the many people with whom we have held discussions, among whom we mention in particular *Stanley Mandelstam* (whose enormous contribution to the subject will be evident form the following pages), *Geoffrey Chew* (who introduced us to this subject), *Elliot Leader, Roger Phillips* and *Alan Martin*; also to our Editor, *G. Höhler*, who has helped in many ways, to our several patient typists, and to the publishers who have produced this volume so speedily.

Durham, England, Oct. 1967

P. D. B. Collins

E. J. Squires

Chapter I. The S-matrix

I.1. Introduction

There is as yet no generally accepted basis for discussing the phenomena of strong interactions, though in the past decade very considerable progress has been made in the development of a dispersion theory for the analytic scattering-matrix (or S-matrix for short). This theory is not yet completely formulated, and indeed it is not generally agreed whether the required analyticity properties should be simply postulated, or whether they can (and should) be deduced from some more general notions of causality.

We shall not find it necessary to enter this controversy directly, though our prejudices will probably be evident, but it was within the framework of S-matrix theory that Regge poles were introduced into high energy physics, and if we are to give more than a purely phenomenological discussion of Regge poles, we shall certainly need to make use of this framework, often conjecturing results for which no formal proof yet exists. This is evidently a much less satisfactory situation than potential scattering, where the foundations are secure, and the properties of Regge poles can be formally demonstrated, but on the other hand it is only in the relativistic theory that Regge poles are truly useful. In potential scattering they are simply a mathematical curiosity.

In this chapter we shall briefly review S-matrix theory in its present state of development, laying special emphasis on those features which will be useful later in the book. We begin with an account of the postulates on which S-matrix theory is based, and of the use of the "bubble" notation to write unitarity equations. The singularity structure which follows from the postulate of "Maximal Analyticity of the First Kind" is briefly reviewed, leading to a more detailed account of the kinematics and singularities of the four-line connected part (the scattering amplitude for a collision of two particles, in which there are only two particles in the final state). Single and double dispersion relations are written down for this amplitude, and the ambiguities stemming from the need for "subtractions" in some of the integrals are demonstrated, paving the way for the introduction of "Maximal Analyticity of the Second Kind" in Chapter II. Finally we show how the elastic part of the double spectral functions may be calculated.

A much more comprehensive account of the axioms, and of diagram analysis, with a fuller list of references, is to be found in the book by *Eden* et al. (1966), while a review of the use of the theory for dynamical calculations is given in several works by *Chew* [(1962a), (1964), (1965a) and (1966)]. Regge poles in potential scattering are discussed comprehensively by *Squires* (1963a), *Newton* (1964), and *De Alfaro* and *Regge* (1965).

I.2. The Postulates of S-matrix Theory

There are various ways of expressing the postulates of S-matrix theory [see for example *Stapp* (1962) and *Olive* (1964)], but the following six seem satisfactory:

(i) The superposition principle of quantum mechanics.

(ii) The existence of a unitary S-matrix.

(iii) The Lorentz invariance of the S-matrix.

(iv) The disconnectedness of the S-matrix, due to the finite range of the strong interaction forces.

(v) Maximal analyticity of the first kind. This is the requirement that the only singularities of the S-matrix should be the poles corresponding to stable or unstable particles, and the further singularities which are generated from those poles by unitarity.

(vi) Maximal analyticity of the second kind. This is the requirement that the S-matrix should be continuable in angular momentum throughout the complex angular-momentum plane, with only isolated singularities.

Of these the first four are relatively uncontroversial, and will almost certainly be required of any successful theory of strong interactions, while (v) and (vi) are very controversial, but form the basis of dynamical S-matrix theory [*Chew* (1964), (1965a) and (1966)]. It should be noted that (vi) follows from (v), at least in part, so it may not be necessary to postulate it separately. It is convenient to treat it independently, however. In this chapter we shall discuss the consequences, or conjectured consequences, of postulates (i) to (v), and leave (vi), which concerns Regge poles, for the next chapter.

Postulate (i) simply requires that if $|\psi_1\rangle$ and $|\psi_2\rangle$ are physical states, then $|\psi_3\rangle = a\,|\psi_1\rangle + b\,|\psi_2\rangle$, a and b being any complex numbers, is also a physical state. In fact this is not always true because of "superselection rules" such as the conservation of charge or baryon number, but these complications can easily be dealt with (*Stapp*, 1963), and we ignore them here.

An experiment in strong interactions consists of observing the initial state $|i\rangle$ of two (or more, in principle, if not in practice) particles, allowing them to interact, and then observing the final state of the arbitrary number of particles resulting, $|f\rangle$. We define the S-matrix element, $\langle f|S|i\rangle$, such that

$$P_{fi} = |\langle f|S|i\rangle|^2 = \langle i\,|S^\dagger|\,f\rangle\,\langle f\,|S|\,i\rangle \qquad (2.1)$$

represents the probability of $|f\rangle$ being the final state, given $|i\rangle$. Because of postulate (iv), and the fact that we are neglecting weak but long-range forces such as electromagnetism and gravitation, we can be sure that our states, $|i\rangle$ and $|f\rangle$, will be states of free particles.

Postulate (ii) requires that there be a complete orthonormal set of free particle states $|m\rangle$, $m = 1, 2, \ldots$. Then, starting from $|i\rangle$, the probability that the particles will end up in some final state is unity, so

$$\sum_m |\langle m\,|S|\,i\rangle|^2 = \sum_m \langle i\,|S^\dagger|\,m\rangle\,\langle m\,|S|\,i\rangle = \langle i\,|S^\dagger S|\,i\rangle = 1\,, \qquad (2.2)$$

when we use the completeness relation

$$\sum_m |m\rangle \langle m| = 1 \,. \tag{2.3}$$

Since this must be true for any choice of the basis states $|m\rangle$, we have the matrix equation

$$S^\dagger S = S S^\dagger = 1 \,, \tag{2.4}$$

i.e. S is unitary.

A free particle state of a single particle is specified completely by giving the particle type, t (this includes the spin), the four-momentum, p, and the helicity, λ, the helicity being defined (*Jacob* and *Wick*, 1959) as the component of the spin, $\boldsymbol{\sigma}$, in the direction of motion, i.e.

$$\lambda = \frac{\boldsymbol{p} \cdot \boldsymbol{\sigma}}{|\boldsymbol{p}|} \tag{2.5}$$

p being the three-vector momentum. λ is an integer (or half-odd-integer for fermions), constrained by $|\lambda| \leq |\boldsymbol{\sigma}|$. Since the particles are free (we are never concerned with interacting states, but only with initial and final states), we have the mass-shell constraint

$$p^2 = m^2 = p_0^2 - \boldsymbol{p}^2 \,, \tag{2.6}$$

in units such that the velocity of light $c = 1$, m being the particle mass, and p_0 its energy. The states may thus be written $|p, t, \lambda\rangle$.

We adopt the relativistic normalisation

$$\langle p', t', \lambda' \mid p, t, \lambda \rangle = (2\pi)^3 \, 2p_0 \, \delta^3(\boldsymbol{p}' - \boldsymbol{p}) \, \delta_{tt'} \, \delta_{\lambda\lambda'} \,, \tag{2.7}$$

and the completeness relation (2.3), for single particle states, becomes

$$\sum_{t'} \sum_{\lambda'} \int |p', t', \lambda'\rangle \langle p', t', \lambda'| \, (2\pi)^{-3} \, \delta^+(p'^2 - m^2) \, \mathrm{d}^4 p' = 1 \,. \tag{2.8}$$

Using P to denote the set of state labels $\{p, t, \lambda\}$, we can denote an N-particle state by $|P_1, P_2, \ldots P_N\rangle$, with orthonormality

$$\langle P_1' \ldots P_N' \mid P_1 \ldots P_N \rangle = \Pi_N \, (2\pi)^3 \, 2p_{0N} \, \delta^3(\boldsymbol{p}_N' - \boldsymbol{p}_N) \, \delta_{NN'} \,, \tag{2.9}$$

and completeness

$$\sum_{N=0}^{\infty} (2\pi)^{-3N} \int \mathrm{d}^4 p_1 \ldots \mathrm{d}^4 p_N \, \delta^+(p_1^2 - m_1^2) \ldots \delta^+(p_N^2 - m_N^2) \, | P_1 \ldots P_N \rangle \times$$
$$\times \, \langle P_1 \ldots P_N | = 1 \,. \tag{2.10}$$

Then the unitarity relation (2.2) becomes

$$\sum_N (2\pi)^{-3N} \int \prod_{i=1}^N \mathrm{d}^4 q_i \, \delta^+(q_i^2 - m_i^2) \, \langle P_1' \ldots P_M' |S| Q_1 \ldots Q_N \rangle \times$$
$$\times \, \langle Q_1 \ldots Q_N |S^\dagger| P_1 \ldots P_M \rangle \tag{2.11}$$
$$= \langle P_1' \ldots P_M' \mid P_1 \ldots P_M \rangle \,,$$

where we have used $Q \equiv \{q, t, \lambda\}$ for the intermediate state particles of four-momentum q.

These unitarity equations are crucial to the development of the theory, but evidently rather cumbersome as they stand. A much simpler way of expressing them has been devised (*Gunson*, 1965): the bubble diagrams.

I.3. Bubble Notation

In this section, for ease of writing, we shall cease to refer to the particle type, t, and helicity, λ. No points of principle are involved, but integrations should be understood to include summations as necessary. Also we shall take all the masses to be equal $(=m)$, and include only stable particles. These restrictions will be relaxed later.

Because of Lorentz invariance [postulate (iv)], an S-matrix element will vanish unless energy and momentum are conserved, i.e.

$$\langle p'_1 \ldots p'_N |S| p_1 \ldots p_N \rangle = 0 \quad \text{unless} \quad \Sigma p = \Sigma p' . \tag{3.1}$$

Thus, in the unitary equation (2.11), only intermediate states such that $(Nm)^2 \leqq (\Sigma p)^2$ will contribute. We incorporate this fact into the following notation (*Eden* et al., 1966; *Olive*, 1964, and *Gunson*, 1965), with bubbles for the S-matrix elements,

$$\langle p'_1 \ldots p'_N |S| p_1 \ldots p_M \rangle \equiv \quad \text{} \tag{3.2}$$

and

$$\langle p'_1 \ldots p'_N |S^\dagger| p_1 \ldots p_M \rangle \equiv \quad \text{} \tag{3.3}$$

Allowed intermediate states are written

$$(2\pi)^{-3N} \int \prod_{i=1}^{N} \mathrm{d}^4 q_i \, \delta^+(q_i^2 - m_i^2) \equiv \quad \text{} \tag{3.4}$$

where the bars on the end of the lines indicate that they are to be joined on to bubbles. Finally, the overlap between initial and final states is represented by lines without bars,

$$\langle p'_1 \ldots p'_M | p_1 \ldots p_M \rangle \equiv \quad \text{} \tag{3.5}$$

Thus, for $(2m)^3 < (\Sigma p)^2 < (3m)^2$, energy-momentum conservation reduces (2.11) to only two-particle states, viz.

$$(2\pi)^{-6} \int \prod_{i=1}^{N} \mathrm{d}^4 q_i \, \delta^+(q_i^2 - m_i^2) \, \langle p'_1, p'_2 |S| q_1 q_2 \rangle \, \langle q_1 q_2 |S^\dagger| p_1 p_2 \rangle \\ = \langle p'_1 p'_2 | p_1 p_2 \rangle , \tag{3.6}$$

which in bubble notation is

$$\tag{3.6a}$$

The stability of the particles forbids single particle intermediate states. At a higher energy, $(3m)^2 < (\Sigma p)^2 < (4m)^2$, two and three particle states are possible, and we have

$$(3.7)$$

where the zero on the right hand side of the second and third equations comes from the orthogonality of $\langle p_1' \ldots p_n' | p_1 \ldots p_m \rangle$ if $n \neq m$. Similar sets of diagrams can be drawn for higher energies (see *Eden* et al., 1966, p. 189).

A further simplification of the unitarity equations arises from the disconnectedness postulate, (iv). Because of the short range of the forces, an S-matrix element can often be broken up into disconnected parts. Thus the S-matrix element with four external lines can be decomposed into

$$= \langle p_1' p_2' | p_1 p_2 \rangle + \langle p_1' p_2' |S| p_1 p_2 \rangle_c \, . \qquad (3.8)$$

The reason for the $+$ sign in the bubble will become clear below.

The crucial difference between these two terms lies in the different δ-functions of energy-momentum conservation. In the first term the particles go straight through without interacting, so p is conserved for each particle separately, while in the second term, the so-called "connected part" of the S-matrix (hence the suffix C), the particles do interact, and only $p_1 + p_2$ is conserved. We can thus re-write the equation (3.6), introducing the A-matrix for the connected part, as follows.

$$\begin{aligned} &= (2\pi)^6 \, 4 p_{01} p_{02} \, \delta^3 (\boldsymbol{p}_1' - \boldsymbol{p}_1) \, \delta^3 (\boldsymbol{p}_2' - \boldsymbol{p}_2) + \\ &+ i (2\pi)^4 \, \delta^4 (p_1 + p_2 - p_1' - p_2') \langle p_1' p_2' |A^+| p_1 p_2 \rangle \, . \end{aligned} \qquad (3.9)$$

The factors i and $(2\pi)^4$ are conventional. However, we have défined our A with the usual convention, which has a $-$ sign relative to *Olive*, 1964, and *Eden* et al., 1966. This is compensated by the $-$ sign in (3.16).

With more than four external lines more complex decompositions can be made, such as

$$(3.10)$$

Many examples are given by *Eden* et al., 1966, p. 190.

For ⟩(s+)⟨ we similarly write

$$\rangle\!(s\!+\!)\!\langle \;=\; =\!\!=\; +\; (-1)\; \rangle\!(-)\!\langle \;,\qquad (3.11)$$

where

$$\rangle\!(-)\!\langle \;=\; i\,(2\pi)^4\,\delta^4(p_1+p_2-p_1'-p_2')\,\langle p_1'p_2'\,|A^-|\,p_1p_2\rangle,\qquad (3.12)$$

and so on for more complex diagrams, a conventional $(-1)^I$ being multiplied onto each term with $I \ominus$ bubbles.

With this decomposition, a unitarity equation such as (3.6a) becomes

$$\Big(=\!\!=\; +\; \rangle\!(+)\!\langle\Big)\Big(=\!\!=\; -\; \rangle\!(-)\!\langle\Big) \;=\; =\!\!=\;,\qquad (3.13)$$

which, on multiplying out and taking terms of the same connectedness (same sets of δ-functions), becomes

$$\rangle\!(+)\!\langle \;-\; \rangle\!(-)\!\langle \;=\; \rangle\!(+)\!=\!\!=\!(-)\!\langle \;.\qquad (3.14)$$

Similarly, at energies corresponding to (3.7),

$$\overline{(s)}\;\;\overline{(s\!+\!)} \;+\; \overline{(s)\!-\!(s\!+\!)} \;=\; =\!\!=$$

gives

$$\overline{(+)} \;-\; \overline{(-)} \;=\; \overline{(+)}\;\;\overline{(-)} \;+\; \overline{(+)\!-\!(-)} \;.\qquad (3.15)$$

In writing out the equations corresponding to these diagrams we find that for each closed loop such as Fig. (I.1) there is only one free momentum,

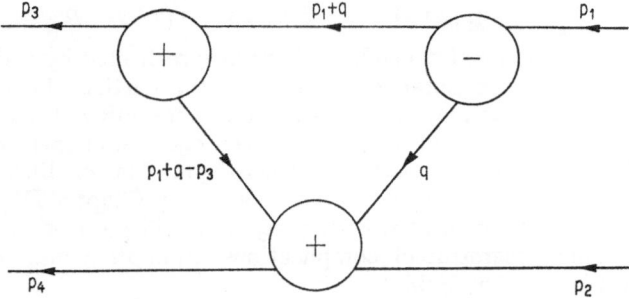

Fig. I.1. A loop diagram. Because of energy-momentum conservation at each bubble, the only free four-momentum is q. For the whole diagram we have the constraint $p_4 = p_1 + p_2 - p_3$.

q, and that δ-functions corresponding to over all p conservation, and p conservation at each vertex, with various factors of i, (2π), etc., can be cancelled out of both sides. Thus we achieve, for each diagram, the simpler set of rules (*Olive, 1964*): —

Each bubble: —

$$\overline{\cdots\underbrace{\pm}\cdots} \equiv (-1) \langle p_1' \ldots p_M' | A^\pm | p_{M+1} \ldots p_N \rangle \tag{3.16}$$

$$\equiv (-1) A^\pm (p_1' \ldots p_N') .$$

Each internal line of momentum q: —

$$\vdash\!\!\!\!\!\underset{q}{\rule{2cm}{0.4pt}}\!\!\!\!\!\dashv \equiv -2\pi i \, \delta^+(q^2 - m^2) . \tag{3.17}$$

Each loop, with q the free momentum:

$$= \frac{i}{(2\pi)^4} \int d^4 q . \tag{3.18}$$

Applying these rules to the particularly important case of the two particle unitarity equation for the four line S-matrix, we get from (3.14)

or

$$A^+ (p_1, p_2, p_1', p_2') - A^- (p_1, p_2, p_1', p_2')$$

$$= \frac{-i}{(2\pi)^4} \int d^4 q \, (-2\pi i)^2 \, \delta^+ [(p_1 + q)^2 - m^2] \, \delta^+ [(p_1 - q)^2 - m^2] \times \tag{3.19}$$

$$\times A^+ (p_1 + q, p_2 - q, p_1', p_2') \, A^- (p_1, p_2, p_1 + q, p_2 - q) .$$

The point of working with the A-matrices, rather than the S-matrix itself, is that we have now removed all the necessary kinematical δ-functions. Each such A-matrix will be a function of the set of variables describing its incoming and outgoing particles;

$$\langle P_1' \ldots P_M' | A^+ | P_{M+1} \ldots P_N \rangle = A^+ (P_1' \ldots P_N) . \tag{3.20}$$

However, the functional dependence is greatly restricted by the requirement that the A's be Lorentz invariant. If the particles have no spin this requirement is simple; the A's can be functions only of the invariants which can be formed out of the p's. For the remainder of this chapter we shall restrict ourselves to particles of zero spin. The modifications required in the presence of spin will be taken up in Chapter IV, where it will be found that much of what we say here will go over unchanged, but with some (regrettable) complications. Spin in S-matrix theory is discussed by *Taylor*, 1966.

If we define a set of invariants of the form

$$s_{ijk \ldots} = (\pm p_i \pm p_j \pm p_k \pm \cdots)^2 ,$$

then we have $A = A (s_1, s_2, s_{12}, \ldots, s_{1 \ldots N})$, but of course these are not all independent variables. For each N-line A matrix there are $4N$ variables, the 4 components of each of the N 4-vectors. But with N mass shell constraints, 4 constraints for over all energy momentum conservation, and 6 constraints for rotational invariance in 4-space, we get only $(3N - 10)$ free variables. A simple way to check this is to note that for a single particle we have $\overset{p_1}{\underset{}{\multimap\!\!\circ\!\!\multimap}}\overset{p_2}{}$, i.e. $(3 \times 2 - 10) = -4$ degrees of freedom, i.e. four constraints; which are that $p_{1\mu} = p_{2\mu}$ for the components $\mu = 1 \ldots 4$.

A collection of particles entering (or leaving) an A-matrix element is referred to as a "channel", and of course in general the various internal quantum numbers (charge, isospin, baryon number, etc.), associated with the particle types, restrict the number of exit channels available from a given entrance channel. Channels with the same internal quantum numbers are said to be "communicating channels" (*Chew*, 1965a).

I.4. The Singularity Structure of Scattering Amplitudes

On the basis of postulates (i) to (iv) it is possible to show that, because of the unitarity equations, the A-matrices must have singularities in the various invariants on which they depend. The deduction of these singularities is discussed in *Eden* et al., 1966, and here we shall content ourselves with quoting some of the results, often without certain qualifying statements as to the rigour of the proofs available.

Firstly we have poles corresponding to single particle internal lines. Thus $A^+ = $ ⟩+⟨ has a pole corresponding to ⟩+⟨—⊕—⟩+⟨ $\overset{p_1}{\underset{m \quad p_2}{}}$, of the form $\dfrac{1}{s_{12} - m^2}$. [The —⊕— bubble on the line simply indicates that it is part of the ⟩+⟨ bubble, and not an internal line of a unitarity integral, which we have seen, (3.17), corresponds to $-2\pi i\,\delta^+(q^2 - m^2)$.] The residue of this pole can be shown to be given by

$$\overset{}{\underset{1 \quad q \quad 2}{\,\rangle\!+\!\langle\,\overset{m}{\frac{}{}}\,\rangle\!-\!\langle\,}}\overset{p_1}{\underset{p_2}{}} = 2\pi i\, A_1^+\, A_2^-\, \delta^+(s_{12} - m^2). \qquad (4.1)$$

The important point to note about this residue is that it can be factorised into the product of the two amplitudes, A_1^+, and A_2^-. Thus the contribution of the pole to the amplitude A^+ is

$$A^+ \approx \frac{A_1^+\, A_2^-}{s_{12} - m^2} \quad \text{for} \quad s_{12} \text{ near } m^2. \qquad (4.2)$$

This remains true whether the particle is stable or unstable, though of course (4.1) can only be a physical process if m is unstable. Otherwise it is a bound-state pole.

Another type of singularity is the branch point which occurs at the threshold for each group of particles which can occur as an intermediate state. Thus our amplitude ⟩+⟨ has a threshold branch point corresponding to ⟩+⟨—⊕—⊕—⟩+⟨ , whose cut discontinuity can be shown to be given by 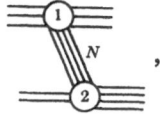 . This will be calculated explicitly below [section (I.9)]. In fact the result generalizes, and the discontinuity across the cut associated with an N particle intermediate state, such as

is given by the Cutkosky rules (*Cutkosky*, 1960, 1961) in the form

$$\text{Disc}_N\,[A] = \int \prod_{l=1}^{N-1} \left[\frac{i\,\mathrm{d}^4 k_l}{(2\pi)^4}\right] \prod_{i=1}^{N} \left[-2\pi i\,\delta^+(q_i^2 - m_i^2)\right] A_1^+ A_2^-\,, \quad (4.3)$$

where the integration is over the $N - 1$ independent loops, l, which are formed by the N intermediate lines. Using the identity

$$\frac{1}{(q_i^2 - m_i^2 \mp i\,\varepsilon)} = \mathrm{P}\,\frac{1}{(q_i^2 - m_i^2)} \pm 2\pi i\,\delta^+(q_i^2 - m_i^2)$$

[P implies principal value],

we can re-write this as

$$\text{Disc}_N\,[A] = \text{Disc}\left\{ \int \prod_{l=1}^{N-1} \left[\frac{i\,\mathrm{d}^4 k_l}{(2\pi)^4}\right] \prod_{i=1}^{N} \frac{1}{(q_i^2 - m_i^2)} A_1^+ A_2^- \right\}.$$

This is similar to the Feynman integral for the corresponding diagram, except that the vertices contain the amplitudes, A^+, A^-, rather than coupling constants. The positions of the singularities of such integrals are given by the Landau rules (*Landau*, 1959): —

$$q_i^2 = m_i^2 \quad \text{for all}\quad i\,,$$

and

$$\sum_{\text{loop}\,j} \alpha_j q_j = 0 \qquad\qquad\qquad (4.4)$$

for some constants α_j, with j summed round each loop, but $\alpha_j \neq 0$. Note that this is only the leading singularity of the corresponding Feynman diagram; others already appear in simpler diagrams.

Thus for a given amplitude we can find a *minimum* set of singularities, by drawing all the (infinite number of) possible diagrams with these external lines, such as

etc., and use the Landau and Cutkosky rules to find the positions and discontinuities associated with these diagrams. Then postulate (v), the postulate of maximal analyticity of the first kind, states that this minimum set of singularities constitutes the complete set of singularities.

There are no other singularities not given by the unitarity equations. If this is true it means that given the complete set of poles, corresponding to all the particles, both stable and unstable, we can, in principle, find all the other singularities by application of the Landau and Cutkosky rules, since (4.3) and (4.4) depend only on the particle masses. However, the number of particles in the theory, and their masses, is completely undetermined. Postulating a new particle simply gives rise to a new (infinite) set of singularities.

If the masses of the particles are real (i.e. for stable particles) the singularities given by the Landau-Cutkosky rules occur for real values of the invariants. For example we find that there is a branch point at $s_{12} \equiv (p_1 + p_2)^2$

$= (m_1 + m_2)^2$ corresponding to the diagram  . Now the

physical amplitude for the process $1 + 2 \to 3 + 4$ is of course to be evaluated with s_{12} real, but we have a choice of reaching the real axis either from above or below. We adopt the $+ i\varepsilon$ prescription, to the effect that in the physical s_α region (for any channel α) the physical amplitude is given by

$$\text{Physical} \quad A^+(s_\alpha, \ldots) = \lim_{\varepsilon \to 0} A^+(s_\alpha + i\varepsilon, \ldots), \quad s_\alpha \text{ real}, \qquad (4.5)$$

i.e. we always approach the real axis from above, as in Fig. (I.2). Clearly if we always adopted the opposite prescription $(- i\varepsilon)$ we should simply be working with complex conjugate amplitudes, and this would make no difference, but we can not be sure that a mixed prescription, approaching some cuts from above, and others from below, should not be used. We adopt the hypothesis that there is a "physical sheet" (*Eden* et. al.,

$\underline{\text{IS}}$

$(m_1+m_2)^2$

Fig. I.2. The direction in which the real s axis is to be approached to obtain the physical amplitude. A cut from the branch point at $s = (m_1 + m_2)^2$ is drawn along the positive real axis.

1966, *Chew*, 1965a) of a given invariant, such that on this sheet continuation to the real axis from above leads to the physical amplitudes. Hopefully it will eventually be possible to prove that this is the case.

Since A^- is the complex conjugate of A^+, we have

$$\text{Physical} \quad A^-(s_\alpha, \ldots) = \lim_{\varepsilon \to 0} A^-(s_\alpha - i\varepsilon, \ldots). \qquad (4.6)$$

Unstable particles have a negative imaginary part to their mass and hence do not lie on the physical sheet. We expect this also to be true of the other singularities generated from these unstable particle poles by unitarity [see Section (I.8)].

I.5. Crossing and the TCP-Theorem

One of the most important results which has been demonstrated with these principles is "crossing". Unfortunately the demonstration does not amount to a proof, because of the impossibility, at the present time, of determining whether the analytic continuation needed, from one region of the variables to another, can be made without leaving the physical sheet. We shall see in detail how this works in Section (I.8). If the con-

tinuation can be made, then, for example,the amplitude

can be analytically continued to the amplitude , where $\bar{5}$ is the anti-particle of 5. It is evident that $\bar{5}$ must have the opposite quantum numbers to 5. Of course these processes occur for different regions of the variables, since in ① we need $\sqrt{s_{12}} \geq \sqrt{s_{34}} + m_5$, while ② needs $\sqrt{s_{34}} \geq \sqrt{s_{12}} + m_{\bar{5}}$ (remember $m_{\bar{5}} = m_5$), but the fact that amplitudes ① and ② correspond to a single function, with an analytic interpolation between the two regions of the variables, is a very strong restriction on their behaviour.

If we move all the legs round,

we are back to the same region of the variables, and so we find that the amplitude for $\bar{3} + \bar{4} + \bar{5} \to \bar{1} + \bar{2}$ is the same as that for $1 + 2 \to 3 + 4 + 5$. This is an example of the TCP theorem, which states that the S-matrix is invariant under the combined operations of time reversal, charge conjugation, and parity inversion.

I.6. The Four-line Connected Part

In this book we shall often be concerned with the simplest possible scattering process, in which two particles enter the scattering region, and two (not necessarily the same) emerge, i.e. the four-line connected part. In the remaining sections of this chapter we shall apply some of the general ideas we have introduced so far to this amplitude (Fig. I.3).

From our discussion in Section I.3 we know that there are just two independent variables among the invariants which can be constructed from the four-vectors p_1, p_2, p_3, and p_4. It is a convenient convention to label all the four particles as ingoing particles, as in Fig. (I.3). Any actual process will have two of the antiparticles of these outgoing.

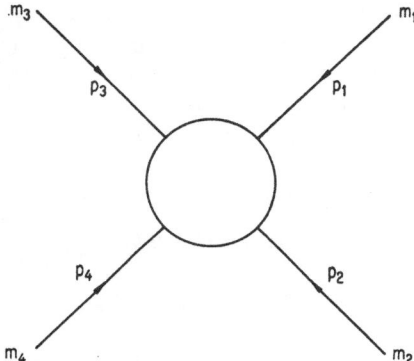

Fig. I.3. The four-line connected part.

Because of the TCP theorem, and crossing, the amplitude will describe the six processes:

$$1 + 2 \to \bar{3} + \bar{4}; \quad 3 + 4 \to \bar{1} + \bar{2}; \quad - \textcircled{1}$$
$$1 + 3 \to \bar{2} + \bar{4}; \quad 2 + 4 \to \bar{1} + \bar{3}; \quad - \textcircled{2}$$
$$1 + 4 \to \bar{2} + \bar{3}; \quad 2 + 3 \to \bar{1} + \bar{4}; \quad - \textcircled{3}$$

but with $\textcircled{1}$, $\textcircled{2}$, and $\textcircled{3}$ in different regions of the variables.

We define the following three invariants:

$$s = (p_1 + p_2)^2 = (p_3 + p_4)^2 , \tag{6.1a}$$
$$t = (p_1 + p_3)^2 = (p_2 + p_4)^2 , \tag{6.1b}$$
$$u = (p_1 + p_4)^2 = (p_3 + p_2)^2 , \tag{6.1c}$$

where the second equality follows from energy-momentum conservation.

In the centre-of-mass system for particles 1 and 2 we can write out the four-vectors explicitly, viz.

$$p_1 = (E_1, \boldsymbol{q}_{s12}) , \quad p_2 = (E_2, - \boldsymbol{q}_{s12}) , \tag{6.2a}$$

where $\boldsymbol{q}_{s12}$ is the three-momentum of particle 1 in this system, and E_1 is its energy, etc. Likewise, for particles 3 and 4 we have

$$p_3 = (E_3, \boldsymbol{q}_{s34}) , \quad p_4 = (E_4, - \boldsymbol{q}_{s34}) . \tag{6.2b}$$

The mass-shell constraints require

$$p_1^2 = E_1^2 - q_{s12}^2 = m_1^2 , \tag{6.3a}$$
$$p_2^2 = E_2^2 - q_{s12}^2 = m_2^2 , \tag{6.3b}$$
$$p_3^2 = E_3^2 - q_{s34}^2 = m_3^2 , \tag{6.3c}$$

and

$$p_4^2 = E_4^2 - q_{s34}^2 = m_4^2 , \qquad (6.3\,\mathrm{d})$$

while

$$s = (E_1 + E_2)^2 = (E_3 + E_4)^2 \qquad (6.4)$$

is the square of the total centre-of-mass energy of particles 1 and 2, or 3 and 4; that is of process ①. We thus call process ① the s-channel, while ② is the t-channel, and ③ the u-channel.

Also, from (6.1)

$$s = p_1^2 + p_2^2 + 2p_1 \cdot p_2 ,$$

which from (6.3)

$$= m_1^2 + m_2^2 + 2p_1 \cdot p_2 , \qquad (6.5\,\mathrm{a})$$

and similarly

$$t = m_1^2 + m_3^2 + 2p_1 \cdot p_3 , \qquad (6.5\,\mathrm{b})$$

and

$$u = m_1^2 + m_4^2 + 2p_1 \cdot p_4 . \qquad (6.5\,\mathrm{c})$$

Adding

$$s + t + u = m_1^2 + m_2^2 + m_3^2 + m_4^2 + 2m_1^2 + 2p_1 \cdot (p_2 + p_3 + p_4) . \qquad (6.6)$$

By four-momentum conservation

$$p_2 + p_3 + p_4 = -p_1 , \qquad (6.7)$$

and using (6.3) we get

$$s + t + u = m_1^2 + m_2^2 + m_3^2 + m_4^2 \equiv \Sigma . \qquad (6.8)$$

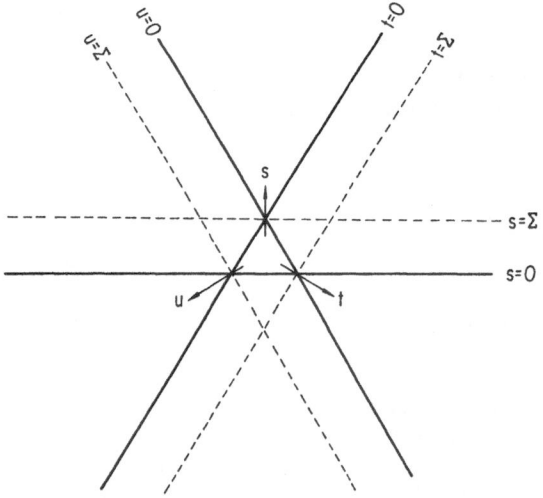

Fig. I.4. The Mandelstam plot, showing the variables s, t and u.

So, as expected, there is a constraint on our three invariants, only two of them being independent. But it is often convenient to work with all three, plotting them on a Mandelstam diagram (*Mandelstam*, 1958) as in Fig. (I.4). We wish to express the other physical variables in terms of these.

From (6.5a)

$$p_1 \cdot p_2 = \frac{1}{2}(s - m_1^2 - m_2^2). \qquad (6.9)$$

Also

$$m_1^2 + p_1 \cdot p_2 = p_1 \cdot (p_1 + p_2)$$
$$= E_1 \sqrt{s}, \qquad (6.10)$$

using (6.2a).

Eliminating $p_1 \cdot p_2$ between (6.9) and (6.10) gives

$$E_1 = \frac{1}{2\sqrt{s}}[s + m_1^2 - m_2^2], \qquad (6.11a)$$

and similarly we can obtain

$$E_2 = \frac{1}{2\sqrt{s}}[s + m_2^2 - m_1^2], \qquad (6.11b)$$

$$E_3 = \frac{1}{2\sqrt{s}}[s + m_3^2 - m_4^2], \qquad (6.11c)$$

$$E_4 = \frac{1}{2\sqrt{s}}[s + m_4^2 - m_3^2]. \qquad (6.11d)$$

Thus s determines the centre of mass energies of each of the particles. The corresponding 3-momenta can be found by combining (6.3a) with (6.11a), giving

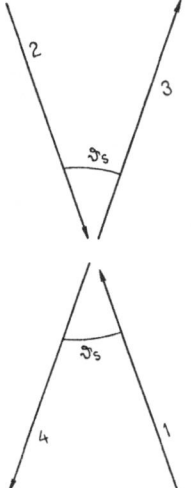

Fig. I.5. The centre-of-mass scattering angle, ϑ_s.

$$q_{s12}^2 = \frac{1}{4s}[s - (m_1 + m_2)^2][s - (m_1 - m_2)^2], \qquad (6.12a)$$

and similarly

$$q_{s34}^2 = \frac{1}{4s}[s - (m_3 + m_4)^2][s - (m_3 - m_4)^2]. \qquad (6.12b)$$

From (6.5b) and (6.2a), (6.2b), we get

$$t = m_1^2 + m_3^2 + 2E_1E_3 - 2q_{s12} \cdot q_{s34},$$
$$= m_1^2 + m_3^2 + 2E_1E_3 - 2q_{s12}q_{s34}\cos\vartheta_s, \qquad (6.13)$$

where ϑ_s is the angle between the directions of motion of particles 1 and 3 in the centre-of-mass system, generally called the centre-of-mass scattering angle [Fig. (I.5)]. Using (6.11a) and (6.11c) we find

$$\cos\vartheta_s = \frac{s^2 + s(2t - \Sigma) + (m_1^2 - m_2^2)(m_3^2 - m_4^2)}{4s\,q_{s12}\,q_{s34}}, \qquad (6.14)$$

so, at a given energy (s), the scattering angle is determined by t. We can find corresponding expressions for $\cos\vartheta_t$ and $\cos\vartheta_u$, the scattering angles in the t and u channels. It is worth noting that, if all the masses are equal, (6.12) and (6.14) greatly simplify to

$$q_s^2 = \frac{s - 4m^2}{4}, \qquad (6.15)$$

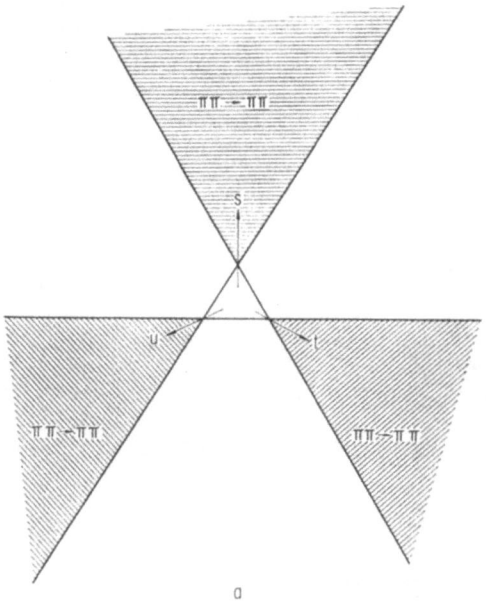

a

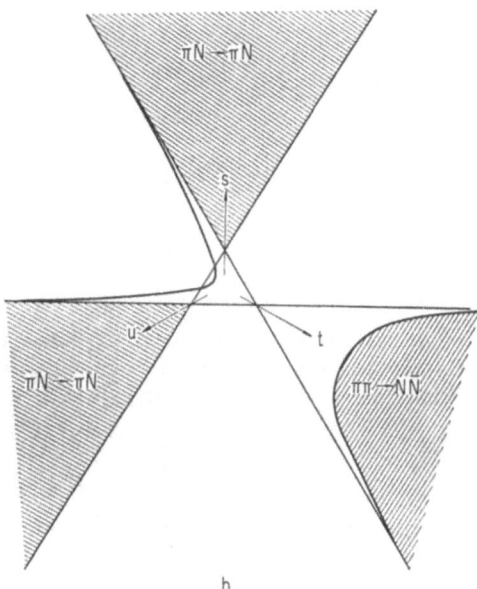

b

Fig. I.6a and b. The physical regions for (a) equal mass scattering such as $\pi\pi \to \pi\pi$, and (b) two different masses such as $\pi N \to \pi N$.

and

$$\cos\vartheta_s = 1 + \frac{t}{2q_s^2} = 1 + \frac{2t}{s - 4m^2} .$$ (6.16)

The physical region for the s channel is evidently given by

$$s > (m_1 + m_2)^2 \quad [\text{i.e. } q_{s12}^2 > 0] ,$$

and

$$-1 \leq \cos\vartheta_s \leq 1 .$$

In terms of s and t this latter condition can be written (*Kibble*, 1960; *Eden* et al., 1966)

$$\begin{vmatrix} 0 & 1 & 1 & 1 & 1 \\ 1 & 0 & m_2^2 & t & m_1^2 \\ 1 & m_2^2 & 0 & m_3^2 & s \\ 1 & t & m_3^2 & 0 & m_4^2 \\ 1 & m_1^2 & s & m_4^2 & 0 \end{vmatrix} \geq 0 ,$$

which, for equal masses, becomes $-(s - 4m^2) \leq t \leq 0$, and the line $t = -(s - 4m^2)$ is, from (6.8), just $u = 0$. Corresponding conditions are obtained for the t and u channel physical regions. In Fig. (I.6) we show the physical regions of the various channels for two different combinations of masses.

To conclude, we expect our two-particle scattering amplitude $A(s, t)$ to be an analytic function of the variables s, t, u, subject to the restraint $s + t + u = \Sigma$, and equal to the physical amplitudes for processes ① — ③ in the appropriate regions.

I.7. The Relation Between the Scattering Amplitude and Measurable Quantities

In order to compare the theory with experiment, we need to know the relation between the invariant scattering amplitudes which we have defined, and the cross-sections measured by experimenters.

The S-matrix element connecting an initial state $|i\rangle$, of total four-momentum p_i, with a final state $|f\rangle$, total four-momentum p_f, may be written

$$S_{fi} = \delta_{fi} + i(2\pi)^4 \delta^4(p_f - p_i) \langle f |A^+| i\rangle .$$ (7.1)

The probability of a transition from $|i\rangle$ to $|f\rangle$, in unit time, is given by

$$\begin{aligned} P_{fi} &= |(2\pi)^4 \delta^4(p_f - p_i) \langle f |A^+| i\rangle|^2 \\ &= (2\pi)^4 \delta^4(p_f - p_i) |\langle f |A^+| i\rangle|^2 , \end{aligned}$$ (7.2)

with the usual rule for squaring a δ-function. If our initial state contains two particles, of four-momenta p_1 and p_2, and the final state an arbitrary number of particles, N, of momenta p_j, this becomes, with the normalisation of (2.7)

$$P_{fi} = (2\pi)^4 \int \Pi_j \left[\frac{d^3 p'_j}{2 p'_{0j} (2\pi)^3}\right] |\langle p'_1 \ldots p'_N |A^+| p_1 p_2 \rangle|^2 \times$$
$$\times \delta^4(p_f - p_1 - p_2) . \tag{7.3}$$

The initial flux of particles, as measured in the centre of mass system, is given by the modulus of the relative velocity with which the two particles approach each other, divided by the normalization volume of (2.7) [volume $= (2E_1 2E_2)^{-1}$].

$$\text{Flux} = 4E_1 E_2 |v_1 - v_2|_{\text{c. of m.}}$$
$$= 4E_1 E_2 \left|\frac{q_{s12}}{E_1} + \frac{q_{s12}}{E_2}\right| \tag{7.4}$$
$$= 4 q_{s12} \sqrt{s} .$$

Hence the cross-section for scattering from $|i\rangle$ to $|f\rangle$ is

$$\sigma = \text{Transition Probability/unit flux}$$
$$= \frac{(2\pi)^4}{4 q_{s12} \sqrt{s}} \int \Pi_j \left[\frac{d^3 p'_j}{2 p'_{0j} (2\pi)^3}\right] |\langle p'_1 \ldots p'_N | A^+ |p_1 p_2 \rangle|^2 \times \tag{7.5}$$
$$\times \delta^4(p_f - p_1 - p_2) .$$

If there are only two particles in the final state, whose 4-vectors in the centre-of-mass system are $p_3 = (E_3, q_{s12})$ and $p_4 = (E_4, -q_{s34})$, this gives

$$\sigma = \frac{1}{(8\pi)^2 q_{s12} \sqrt{s}} \int \frac{d^3 p_3 \, d^3 p_4}{E_3 E_4} |\langle p_3 p_4 |A^+| p_1 p_2 \rangle|^2 \delta^4(p_3 + p_4 - p_1 - p_2) ,$$
$$\tag{7.6}$$

and we can use the δ-function to perform one of the integrations, obtaining

$$\sigma = \frac{1}{(8\pi)^2 q_{s12} \sqrt{s}} \int \frac{d^3 q_{s34}}{E_3 E_4} \delta(\sqrt{s} - E_3 - E_4) |\langle p_3 p_4 |A^+| p_1 p_2 \rangle|^2 \tag{7.7}$$

Then with

$$\int d^3 q_{s34} = \int q_{s34}^2 \cdot dq_{s34} \cdot d\Omega , \tag{7.8}$$

where $\int d\Omega$ is the angular integral over the direction of motion of particle 3, we get the cross-section

$$\sigma = \frac{q_{s34}}{(8\pi)^2 q_{s12} s} \int |\langle p_3 p_4 |A^+| p_1 p_2 \rangle|^2 \, d\Omega . \tag{7.9}$$

The differential cross section is thus

$$\frac{d\sigma}{d\Omega} = \frac{q_{s34}}{q_{s12}} \frac{1}{(8\pi)^2 s} |\langle p_3 p_4 |A^+| p_1 p_2 \rangle|^2 , \tag{7.10}$$

or, in terms of our invariants,

$$d\Omega = \frac{dt}{2 q_{s12} q_{s34}} d\phi \tag{7.11}$$

and

$$\frac{d\sigma}{dt} = \frac{1}{64\pi s\, q_{s12}^2} |A\,(s,\,t)|^2 \,. \tag{7.12}$$

We shall also need the familiar relation between the total cross section and the imaginary part of the elastic scattering amplitude, the optical theorem. The unitarity relation (2.2) gives

$$(SS^\dagger)_{fi} = \sum_m S_{fm}\, S_{mi}^\dagger = \delta_{fi}\,. \tag{7.13}$$

Substituting (7.1), we get

$$i\,[\langle f\,|A^+|\,i\rangle - \langle f\,|A^-|\,i\rangle] = -\,(2\pi)^4 \sum_m \delta(p_i - p_f)\times$$
$$\times \langle f\,|A^+|\,m\rangle \langle m\,|A^-|\,i\rangle\,, \tag{7.14}$$

so, if the initial and final states are identical,

$$2\,\mathrm{Im}\{\langle i\,|A|\,i\rangle\} = (2\pi)^4 \sum_m \delta(p_i - p_m)\,|\langle m\,|A^+|\,i\rangle|^2\,. \tag{7.15}$$

Comparing this with (7.5), we see that the right hand side of (7.15) is proportional to the sum of the cross-sections for the transitions from the two particle state $|i\rangle = |p_1 p_2\rangle$ to all the other states which are available given the need to conserve energy and momentum.

Thus we have

$$\sigma_{i\to \mathrm{all}}^{\mathrm{tot}} = \frac{1}{2q_{s12}\sqrt{s}}\,\mathrm{Im}\,\{\langle i\,|A|\,i\rangle\}\,. \tag{7.16}$$

The amplitude $\langle i\,|A|\,i\rangle$ has its final state identical with the initial state, so the final state particles must be in the forward direction, i.e. $t = 0$, so

$$\sigma^{\mathrm{tot}} = \frac{1}{2q_{s12}\sqrt{s}}\,\mathrm{Im}\,\{A\,(s,\,0)\}\,. \tag{7.17}$$

Equations (7.10), (7.12) and (7.17) will be important for future reference.

I.8. The Singularities of the Four-line Connected Part

In each channel there will be the singularities demanded by unitarity. Thus, if there is a particle of mass M which has the same internal quantum numbers as the two particle system $1 + 2$, and hence communicates with the s channel, we have a pole

at $s = M^2$.

Then there will be threshold branch points corresponding to the various production thresholds,

etc.,

for all the sets of particles $\{a, b\}$, $\{c, d, e\}$, etc., communicating with the channel. The positions of these thresholds are at $s = (m_a + m_b)^2$, $s = (m_c + m_d + m_e)^2$ etc. For stable particles (m real) the branch points lie on the real s axis, and we can draw branch cuts from the branch points, running along the real axis to $s = +\infty$. This is shown in Fig. (I.7), where we have assumed that $1 + 2$ is the lowest threshold communicat-

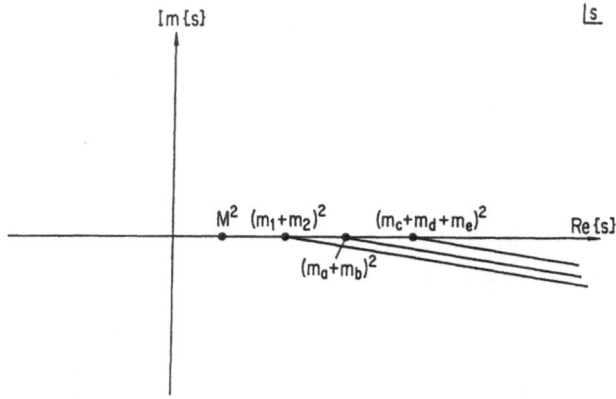

Fig. I.7. The s physical sheet, showing a bound state pole at M^2, and some of the threshold branch cuts.

ing with the channel. The cuts have been drawn slightly displaced downwards so as not to obscure one another.

Of course we are free to draw them in any direction, but this is the most convenient choice as it makes the amplitude hermitian analytic.

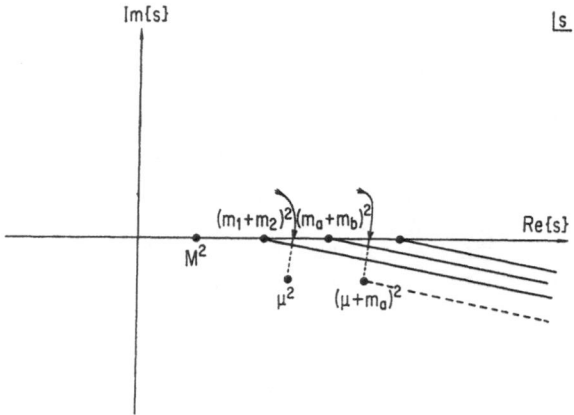

Fig. I.8. The physical sheet, showing an unstable particle pole μ, and a threshold involving this particle.

In accordance with the $i\varepsilon$ prescription of section I.4, to obtain the physical amplitude for $s > (m_1 + m_2)^2$, we must go above the cuts, that is

$$\text{Physical (s channel)} \quad A^+(s, t) = \lim_{\varepsilon \to 0} A(s + i\varepsilon, t).$$

The sheet for which this is the correct continuation, the one actually exhibited in Fig. (I.6), is the physical sheet. Similarly for $A^-(s, t)$, the correct boundary value is $\lim_{\varepsilon \to 0} A(s - i\varepsilon, t)$. Below the lowest threshold $A^+ = A^-$, so A is hermitian analytic.

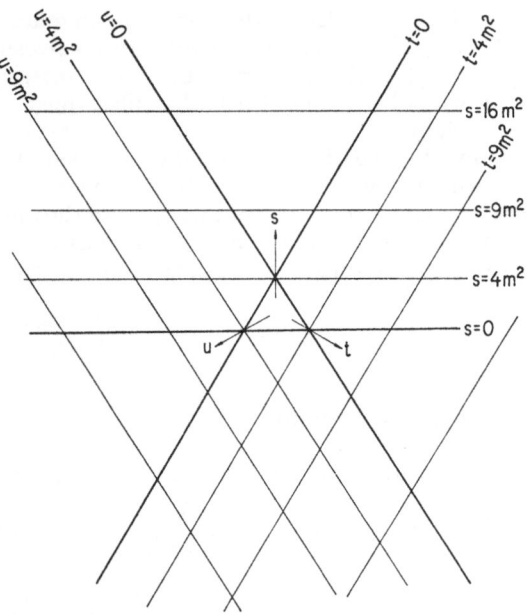

Fig. I.9. Singularities on the Mandelstam plot for equal mass scattering, showing thresholds at $(2m)^2$, $(3m)^2$, and $(4m)^2$ in each channel.

If we have an unstable particle of mass μ, $\mathrm{Re}\{\mu\} > (m_1 + m_2)$ it will have a negative imaginary part determined by the width corresponding to the Breit-Wigner formula,

$$A(s, t) = \frac{g}{E - E_R + \dfrac{i\Gamma}{2}} = \frac{g}{E - \mu}, \qquad (8.1)$$

where $E = \sqrt{s}$, is the centre of mass energy, $E_R = \mathrm{Re}\{\mu\}$, is the resonance position, and Γ its width. g is the residue of the pole. For $s \approx s_R \equiv E_R^2$ we have

$$A(s, t) \approx \frac{2 E_R g}{s - s_R + i E_R \Gamma} \approx \frac{2 E_R g}{s - \mu^2} \qquad (8.2)$$

in the approximation that $\Gamma \ll \mathrm{Re}\{\mu\}$. This puts the pole on the first unphysical sheet, as shown in Fig. (I.8), and similarly any threshold involving the production of μ, such as $(\mu + m_a)$, will be on an unphysical sheet.

It is believed that stable particle poles, and thresholds involving only stable particles, are the only singularities on the physical sheet (we are neglecting anomalous thresholds which result if certain inequalities between the various particle masses are satisfied (see *Eden* et al., 1966)), and that all the other, more complex, singularities lie on unphysical sheets. These poles and thresholds will occur for each of three channels s, t, and u, so in the Mandelstam diagram the singularities will occur as in Fig. (I.9). More complicated singularities correspond to more complicated diagrams, with larger masses in the intermediate states, and so, in accord with the Landau-Cutkosky rules, must occur for larger values of the invariants.

The low energy singularities will usually take simple forms. Apart from the poles, which have the form (4.2), the lowest singularities will often be the two particle thresholds, and we can use the unitarity relation (3.9) to calculate the discontinuity across the corresponding cuts. As this will be important for future discussions we give the derivation explicitly.

I.9. The Two-particle Discontinuity

We define the discontinuity across the cuts in s to be

$$D_s(s, t) = \frac{1}{2i} [A^+(s, t) - A^-(s, t)] = \frac{1}{2i} [A(s_+, t) - A(s^-, t)] , \quad (9.1)$$

where $s_+ \equiv s + i\varepsilon$, is evaluated just above the cut, and s_- just below. If we consider a system where all the particles have the same mass, and take

$$(2m)^2 < s < (3m)^2$$

so that only the two particle state is possible, we have, from (3.19),

$$D_s(s, t) = \frac{1}{2(2\pi)^2} \int d^4q \, \delta^+[(p_1 + q)^2 - m^2] \, \delta^+[(p_2 - q)^2 - m^2] \, A^+ A^- . \tag{9.2}$$

It is convenient to transform the integration variable

$$q \to q - p_1 ,$$

giving

$$D_s(s, t) = \frac{1}{2(2\pi)^2} \int d^4q \, \delta^+(q^2 - m^2) \, \delta^+[(p_1 + p_2 - q)^2 - m^2] \, A^+ A^- . \tag{9.3}$$

In the centre of mass system

$$p_1 = (p_{01}, \boldsymbol{p}), \quad \text{and} \quad p_2 = (p_{02}, -\boldsymbol{p}),$$

so that

$$(p_1 + p_2) = (p_{01} + p_{02}, 0) = (\sqrt{s}, 0) .$$

Putting (say)
$$q \equiv (q_0, \boldsymbol{q}) ,$$
we have for the argument of the second δ-function in (9.3)

$$(\not p_1 + \not p_2 - q)^2 - m^2 = s - 2\sqrt{s}\, q_0 + q^2 - m^2$$
$$= s - 2\sqrt{s}\, q_0 ,$$

from the first δ-function in (9.3). So

$$D_s(s, t) = \frac{1}{2(2\pi)^2} \int d^4q \; \delta^+(q^2 - m^2) \cdot \delta^+(s - 2\sqrt{s}\, q_0) A^+ A^-$$

$$= \frac{1}{2(2\pi)^2 \, 2\sqrt{s}} \int dq_0 \cdot d^3\boldsymbol{q} \cdot \delta^+(q_0^2 - |\boldsymbol{q}|^2 - m^2) \cdot \delta^+\left(\frac{\sqrt{s}}{2} - q_0\right) A^+ A^-$$

$$= \frac{1}{(4\pi)^2 \sqrt{s}} \int d^3\boldsymbol{q} \cdot \delta^+\left(\frac{s}{4} - |\boldsymbol{q}|^2 - m^2\right) A^+ A^- , \qquad (9.4)$$

and using

$$\int d^3\boldsymbol{q} = \frac{1}{2} \int |\boldsymbol{q}| \cdot d|\boldsymbol{q}|^2 \cdot d\Omega ,$$

$$D_s(s, t) = \frac{\sqrt{\frac{s}{4} - m^2}}{(4\pi)^2 \, 2\sqrt{s}} \int d\Omega \, A^+(s, t') \, A^-(s, t'') , \qquad (9.5)$$

$$= \frac{q_s}{32\pi^2 \sqrt{s}} \int d\Omega \, A^+(s, t') \, A^-(s, t'') .$$

The relation between t', t'', and t is given in section (I.12).

The more general result is that in the amplitude for a transition from channel a to channel b, such as

the discontinuity across the cut associated with channel n is, by a generalization of (9.5),

$$A^{ab}(s_+, t) - A^{ab}(s_-, t) = 2i \frac{q_{sn}}{32\pi^2 \sqrt{s}} \int d\Omega_n \, A^{bn}(s_+, t') \, A^{na}(s_-, t'') , \qquad (9.6)$$

q_{sn} being the momentum in state n.

I.10. Single Variable Dispersion Relations

Since the poles and thresholds in each channel are the only singularities on the physical sheet, the physical sheet singularities in any one of the invariants s, t, or u, can easily be deduced from the Mandelstam plot, Fig. (I.9). If we hold s fixed, real, and positive, then in the t-plane we shall find the singularities shown in Fig. (I.10).

For positive t there will be (say) a pole at $t = M_t^2$, corresponding to a bound state of mass M_t in the t-channel. Then there will be a branch cut beginning at the lowest t threshold, $t_0 = (m_1 + m_3)^2$, and extending to the right. Similarly on the left we may find a pole at $u = M_u^2$, and the

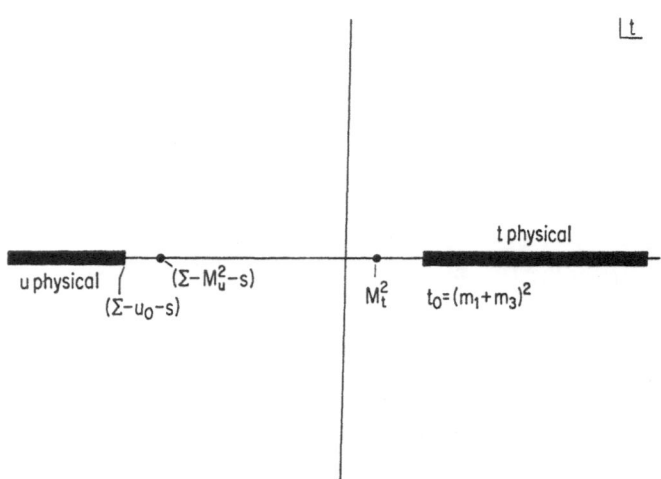

Fig. I.10. Singularities in the complex t plane at fixed s.

u-channel threshold at $u_0 = (m_1 + m_4)^2$, with a cut extending to $u = \infty$. Since $t = \Sigma - s - u$ from (6.8), these singularities are at $t = \Sigma - s - M_u^2$, and $t = \Sigma - s - u_0$, respectively. Note also that though the physical u-channel amplitude is $\lim_{\varepsilon \to 0} A(s, t, u + i\varepsilon)$, in the t-plane this corresponds to the negative real t axis approached from *below*, because of (6.8). (In this section we include u in the variables determining $A(s, t, u)$, but of course (6.8) still means that only two are independent).

We define the discontinuity functions at fixed s

$$D_t(s, t) = \frac{1}{2i} \left[A(s, t_+, u) - A(s, t_-, u) \right],$$

$$D_u(s, u) = \frac{1}{2i} \left[A(s, t, u_+) - A(s, t, u_-) \right], \tag{10.1}$$

where $t_\pm \equiv \lim_{\varepsilon \to 0} t \pm i\varepsilon$, the discontinuity being taken across all the cuts in t (or u) at fixed s. We have suppressed the third, dependent, variable in D_t and D_u. By hermitian analyticity

$$A(s, t^*, u) = A^*(s, t, u) \tag{10.2}$$

so we have

$$D_t(s, t) = \Delta_t\{A(s, t, u)\}, \quad t > t_0,$$

and
$$D_u(s, u) = \Delta_u\{A(s, t, u)\}, \qquad u > u_0, \tag{10.3}$$

(where Δ_t implies the discontinuity across the t cut divided by $2i$, etc.)

Then, if these are the only physical sheet singularities, we can use Cauchy's theorem, and find

$$A(s, t, u) = \frac{g_t}{M_t^2 - t} + \frac{g_u}{M_u^2 - u} + \frac{1}{2\pi i} \int_C \frac{dt'}{t' - t} A(s, t', u'), \tag{10.4}$$

where the contour C is as shown in Fig. (I.11).

If $A(s, t, u)$ vanishes sufficiently fast as $t \to \infty$, we get from (10.4)

$$A(s, t, u) = \frac{g_t}{M_t^2 - t} + \frac{g_u}{M_u^2 - u} + \frac{1}{\pi} \int_{t_0}^{\infty} \frac{dt'}{t' - t} D_t(s, t') + \frac{1}{\pi} \int_{u_0}^{\infty} \frac{du'}{u' - u} D(s, u'),$$
$$\tag{10.5}$$

where $u' + t' + s = \Sigma$.

The validity of this sort of dispersion relation is a test of our maximal analyticity postulate (v), and for some processes such relations have been checked with experiment, as far as the available data permits. (See, for example, *Hamilton*, 1964, for review of such comparisons in $\pi - N$ scattering).

In general the behaviour of the amplitude at infinity is not sufficiently convergent for the dispersion relation, as we have written it, to be meaningful. The integrals will diverge, and the contributions from the circle at infinity will not vanish. When this happens we resort to "subtractions", that is we write a dispersion relation for $A(s, t, u) [(t - t_1)(t - t_2) \ldots$ $\ldots (t - t_N)]^{-1}$, instead of for $A(s, t, u)$. Let us for the moment neglect the

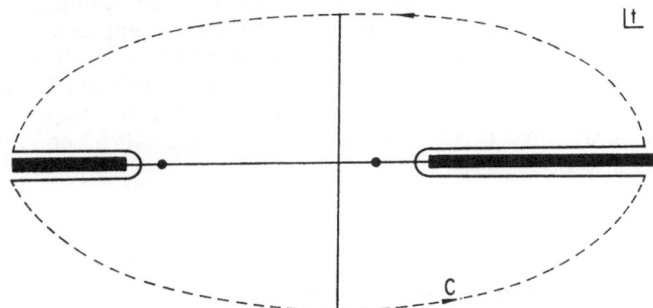

Fig. I.11. Contour of integration in the complex t plane.

bound state pole terms in (10.5), and suppose that in some formal sense we have the representation

$$A(s, t, u) = \frac{1}{\pi} \int_{t_0}^{\infty} \frac{dt'}{t' - t} D_t(s, t') + \frac{1}{\pi} \int_{u_0}^{\infty} \frac{du'}{u' - u} D_u(s, u'), \tag{10.6}$$

but that the integrals diverge. Then we can write

$$A\,(s,\,t,\,u)\,\prod_{i=1}^{N}\,(t-t_i)^{-1} = \sum_{j=1}^{N}\frac{A\,(s,\,t_j,\,u)}{(t-t_j)}\,\prod_{\substack{i=1\\i\neq j}}^{N}(t_j-t_i)^{-1}+$$

$$(10.7)$$

$$+\frac{1}{\pi}\int_{t_0}^{\infty}\frac{D_t(s,\,t')\;\mathrm{d}t'}{(t'-t_1)\ldots(t'-t_N)\,(t'-t)}+(u\text{ term})\,,$$

where we have picked up a contribution from the poles at $t = t_i$. So

$$A\,(s,\,t,\,u) = F^{N-1}(s,\,t)+\frac{1}{\pi}\,\prod_{i=1}^{N}(t-t_i)\int_{t_0}^{\infty}\frac{D_t(s,\,t')\;\mathrm{d}t'}{(t'-t_1)\ldots(t'-t_N)\,(t'-t)}+$$

$$+\,(u\text{ term})\qquad\qquad\qquad\qquad\qquad (10.8)$$

where $F^{N-1}(s,\,t)$ is a function of s multiplied by a polynomial of degree $N-1$ in t. Thus if

$$D_t(s,\,t)\underset{t\to\infty}{\sim} t^{N-\varepsilon}\quad\text{at fixed }s,\quad \varepsilon > 0\,,$$

the integral converges, and the dispersion relation is well defined. (We neglect any subtractions which may also be needed for the D_u term.) If $D_t(s,\,t)$ is bounded by some power of t it will always be possible to do this. However, unlike the previous expression, (10.6), $A\,(s,\,t,\,u)$ is no longer completely determined by $D_t(s,\,t)$ [and D_u]. We now also require to know N further pieces of information, the values of $A\,(s,\,t_j,\,u)$ for $j = 1 \ldots N$.

   Thus, requiring that the amplitude satisfy maximal analyticity of the first kind, with all its singularities given by the Landau-Cutkosky rules, is not necessarily sufficient to determine it completely. It would be sufficient if the amplitude were known to vanish suitably at infinity, but otherwise subtractions, which may introduce arbitrary parameters, are needed. We shall see in the next chapter how the requirement of maximal analyticity of the second kind [postulate (vi)] enables these ambiguities to be resolved.

I.11. The Mandelstam Representation

   In the last section we defined the discontinuities D_s, D_t, D_u, of the four-line connected part, in the various invariants s, t and u. However, maximal analyticity, (v), also requires that these discontinuities themselves have branch points, so that, for example, $D_t(s,\,t)$ is cut in s, and also in u [remember (6.8)]. It is hoped, but unfortunately not yet proved, that all such branch points lie on the real axis, so that again we can draw the cuts along the real axis, and write simple dispersion relations for D_t etc.

We define the discontinuity in s of D_t to be

$$\varrho_{st}(s, t) = \frac{1}{2i}\left[D_t(s_+, t) - D_t(s_-, t)\right], \quad s > b_1(t) > 0, \quad (11.1)$$

and

$$\varrho_{tu}(t, u) = \frac{1}{2i}\left[D_t(u_+, t) - D_t(u_-, t)\right], \quad u > b_2(t) > 0, \quad (11.2)$$

so that

$$D_t(s, t) = \frac{1}{\pi}\int_{b_1(t)}^{\infty}\frac{\varrho_{st}(s'', t)}{s'' - s}\,ds' + \frac{1}{\pi}\int_{b_2(t)}^{\infty}\frac{\varrho_{tu}(t, u'')}{u'' - u}\,du''. \quad (11.3)$$

These integrals are to be taken over the regions of s'' and u'' for which ϱ_{st} and ϱ_{tu} exist for a given value of t. The boundaries $s = b_{1,2}(t)$ etc. will be discussed in the next section.

Similarly we can write

$$D_u(s, u) = \frac{1}{\pi}\int^{\infty}\frac{\varrho_{su}(s'', u)}{s'' - s}\,ds'' + \frac{1}{\pi}\int^{\infty}\frac{\varrho_{tu}(t'', u)}{t'' - t}\,dt''. \quad (11.4)$$

Substituting (11.3) and (11.4) into (10.6), we find

$$A(s, t) = \frac{1}{\pi^2}\int\int^{\infty}\frac{\varrho_{st}(s'', t')}{(s'' - s)(t' - t)}\,ds''\,dt' + \frac{1}{\pi^2}\int\int^{\infty}\frac{\varrho_{tu}(t', u'')}{(u'' - u')(t' - t)}\,dt'\,du'' +$$

$$+ \frac{1}{\pi^2}\int\int^{\infty}\frac{\varrho_{su}(s'', u')}{(s'' - s)(u' - u)}\,ds''\,du' + \quad (11.5)$$

$$+ \frac{1}{\pi^2}\int\int^{\infty}\frac{\varrho_{tu}(t'', u')}{(t'' - t')(u' - u)}\,dt''\,du'.$$

Note that this relation, like (10.6), is written at fixed s, so that in the second and fourth terms on the right-hand side we need a prime on the variable which comes from the denominator in (11.3) and (11.4),

$$\text{i.e.} \quad s + t + u = s + t' + u' = \Sigma \quad (11.6)$$

from (6.8). Interchanging the primes on the fourth term, and adding the 2nd and 4th terms, we get,

$$\int\int \varrho_{tu}(t', u'')\left[\frac{1}{(u'' - u')(t' - t)} + \frac{1}{(t' - t'')(u'' - u)}\right]dt'\,du'',$$

which, using (11.6), simplifies to

$$\int\int \varrho_{tu}(t', u'')\frac{1}{(t' - t)(u'' - u)}\,dt'\,du'',$$

so from (11.5) we finally obtain

$$A(s, t) = \frac{1}{\pi^2}\int\int^{\infty}\frac{\varrho_{st}(s'', t')}{(s'' - s)(t' - t)}\,ds''\,dt' +$$

$$+ \frac{1}{\pi^2}\int\int^{\infty}\frac{\varrho_{su}(s'', u')}{(s'' - s)(u' - u)}\,ds''\,du' + \quad (11.7)$$

$$+ \frac{1}{\pi^2}\int\int\frac{\varrho_{tu}(t', u'')}{(t' - t)(u'' - u)}\,dt'\,du''.$$

This is the Mandelstam representation (*Mandelstam*, 1958, 1959 a, 1959 b). Though we have derived (11.7) from the fixed s dispersion relation (10.6), it is clear, from the symmetry in s, t and u, that the same result would be obtained whichever variable we dispersed first. This is because, from (11.1) and (10.1) (suppressing the dependent variable u),

$$\varrho_{st}(s,t) = \frac{1}{2i} \left\{ \frac{1}{2i} \left[A\left(s_+, t_+\right) - A\left(s_-, t_+\right) \right] - \frac{1}{2i} \left[A\left(s_+, t_-\right) - A\left(s_-, t_-\right) \right] \right\}$$

$$= \frac{-1}{4} \left\{ A\left(s_+, t_+\right) + A\left(s_-, t_-\right) - A\left(s_-, t_+\right) - A\left(s_+, t_-\right) \right\},$$

which can be expressed as

$$(1/2i) \left[(D_t(s_+, t) - D_t(s_-, t) \right] \quad \text{or} \quad (1/2i) \left[(D_s(s, t_+) - D_s(s, t_-) \right].$$

General proofs that the amplitudes can be written in this form are lacking, and we refer the reader to *Eden* et al., 1966, for a discussion of the progress which has been made so far.

In the next section we shall show how to calculate the "double spectral functions" ϱ_{st}, ϱ_{su}, ϱ_{tu}, at least for limited regions of the variables. In particular we shall show how the boundaries of the double spectral functions can be calculated given the thresholds in the three channels s_0, t_0 and u_0. These are the lowest values of s, t and u for which D_s, D_t and D_u, respectively, exist. For example in the amplitude for $\pi - \pi$ elastic scattering we find that $\varrho_{st}(s, t)$ is bounded by the two curves $t = 4 t_0 + t_0/q_s^2$ and $s = 4 s_0 + s_0/q_t^2$. With $s_0 = t_0 = 4 m_\pi^2$, and $q_s^2 = s/4 - m_\pi^2$, $q_t^2 = t/4 - m_\pi^2$, we obtain the boundaries shown in Fig. (I.15). The double spectral functions can only exist in unphysical regions of the variables.

If we know the double spectral functions, then (11.7) enables us to determine the amplitude for all s and t. But again we have the problem of subtractions noted in the last section. It can be seen from (11.1) that if $D_t(s, t) \underset{t \to \infty}{\sim} t^N$, then, usually, $\varrho_{st}(s, t) \sim t^N$ also. So, in general, the Mandelstam representation only defines the amplitude up to an unspecified number of arbitrary subtractions in each of the variables s, t, and u.

It appears however that there must be at least some connection between this asymptotic behaviour, and the nature of the bound-state and resonance poles which are contained in the theory. For example, a resonance of spin 1 in the s channel, of mass $\sqrt{s_R}$ and width Γ, contributes to the amplitude a pole of the form [c.f. (8.2)]

$$A(s, t) \approx \frac{g(s) \, P_l(\cos\vartheta_s)}{s - s_R + i \, \Gamma \sqrt{s_R}},$$

where $g(s)$ is the residue of the pole. For small Γ we have approximately

$$D_s(s, t) \approx g(s) \, P_l(\cos\vartheta_s) \, \delta(s - s_R),$$

Now [B 1, 3.9 (19)]

$$P_l(\cos\vartheta_s) \underset{\cos\vartheta_s \to \infty}{\sim} (\cos\vartheta_s)^l, \quad l > -\frac{1}{2},$$

and from (6.14)

$$\cos\vartheta_s \xrightarrow[t \to \infty]{} \frac{t}{2 q_{s12} \, q_{s34}},$$

so we have
$$D_s(s, t) \underset{t \to \infty}{\sim} t^l .$$

Hence, if the asymptotic t behaviour for all s is bounded by t^l, then l is the highest spin which an s-channel pole can have. But of course there may be an arbitrarily large number of particles of all spins $\leq l$. This should not surprise us as maximal analyticity of the first kind does not limit the number of poles which can be present, it only allows us to find all the other singularities given the poles.

On the other hand it would be surprising if all the poles really could be specified arbitrarily. For instance suppose we include the neutron and proton poles in the S-matrix. We would expect the deuteron pole to be generated by the "force" between these two particles, so there should be no need to put it in beforehand. Our expectation about this is clearly based on the feeling that the deuteron is a composite particle, and that composites should be consequences of the theory, not part of the postulates. In quantum-electrodynamics one has to specify the masses and charges of the electron and positron, but not those of positronium, which can be calculated. To add to the theory the requirement that the positronium mass take some particular value other than the experimental one would certainly be inconsistent. A theory of strong interactions which enables one to specify the masses and couplings of all the particles arbitrarity is almost certainly similarly contradictory.

So the basic difficulty of using an S-matrix theory based on postulates (i) to (v) is just this; that we do not know how much information, how many particle masses and couplings, if any, we need to put into the theory at the start. In non-relativistic physics, and probably also in quantum electrodynamics, we can distinguish between the "elementary" particles, such as the electron and proton, and the composite particles, such as the hydrogen atom or positronium, but with strong interactions this distinction is harder, perhaps impossible, to make. But evidently the answer to the problem is bound up with the angular momentum properties of the S-matrix, and these are the subject of the next chapter.

In the final section we shall show how the double spectral functions can be calculated when two particle unitarity holds.

I.12. The Elastic Double Spectral Function

In general it is not possible to calculate the double spectral functions of the Mandelstam representation. But, if a given amplitude has a region in which elastic unitarity holds, we can make use of (9.5) to find $D_s(s, t)$, and, by taking the t discontinuity of D_s, can find the double spectral function, at least for that region of s where elastic unitarity is valid (*Mandelstam*, 1958, 1959a, 1959b). In principle we could use (9.6) for a set of coupled two-particle channels, but this would involve treating several amplitudes at once, which would be prohibitively difficult from a computational point of view. Once multi-particle thresholds are reached the problem becomes quite intractable. It is worth persuing the question of calculating the elastic double spectral functions, however, because, for most of the amplitudes with which we shall be concerned, the elastic thresholds are the lowest communicating thresholds, which means that

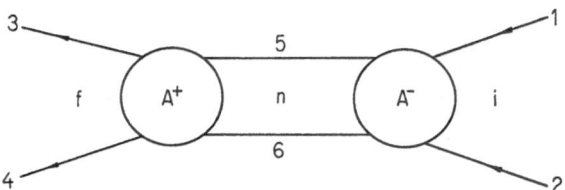

Fig. I.12. Two particle unitarity with intermediate state n.

the elastic double spectral function boundary is the boundary of the whole double spectral function. We are thus able to determine the boundary. Also for some purposes it is a worthwhile approximation to simply neglect inelastic unitarity, and we shall often make use of this in Chapter VI.

We first re-write (9.5) for the discontinuity corresponding to Fig. (I.12),

$$D_s(s, t) = \frac{q_s}{32\pi^2 \sqrt{s}} \int d\Omega_s \, A^+(s, t') \, A^-(s, t'') \, . \qquad (12.1)$$

Here $t' = t(z', s)$, where $z' \equiv \cos\vartheta_{nf}$, and corresponds to the angle between the direction of motion of the particles in state f and in state n (in the

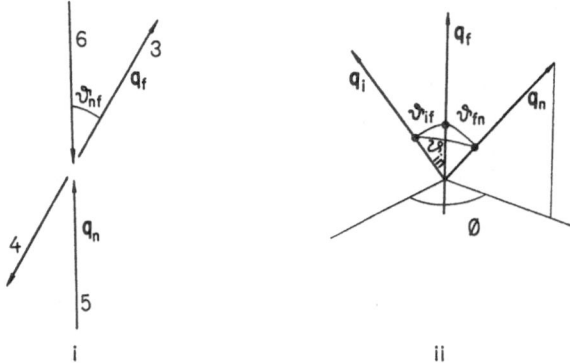

Fig. I.13. (i) The centre-of-mass 3-momenta, and the scattering angle ϑ_{nf}. (ii) The relation between the three angles ϑ_{if}, ϑ_{in}, ϑ_{nf}. ϑ is measured about the q_f axis, with its origin in the plane containing q_f and q_i.

centre of mass system), while $t'' = t(z'', s)$, where $z'' \equiv \cos\vartheta_{in}$, gives the angle between i and n. These angles are related to that corresponding to t [$= t(z_s, s)$ where $z_s = \cos\delta_{if}$] by the addition theorem for cosines (see Fig. I.13),

$$\cos\vartheta_{in} = \cos\vartheta_{if} \cos\vartheta_{fn} + \sin\vartheta_{if} \sin\vartheta_{fn} \cos\phi \, ,$$

or

$$z'' = z_s \, z' + \sqrt{1 - z_s^2} \, \sqrt{1 - z'^2} \cos\phi \, . \tag{12.2}$$

If we now substitute (10.6) for each of the amplitudes in (12.6) [Equation (10.6) supposes there are no bound-state poles; we shall include them later]. We have

$$D_s(s, t) = \frac{q_s}{32\pi^2 \sqrt{s}} \int d\Omega_s \left[\frac{1}{\pi} \int\limits_{t_0}^{\infty} \frac{D_t(s_t, t_1)}{t_1 - t''} \, dt_1 + \frac{1}{\pi} \int\limits_{u_0}^{\infty} \frac{D_u(s_t, u_1)}{u_1 - u''} \, du_1 \right] \times$$

$$\times \left[\frac{1}{\pi} \int\limits_{t_0}^{\infty} \frac{D_t(s_-, t_2)}{t_2 - t'} \, dt_2 + \frac{1}{\pi} \int\limits_{u_0}^{\infty} \frac{D_u(s_-, u_2)}{u_2 - u'} \, du_2 \right], \tag{12.3}$$

where

$$s + t + u = s + t_1 + u_1 = s + t' + u' = s + t'' + u'' = \Sigma. \tag{12.4}$$

If we now replace the t and u variables by the corresponding cosines we get

$$D_s(s, t) = \frac{q_s}{32\pi^2 \sqrt{s}} \frac{1}{\pi^2} \int\limits_{-1}^{1} dz' \int\limits_{0}^{2\pi} d\phi \left[\int\limits_{z_{t0}}^{\infty} \frac{D_t(s_+, t_1)}{z_1 - z''} \, dz_1 + \int\limits_{z_{u0}}^{-\infty} \frac{D_u(s_+, u_1)}{z_1 - z''} \, dz_1 \right] \times$$

$$\times \left[\int\limits_{z_{t0}}^{\infty} \frac{D_t(s_-, t_2)}{z_2 - z'} \, dz_2 + \int\limits_{z_{u0}}^{-\infty} \frac{D_u(s_-, u_2)}{z_2 - z'} \, dz_2 \right], \tag{12.5}$$

where

$$z_{t0} = z_s(s, t_0) \quad \text{and} \quad z_{u0} = z_s(s, \Sigma - s - u_0) \, .$$

Interchanging the order of integration in (12.5), we get the angular integration

$$\int\limits_{-1}^{1} dz' \int\limits_{0}^{2\pi} d\phi \frac{1}{(z_1 - z'')(z_2 - z')} \, ,$$

which, using (12.2), can be shown (with some effort) to give

$$\frac{2\pi}{k^{\frac{1}{2}}} \log \left[\frac{z_s - z_1 z_2 + k^{\frac{1}{2}}}{z_s - z_1 z_2 - k^{\frac{1}{2}}} \right],$$

where

$$k \equiv [z_s^2 + z_1^2 + z_2^2 - 1 - z_s z_1 z_2] \, ,$$

and we must take the branch of the logarithm which is real for $-1 < z_s < 1$. If we now convert the result back to the t variables, using (6.14) and the fact that for elastic scattering $q_{s12} = q_{s34} \equiv q_s$ (say), we end up with

$$D_s(s, t) = \frac{1}{16\pi^3} \frac{q_s}{\sqrt{s}} \int\limits_{t_0}^{\infty} \frac{dt_1}{2q_s^2} \int\limits_{t_0}^{\infty} \frac{dt_2}{2q_s^2} \, [D_t(s_+, t_1) + D_u(s_+, t_1)] \times$$

$$\times [D_t(s_-, t_2) + D_u(s_-, t_2)] \, 2q_s^2 \, K^{-\frac{1}{2}} \, (t, t_1, t_2, s) \times$$

$$\times \log \left[\frac{t - t_1 - t_2 - t_1 t_2 / 2q_s^2 + K^{\frac{1}{2}}}{t - t_1 - t_2 - t_1 t_2 / 2q_s^2 - K^{\frac{1}{2}}} \right] \tag{12.6}$$

3*

where

$$K\left(t, t_1, t_2, s\right) = \left[t^2 + t_1^2 + t_2^2 - 2\left(tt_1 + tt_2 + t_1 t_2\right) - \frac{tt_1 t_2}{q_s^2}\right], \quad (12.7)$$

and t_0 is the lower of t_0 and u_0.

To find the double spectral functions we simply have to find the discontinuities of (12.6). It will simplify the discussion if for the moment we

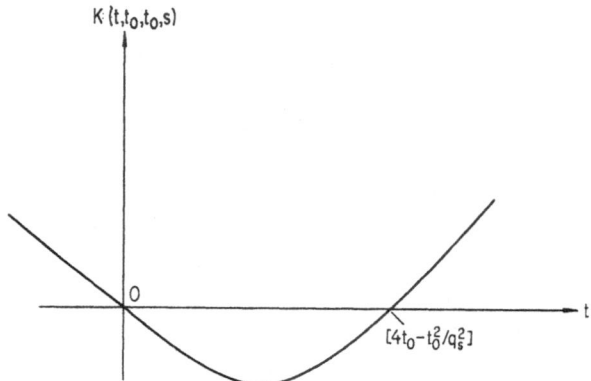

Fig. I.14. The variation of $K\left(t, t_0, t_0, s\right)$ with t, at fixed s.

ignore the u channel discontinuity D_u, so that $D_s(s, t)$ has only one t discontinuity, which is given by the double spectral function $\varrho_{st}(s, t)$.

So we have

$$
\begin{aligned}
D_s(s, t) &= \frac{1}{\pi} \int_{b(s)}^{\infty} \frac{\varrho_{st}(s, t')}{t' - t}\, \mathrm{d}t' \\
&= \frac{1}{16\pi^3} \frac{q_s}{\sqrt{s}} \int_{t_0}^{\infty} \frac{\mathrm{d}t_1}{2q_s^2} \int_{t_0}^{\infty} \frac{\mathrm{d}t_2}{2q_s^2} \frac{D_t(s_+, t_1)\, D_t(s_-, t_2)}{(1/2q_s^2)\, K^{\frac{1}{2}}\,(t, t_1, t_2, s)} \log\left[\cdots\right]. \quad (12.8)
\end{aligned}
$$

Then $\varrho_{st}(s, t)$ can be found by taking the t discontinuity of each side of this equation. On the right-hand side the discontinuity arises from the vanishing $K^{\frac{1}{2}}$. The lowest values of t_1 and t_2 in the integration are

$$t_1 = t_2 = t_0 \, ,$$

whereupon K simplifies to

$$K\left(t, t_0, t_0, s\right) = t\left[t - 4t_0 - \frac{t_0^2}{q_s^2}\right],$$

so $K = 0$ if $t = 0$, or $t = [4t_0 - t_0^2/q_s^2]$, and its variation with t is shown in Fig. (I.14). However $K = 0$ makes the argument of the log in (12.6) equal to 1. Now

$$\log\left(z\right) = \log\left(|z|\right) + i\, am\left(z\right) \quad (12.9)$$

so $\log 1 = 2\pi n i$, with n depending on the branch of the logarithm which is chosen. We have noted that the branch of the logarithm which is real for $-1 \leq z_s \leq 1$ should be taken, and $z_s = 1$ corresponds to $t = 0$, (6.16), so $t = 0$ is not a singular point of $K^{-\frac{1}{2}} \log [\, \cdots \,]$. But as t is increased the phase of the argument of the logarithm goes round to 2π at

$$t = 4t_0 - \frac{t_0^2}{q_s^2} \, ,$$

which is a singular point of $D_s(s, t)$. Higher values of t_1 and t_2 in the integral require a larger value of t for K to vanish (at a given q_s^2), so the boundary of the double spectral function is

$$t = 4t_0 - \frac{t_0^2}{q_s^2} \equiv b(s) \, . \tag{12.10}$$

If we take the t discontinuity of (12.8) for $t > b(s)$, using the fact that from (12.9)

$$\text{Disc} \, \{ \log [\, \cdots \,] \} = 2\pi \quad \text{for} \quad K > 0 \, ,$$

we get

$$\varrho_{st}(s, t) = \frac{1}{8\pi^2} \frac{q_s}{\sqrt{s}} \int_{t_0}^{K(t, t_1, t_2, s) = 0} \frac{dt_1}{2q_s^2} \int_{t_0} \frac{dt_2}{2q_s^2} \frac{D_t(s_+, t_1) \, D_t(s_-, t_2)}{(1/2 \, q_s^2) \, K^{\frac{1}{2}}(t, t_1, t_2, s)} \, ,$$

$$\text{for} \quad t > b(s) \, . \tag{12.11}$$

The integration is only over that range of t_1 and $t_2 > t_0$ for which $K(t, t_1, t_2, s)$ is greater than zero, since for $K < 0$ there is no t discontinuity of (12.8). The fact that these are finite integrals will be very important for the

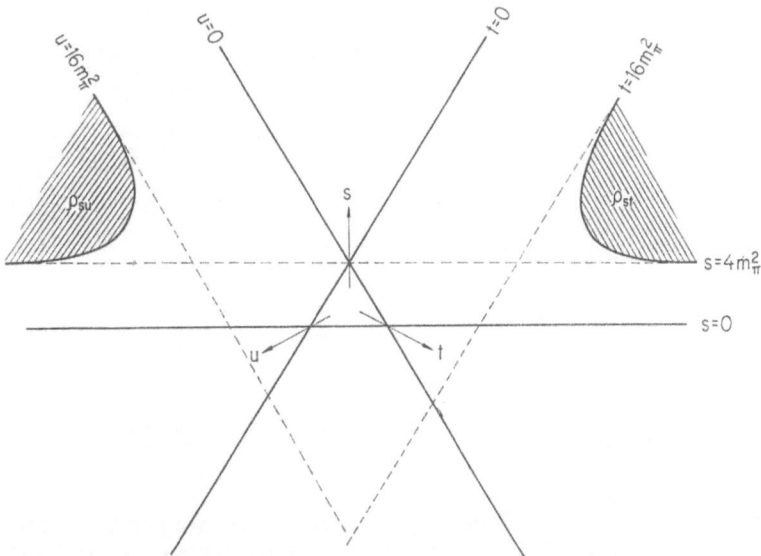

Fig. I.15. The Mandelstam diagram for π-π scattering, showing the elastic s-channel double spectral functions. The boundaries are given by (12.10), and are asymptotic to $s = 4m_\pi^2$, and $t(\text{or } u) = 16m_\pi^2$. They are the complete double spectral functions only up to $s = 16m_\pi^2$.

applications which we will discuss in Chapter VI. Equation (12.11) was calculated on the assumption that there was no u-channel discontinuity. If we return to (12.6) and carry out a similar analysis we find

$$\varrho_{st}^{\text{sel}}(s, t) = \frac{1}{16\pi^2 q_s \sqrt{s}} \int\limits_{t_0}^{K=0}\!\!\int dt_1\, dt_2\, \frac{[D_t(s_+, t_1)\, D_t(s_-, t_2) + D_u(s_+, t_1)\, D_u(s_-, t_2)]}{K^{\frac{1}{2}}(t, t_1, t_2, s)}$$

and (12.12)

$$\varrho_{su}^{\text{sel}}(s, t) = \frac{1}{16\pi^2 q_s \sqrt{s}} \int\limits_{t_0}^{K=0}\!\!\int dt_1\, dt_2\, \frac{[D_t(s_+, t_1)\, D_u(s_-, t_2) + D_u(s_+, t_1)\, D_t(s_-, t_2)]}{K^{\frac{1}{2}}(t, t_1, t_2, s)}\,.$$

(12.13)

These are the elastic s-channel double spectral functions, which exist in the regions shown in Fig. (I.15). If we calculate in a similar manner the

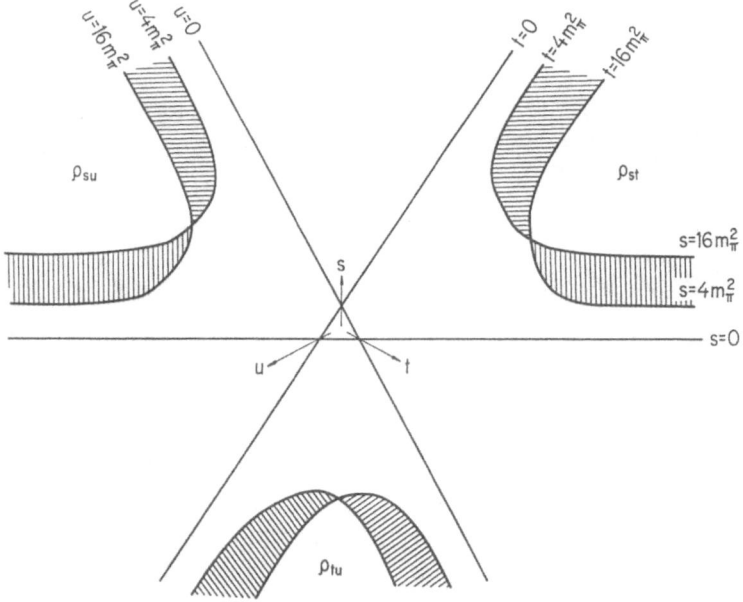

Fig. I.16. The set of elastic double spectral functions for π-π scattering. They are the full double spectral functions in the shaded regions.

elastic t- and u-channel double spectral functions, we get the complete set shown in Fig. (I.16). These are exact only for $s < s_{\text{I}}$ (and $t < t_{\text{I}}$, $u < u_{\text{I}}$), where s_{I} is the lowest s channel inelastic threshold, but of course they can still be calculated for $s > s_{\text{I}}$, and will probably be a good approximation for some greater region of s.

In the dispersion relation (10.6), used in (12.3), we neglected the possibility of bound state poles, such as occur in (10.5). If they are present we get an extra contribution

$$D_t(s, t) = \pi\, g_t\, \delta\,(t - M_f^2)$$

to put into (12.13), and the integrals over t_1 and t_2 must include the points $t_1 = M_f^2$, and $t_2 = M_f^2$. The boundary of the double spectral function then becomes

$$t = 4\,M_f^2 - \frac{(M_t)^4}{q_s^2} \equiv b\,(s) \qquad (12.14)$$

instead of (12.10).

In diagramatic terms the unitarity relation (12.1) corresponds to

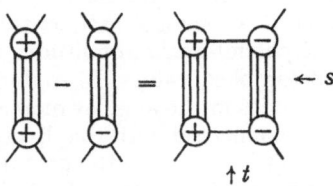

So we obtain all possible t channel exchanges, with two particle unitarity in the s channel. The ability to calculate such "ladders" is one of the most useful features of the Mandelstam representation (*Chew* and *Frautschi* 1960, 1961a; *Chew*, 1961). Of course, in order to use (12.13) we need to know the functions D_t and D_u, but in Chapter VI we shall develop an iteration scheme, based on this equation, which enables us to find both the discontinuities and the double spectral functions.

Chapter II. Partial-Wave Amplitudes and Regge Poles

II.1. Partial-Wave Amplitudes

We indicated in the previous chapter that angular momentum is likely to be important in any discussion of the subtractions required by the Mandelstam representation, because at least some of the divergences must stem from the presence of poles corresponding to particles with spin. In this chapter we shall discuss in some detail the partial-wave decomposition of the scattering amplitude. In the first instance we shall consider only integer values of angular momentum, which are of course the only physically allowed values (for bosons) given the quantization of angular momentum. But we shall subsequently find it interesting to extend our definition of partial-wave amplitudes to include unphysical non-integer, and indeed complex, values. In so doing we shall encounter singularities of the amplitudes in the angular momentum plane, "Regge" poles and cuts, which are connected with the divergences of the Mandelstam representation. In fact if we invoke the postulate of maximal analyticity of the second kind, [postulate (vi) of Section I.2] we shall find that all the divergence problems are solved, and that all the particle poles correspond to "Regge" poles, like those found in potential scattering (*Regge*, 1959, 1960). The scattering amplitude can then be represented in terms of its singularities in the angular-momentum plane. Of course all of this depends on postulate (vi) being true. Our chief justification will lie in the experimental consequences of the Regge hypothesis to be discussed in Chapter VIII

We begin, then, by defining partial-wave amplitudes. Any S-matrix element can be decomposed into "partial-waves", but we shall restrict ourselves here to the four-line connected part, where the external particles have no spin. The modifications needed when spin is included will be dealt with in Chapter IV.

The amplitude $A(s, t)$ of Section (I.6) can be written $A[s, t(z_s, s)]$, where $z_s = z_s(s, t) \equiv \cos\vartheta_s$, since we have seen, (I.6.14), that at fixed s the invariant t depends only on the cosine of the scattering angle [the same is true for u from (I.6.8)]. Then we can define a partial wave amplitude in the s-channel by

$$A_l(s) = \frac{1}{(16\pi)} \frac{1}{2} \int_{-1}^{1} dz_s \, P_l(z_s) \, A[s, t(z_s, s)] \,, \quad \text{for} \quad l = 0, 1, 2, \ldots . \quad (1.1)$$

Here $P_l(z)$ is the Legendre polynomial, of the first kind, of order l. The factor $(16\pi)^{-1}$ is introduced to simplify the two-particle unitarity condition, which we shall discuss in Section (II.5).

Because of the orthogonality of Legendre polynomials [B 1, 3.12 (10)],

$$\int_{-1}^{1} P_l(z) \, P_{l'}(z) \, dz = \frac{2}{2l + 1} \delta_{ll'} \,, \quad (1.2)$$

we can invert (1.1) and write

$$A\left[s, t(z_s, s)\right] = 16\pi \sum_{l=0}^{\infty} (2l + 1)\, A_l(s)\, P_l(z_s)\,.\tag{1.3}$$

Evidently this series cannot converge for all s and t, for $P_l(z_s)$, with l integer ≥ 0, is an entire function of z_s (that is, it is holomorphic, free of singularities for finite z_s), so $A(s, t)$, as defined by (1.3), can have no

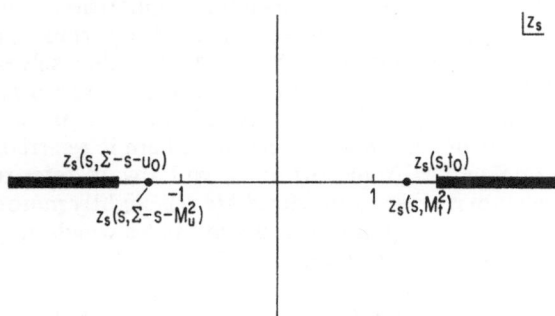

Fig. II.1. The z_s plane singularities corresponding to Fig. I.10.

singularities in t (or u). The series will break down at the nearest t (or u) singularity. In Fig. (I.10) we showed the t-plane singularities at fixed s, and using (I.6.14) we can re-draw these in the z_s plane at fixed s, which we do in Fig. (II.1). The series (1.3) will break down at $z_s = z_s(s, M_t^2)$, or $z_s(s, \Sigma - s - M_u^2)$, whichever is the nearer. The t-channel poles and thresholds occur for positive z_s (> 1), while those of the u-channel appear at negative z_s ($<- 1$).

II.2. Amplitudes of Definite Signature

We have seen that in the z_s-plane we have right-hand singularities corresponding to the t-channel singularities, and left-hand singularities corresponding to those of the u-channel. However we shall find it more convenient (see Section II.8) to work with amplitudes which have only right-hand singularities. These are called "amplitudes of definite signature", and we construct them as follows. Let

$$A(s, t) = A^R(s, t) + A^L(s, t)\tag{2.1}$$

where $A^R(s, t)$ contains only the right-hand singularities of $A(s, t)$, and $A^L(s, t)$ the left-hand ones. From (I.10.5) we can write dispersion relations in z for these functions;

$$A^R(s, t) = \sum_{t\ \text{poles}} \frac{g_{ti}(s)}{z_s(s, t_i) - z_s(s, t)} + \frac{1}{\pi} \int_{z_s(s, t_0)}^{\infty} \frac{D_t(s, t')}{z' - z_s(s, t)}\, \mathrm{d}z'\tag{2.2}$$

$$A^L(s, t) = \sum_{\substack{i \\ u\text{-poles}}} \frac{g_{ui}(s)}{z_s(s, t_i) - z_s(s, t)} + \frac{1}{\pi} \int\limits_{z_s(s, \Sigma - s - u_0)}^{-\infty} \frac{D_u(s, t')}{z' - z_s(s, t)} \, \mathrm{d}z' \quad (2.3)$$

where $t' = t(z', s)$.

We then define amplitudes of definite signature,

$$A^\pm(s, t) = A^R[s, t(z_s, s)] \pm A^L[s, t(-z_s, s)], \qquad (2.4)$$

each of which clearly has only right-hand singularities. Note that with this definition $A^+(s, t)$ contains the even part of $A(s, t)$ in z_s, and $A^-(s, t)$ contains the odd part, though $A^+(s, t)$ and $A^-(s, t)$ themselves are neither even nor odd. [By "even" we mean that $f(x)$ is even in x if $f(x) = f(-x)$, and "odd" if $f(x) = -f(-x)$]. This use of the $\pm$ superscript is quite different from that of the previous chapter where it referred to the connected parts of S or $S^\dagger$. From now on $\pm$ will always refer to signature.

The Mandelstam representation for $A^\pm(s, t)$ is slightly more complicated than for $A(s, t)$. Neglecting any bound-state poles which may be present we have, corresponding to (I.10.6),

$$A^\pm(s, t) = \frac{1}{\pi} \int\limits_{t_0}^{\infty} \frac{D_t(s, t'')}{(t'' - t)} \, \mathrm{d}t'' \pm \frac{1}{\pi} \int\limits_{u_0}^{\infty} \frac{D_u(s, u'')}{(u'' - t)} \, \mathrm{d}u''. \qquad (2.5)$$

Note the t in the denominator of the second term which comes from replacing z_s by $-z_s$. Substituting (I.11.3) and (I.11.4) gives (remembering that $t'' + u'' + s = \Sigma$ and $s' + u' + t = \Sigma$), after some manipulation

$$A^\pm(s, t) = \frac{1}{\pi^2} \int\int\limits^{\infty} \frac{\varrho_{st}(s', t'') \pm \varrho_{su}(s', t'')}{(s' - s)(t'' - t)} \, \mathrm{d}s' \, \mathrm{d}t'' +$$

$$+ \frac{1}{\pi^2} \int\int\limits^{\infty} \frac{\varrho_{tu}(t'', u') \pm \varrho_{tu}(u', t'')}{(u' - u'')(t'' - t)} \, \mathrm{d}u' \, \mathrm{d}t''. \qquad (2.6)$$

Equation (2.5) may also be written

$$A^\pm(s, t) = \frac{1}{\pi} \int\limits_{t_0}^{\infty} \frac{D_t^\pm(s, t'')}{t' - t} \, \mathrm{d}t'', \qquad (2.7)$$

where

$$D_t^\pm(s, t) = D_t(s, t) \pm D_u(s, t)$$

$$= \frac{1}{\pi} \int \frac{\varrho_{st}(s', t)}{(s' - s)} \, \mathrm{d}s' \pm \frac{1}{\pi} \int \frac{\varrho_{su}(s', t)}{(s' - s)} \, \mathrm{d}s' + \int \frac{\varrho_{tu}(t, u')}{(u' - u)} \, \mathrm{d}u' \pm$$

$$\pm \frac{1}{\pi} \int \frac{\varrho_{tu}(u', t)}{(u' - u)} \, \mathrm{d}u'. \qquad (2.8)$$

In (2.7) we have written t_0 for the lower of t_0 and u_0. We can also define the s-discontinuity function

$$D_s^\pm(s, t) = \frac{1}{\pi} \int \frac{\varrho_{st}(s, t'') \pm \varrho_{su}(s, t'')}{(t'' - t)} \, \mathrm{d}t''. \qquad (2.9)$$

It is because we have taken amplitudes of definite signature in the s-channel that our equations cease to be symmetrical in s, t and u.

We shall now use these dispersion relations to obtain an alternative definition of the partial-wave projection (1.1).

II.3. The Froissart-Gribov Projection

Putting back possible t-channel poles into (2.7) we have the dispersion relation

$$A^{\pm}(s, t) = \sum_i \frac{g_{ti}}{M_{ti}^2 - t} \pm \sum_i \frac{g_{ui}}{M_{ui}^2 - t} + \frac{1}{\pi} \int\limits_{t_0}^{\infty} \frac{D_t^{\pm}(s, t')}{t' - t} \, dt' , \quad (3.1)$$

which. using (I.6.14), becomes

$$A^{\pm}(s, t) = \sum_{t\text{-poles}} i \frac{g_{ti}(s)}{z_s(s, M_{ti}^2) - z_s(s, t)} \pm$$

$$\pm \sum_{u\text{-poles}} i \frac{g_{ui}(s)}{z_s(s, \Sigma - s - M_{ui}^2) - z_s(s, t)} + \quad (3.2)$$

$$+ \frac{1}{\pi} \int\limits_{z_s(s, t_0)}^{\infty} \frac{D_t^{\pm}(s, t')}{z' - z_s(s, t)} \, dz' ,$$

where

$$g_{ti}(s) = \frac{g_{ti}}{2q_{s12} \, q_{s34}} , \quad (3.3\,a)$$

and

$$g_{ui}(s) = \frac{g_{ui}}{2q_{s12} \, q_{s34}} , \quad (3.3\,b)$$

and

$$z' = z_s(s, t') .$$

We now substitute (3.2) into (1.1), invert the order of integration, and use Neumann's formula [B 1, 3.6(29)]

$$Q_l(z) = -\frac{1}{2} \int\limits_{-1}^{+1} \frac{dz'}{z' - z} P_l(z') \quad (3.4)$$

to do the integral over z'. The result is

$$A_{l}^{\mp}(s) = (16\pi)^{-1} \sum \{g_{ti}(s) \, Q_l[z_s(s, m_{ti}^2)] \pm g_{ui}(s) \, Q_l[z_s(s, \Sigma - s - M_{ui}^2)]\} +$$

$$+ (16\pi)^{-1} \frac{1}{\pi} \int\limits_{z_s(s, t_0)}^{\infty} D_{l}^{\mp}(s, t') \, Q_l(z') \, dz' , \quad (3.5)$$

provided the integral converges (otherwise we would not be able to invert the order of integration). This equation is known as the Froissart-Gribov projection (*Froissart*, 1961; *Gribov*, 1961).

So (neglecting the poles for ease of writing) we have two expressions for the partial wave amplitudes of definite signature,

$$A_l^\pm(s) = \frac{1}{16\pi^2} \int_{z_s(s,\,t_0)}^{\infty} D_l^\pm(s,t')\, Q_l(z')\, \mathrm{d}z' , \quad \text{from (3.5)} , \qquad (3.6)$$

and

$$A_l^\pm(s) = \frac{1}{32\pi} \int_{-1}^{1} P_l(z')\, A^\pm(s,t')\, \mathrm{d}z', \quad \text{from (1.1)} , \qquad (3.7)$$

both for $l = 0, 1, 2, \ldots$. Now $D_l^\pm(s,t)$ is the discontinuity in t of $A^\pm(s,t)$, which exists only for $z_s > z_s(s,t_0)$, while the discontinuity of $Q_l(z)$ is, from (3.4),

$$\mathrm{Im}\{Q_l(z)\} = -\frac{\pi}{2} P_l(z), \quad -1 < z < 1 , \qquad (3.8)$$
$$= 0 , \quad z < -1, \quad \text{and} \quad z > 1, \quad l = 0, 1, 2, \ldots .$$

Hence we can combine (3.6) and (3.7) in the single expression

$$A_l^\pm(s) = \frac{1}{32\pi^2 i} \int_{c_1 \text{ or } c_2} \mathrm{d}z'\, Q_l(z')\, A^\pm(s,t') , \qquad (3.9)$$

where the contours c_1 and c_2 are shown in Fig. (II.2). The partial wave series for these amplitudes of definite signature, corresponding to (1.3), is

$$A^\pm(s,t) = (16\pi) \sum_{l=0}^{\infty} (2l+1)\, A_l^\pm(s)\, P_l(z_s) . \qquad (3.10)$$

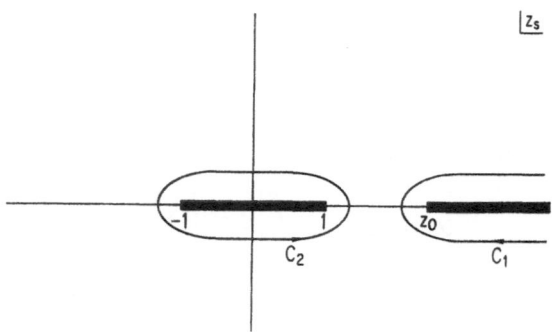

Fig. II.2. The integration contours in the complex z_s-plane, corresponding to (3.9).

We noted in Section (II.2) that $A^+(s,t)$ contains the even part of $A(s,t)$ in z_s, and A^- the odd part. Now [B 1, 3.2 (10)]

$$P_l(-z_s) = (-1)^l P_l(z_s) , \quad \text{for } l \text{ integer} , \qquad (3.11)$$

so $P_l(z)$ is even or odd according as l is even or odd.

Hence

$$A_l^+(s) = A_l(s) , \quad \text{for even } l , \tag{3.12}$$

and

$$A_l^-(s) = A_l(s) , \quad \text{for odd } l . \tag{3.13}$$

We shall often find it convenient to refer to these as the "physical" values of l for the partial-wave amplitudes of definite signature.

II.4. The Singularities of Partial-Wave Amplitudes

It is evident from (3.9) that $A^\pm(s)$ will have the same thresholds in s as $A^\pm(s, t)$, corresponding to all the channels which communicate with particles $1 + 2$, as in Fig. (I.7). It will also have those s-channel bound-state or resonance poles which have spin l. Thus in the s plane $A_l^+(s)$ has the same right-hand singularities as $A^\pm(s, t)$ except that it will not necessarily have all the poles.

In addition to these there will be a set of left-hand singularities generated by the pinching of the t- or u-channel singularities with the branch points of $Q_l(z)$ at $z = \pm 1$. For example, if $A^\pm(s, t)$ has a t-channel bound-state pole at $z_s = z_s(s, t_p)$, so that

$$D_l^\pm(s, t) = g_p(s) \, \pi \, \delta[z_s - z_s(s, t_p)] , \tag{4.1}$$

we get from (3.6) [c.f. (3.5)]

$$A_l^\pm(s) = \frac{1}{16\pi} g_p(s) \, Q_l[z_s(s, t_p)] , \tag{4.2}$$

so that $A_l^\pm(s)$ has branch points at $z_s(s, t_p) = \pm 1$. More generally, for any singularity of $A^\pm(s, t)$ at $t = t_n$, we have a branch point in $A_l^\pm(s)$ at [from (I.6.14)]

$$\frac{s^2 + s(2t_n - \Sigma) + (m_1^2 - m_2^2)(m_3^2 - m_4^2)}{4s \, q_{s12} \, q_{s34}} = \pm 1 . \tag{4.3}$$

Let us examine two simple examples.

For $\pi - \pi$ scattering we know that $A_l^\pm(s)$ will have right-hand branch points at the threshold for two pions ($4m_\pi^2$), four pions ($16m_\pi^2$), and so on. The t-channel singularities at t_n will give branch points at

$$\pm 1 = 1 + \frac{2t_n}{s - 4m_\pi^2} . \tag{4.4}$$

[Remember that amplitudes of definite signature have no u-channel singularities, since they have been "folded over" into the t channel by Eq. (2.4).] These are at $s = \infty$ and $s = 4m_\pi^2 - t_n$. The t channel singularities are of course the same as s channel ones, the various thresholds, so we have $t_n = 4m_\pi^2, 16m_\pi^2, \ldots$. However the $\pi - \pi$ amplitude also contains resonance poles such as the ρ pole with spin 1 at

$$s = m_\rho^2 \approx (29 - i \cdot 0.8) \, m_\pi^2 .$$

This gives a right-hand pole to $A_1^\pm(s)$ only. But the t-channel ρ gives left-hand branch points at $s = \infty$, and $s = 4m_\pi^2 - m_\rho^2$, to $A_l^\pm(s)$, for all l. It is usual to draw cuts from the right-hand branch points along the positive real axis, as for $A^\pm(s, t)$, and from the left-hand branch points along the negative real axis, as shown in Fig. (II.3).

This simplification of equal mass kinematics is not typical. In general there are 4 solutions to (4.3), one at $s = 0$, one at $s = \infty$, and two others. Thus in $\pi - N$ scattering we get, from the N exchange pole at $t = M_N^2$, branch points at $s = -\infty$, 0, $s = (M_N - m_\pi^2/M_N)^2$, and $s = M_N^2 + 2m_\pi^2$, and these are usually connected as shown in Fig. (II.4), so that there is an additional "short cut". If the pion were of equal mass to the nucleon, then the left-hand cut would be continuous from $s = 3M_N^2$ to $-\infty$. Further details of the πN singularities are given by e.g. *Hamilton*, 1964.

With the cuts drawn in this way, we can write dispersion relations for the partial-waves

$$A_l^\pm(s) = \frac{1}{\pi} \int\limits_{\text{L. H. cut}} \frac{\text{Im}[A_l^\pm(s')]}{s' - s}\,ds' + \frac{1}{\pi} \int\limits_{\text{R. H. cut}} \frac{\text{Im}[A_l^\pm(s')]}{s' - s}\,ds'. \quad (4.5)$$

If we restrict ourselves to the "physical" values of l for each signature, we find that the left-hand cut discontinuity is given by the discontinuity of $Q_l(z)$ in Eq. (3.6), [see (3.8)] so, remembering that $D_l^\pm(s, t)$ is zero for $z_s < z_s(s, t_0)$,

$$\text{Im}[A_l^\pm(s)] = \frac{1}{32\pi} \int\limits_{-1}^{z_s(s, t_0)} P_l(z')\,D_l^\pm(s\,t')\,dz', \quad \text{on L.H.cut.} \quad (4.6)$$

Because of the extra pieces of double spectral function [see (2.6)] in amplitudes of definite signature, this is not in general the only contribution to the left-hand cut discontinuity. We shall discuss this in detail in Section (II.11). However, for the physical values of l, (4.6) is correct.

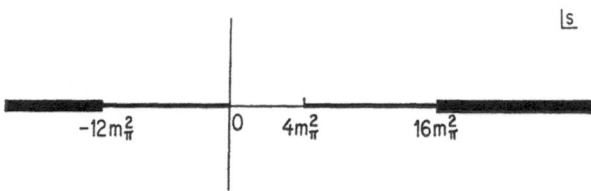

Fig. II.3. Cuts of the π-π partial-wave amplitudes.

On the right the discontinuities will be those of $A^\pm(s, t)$, so from Eq. (3.7) we get

$$\text{Im}[A_l^\pm(s)] = \frac{1}{32\pi} \int\limits_{-1}^{1} P_l(z')\,D_s^\pm(s, t')\,dz', \quad \text{on R.H.cut.} \quad (4.7)$$

The partial-wave projection introduces a kinematical zero in the amplitude at the threshold. From (I.6.14) and (I.6.12) we see that $z_s \to \infty$ when q_{s12} or $q_{s34} \to 0$. This happens at the thresholds of the two channels, $s = s_a \equiv (m_1 + m_2)^2$ and $s = s_b \equiv (m_3 + m_4)^2$, as well as at the unphysical

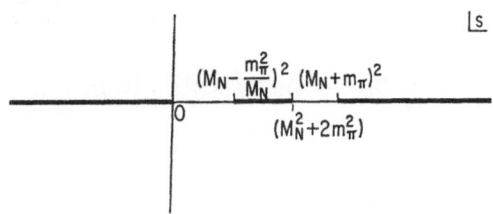

Fig. II.4. Cuts of the π-N partial-wave amplitudes due to the nucleon exchange pole.

points $s = (m_1 - m_2)^2$ and $s = (m_3 - m_4)^2$. We defer discussion of the latter points, which occur only if the masses of the scattering particles are unequal, until Chapter III, but the threshold behaviour is of more immediate concern.

Now [B 1, 3.9 (21)]

$$Q(z) \underset{z \to \infty}{\sim} z^{-(l+1)} \tag{4.8}$$

so

$$Q_l(z) \sim [q_{s12}(s)\, q_{s34}(s)]^{l+1}, \quad \text{as } s \to s_a \text{ or } s_b.$$

Hence the threshold behaviour of the partial wave amplitudes will be [note that there is another factor $(q_{s12}\, q_{s34})$ from $\mathrm{d}z' = \mathrm{d}t'/(2 q_{s12}\, q_{s34})$ in (3.6)]

$$A_l(s) \sim [q_{s12}(s)\, q_{s34}(s)]^l, \quad \text{as } s \to s_a \text{ or } s_b. \tag{4.9}$$

if the integral in (3.6) converges, i.e. if the Froissart-Gribov projection is defined. Where it is not defined one can not prove that this will be the behaviour, but there is a general feeling that it will always be of this form (*Chew*, 1965), as it is in potential scattering with reasonably behaved potentials.

If $s_a \neq s_b$, i.e. $(m_1 + m_2) \neq (m_3 + m_4)$, these zeros correspond to branch points for odd l, and the dispersion relation (4.5) is no longer valid. We shall show how to correct for this in the next section.

II.5. Unitarity of Partial-Wave Amplitudes and Phase Shifts

It is very convenient that the unitarity relations for a partial wave amplitude are simpler than those for the amplitude as a whole. For example if we take the two-particle discontinuity formula (I.9.6), and substitute the partial wave series (1.3) for each of the amplitudes, we obtain

$$16\pi \sum_l (2l+1) \left[A_l^{ab}(s_+) - A_l^{ab}(s_-) \right] P(z_s) = 2i \frac{q_{sn}}{32\pi^2 \sqrt{s}} \times$$

$$\times (16\pi)^2 \int_0^{2\pi} d\phi \int_{-1}^{1} dz' \sum_{l'} (2l'+1) A_{l'}^{bn}(s_+) P_{l'}(z') \times$$

$$\times \sum_{l''} (2l''+1) A_{l''}^{na}(s_-) P_{l''}(z'') . \tag{5.1}$$

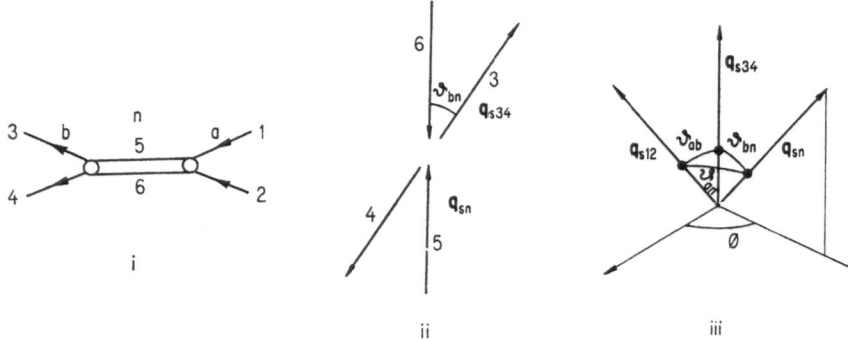

Fig. II.5. (i) The amplitude $1 + 2 \to 3 + 4$ via intermediate state $5 + 6$, i.e. $a \to b$ via n. (ii) The centre of mass 3-momenta, and the scattering angle ϑ_{bn}. (iii) The relation between the three angles ϑ_{ab}, ϑ_{bn}, and ϑ_{an}. ϕ is measured about the axis q_{s34}, with its origin in the plane containing q_{s34} and q_{s12}.

Here $z' = \cos\vartheta_{bn}$ is the cosine of the angle between the direction of motion of the particles in state b and that in state n, in the centre of mass system, while $z'' = \cos\vartheta_{an}$ is the cosine of the angle between the direction in n and in a; see Fig. (II.5). The addition theorem for cosines gives the relation

$$\cos\vartheta_{an} = \cos\vartheta_{ab} \cos\vartheta_{bn} + \sin\vartheta_{ab} \sin\vartheta_{bn} \cos\phi ,$$

or

$$z'' = z_s z' + \sqrt{1 - z_s^2} \sqrt{1 - z'^2} \cos\phi , \tag{5.2}$$

and the addition theorem for Legendre functions [B 1, 3.11 (1)] gives

$$P_l(z'') = P_l(z_s) P_l(z') + 2 \sum_{m=1}^{l} (-1)^m \frac{\Gamma(l-m+1)}{\Gamma(l+m+1)} P_l^m(z_s) \times \tag{5.3}$$

$$\times P_l^m(z') \cos m\phi ,$$

where $P_l^m(z_s)$ is the associated Legendre function of the first kind. Using the orthogonality relations (1.2), and [B 1, 3.12 (19) and 3.12 (21)] we get

$$\int_0^{2\pi} d\phi \int_{-1}^{1} dz' P_{l'}(z') P_{l''}(z'') = \delta_{l'l''} \frac{4\pi}{2l'+1} P_{l'}(z_s) , \tag{5.4}$$

which, substituted in (5.1), gives, with much cancellation,

$$A_l^{ab}(s_+) - A_l^{ab}(s_-) = 4i\,\frac{q_{sn}}{\sqrt{s}}\,A_l^{bn}(s_+)\,A_l^{na}(s_-)\,,\tag{5.5}$$

or

$$\Delta_n[A_l^{ab}(s)] = \frac{2q_{sn}}{\sqrt{s}}\,A_l^{bn}(s_+)\,A_l^{na}(s_-)\tag{5.6}$$

where Δ_n is the discontinuity across the cut associated with the two-particle state n, divided by 2. The absence of the factor 16π stems from our choice of normalization in Eq. (1.1).

In particular, for elastic unitarity

(5.6) becomes

$$\operatorname{Im}[A_l^{aa}(s)] = \frac{2q_{s12}}{\sqrt{s}}\,|A_l^{aa}(s)|^2 = \varrho^a(s)\,|A_l^{aa}(s)|^2\,,\quad \text{for}\quad s < s_I\,,\tag{5.7}$$

where $\varrho^a(s)$ is the two-particle phase-space factor for channel a, $\varrho^a(s) = q_{s12}/\sqrt{s}$, and s_I is the lowest inelastic threshold. Note that, from (I.6.12), $\varrho^a(s) \leqq 1$ for all s, so that

$$|A_l^{aa}(s)| \leqq 1\,.\tag{5.8}$$

In non-relativistic potential scattering (see, for example, Wu and $Ohmura$, 1962) we often write the partial wave amplitude f_l in the form

$$f_l(s) = \frac{e^{i\delta_l(s)}\sin\delta_l(s)}{q_s}\,,\tag{5.9}$$

where $\delta_l(s)$ is the "phase shift". With $\delta_l(s)$ real, this equation embodies non-relativistic unitarity in that

$$\operatorname{Im}\{f_l(s)\} = \frac{\sin^2\delta_l(s)}{q_s} = q_s\,|f_l(s)|^2\,.\tag{5.10}$$

By comparison we write

$$A_l^{aa}(s) = \frac{e^{i\delta_l(s)}\sin\delta_l(s)}{\varrho^a(s)} = \frac{e^{2i\delta_l(s)}-1}{2\,i\,\varrho^a(s)}\,,\tag{5.11}$$

which defines a relativistic phase shift, and embodies the unitarity relation (5.7). The extra factor $\sqrt{s}$, (and the implicit 16π), come from our use of the invariant normalization (I.2.7).

If many channels are open (5.6) gives

$$\operatorname{Im}\{A_l^{ab}(s)\} = \sum_n \varrho^n(s)\,A_l^{bn}(s_+)\,A_l^{na}(s_-)$$
$$+ \text{(many particle channels)}\tag{5.12}$$

where the sum is over all the two particle channels n which are open. This unitarity condition can only be satisfied if $\delta_l(s)$ in (5.11) is non-real.

The optical theorem (I.7.17) gives

$$\sigma_{a\to \text{all}}^{\text{tot}}(s) = \frac{1}{2a_{s12}\sqrt{s}}\operatorname{Im}\{A^{aa}(s,0)\}\,,\tag{5.13}$$

which, using (1.3), becomes

$$\sigma^{\text{tot}}(s) = \frac{8\pi}{q_{s12}\sqrt{s}} \sum_l (2l+1)\,\text{Im}\{A_l^{aa}(s)\}, \tag{5.14}$$

since $P_l(\cos\vartheta = 1) = 1$ for all l. So we can define the partial-wave total cross-section to be

$$\sigma_l^{\text{tot}}(s) = \frac{8\pi}{q_{s12}\sqrt{s}} (2l+1)\,\text{Im}\{A_l^{aa}(s)\}. \tag{5.15}$$

On the other hand the partial-wave elastic cross section, obtained by substituting (1.3) in (I.7.9) and using the relation (1.2), is

$$\sigma_l^{\text{elaa}}(s) = (2l+1)\frac{16\pi}{s}|A_l^{aa}(s)|^2. \tag{5.16}$$

Below the inelastic thresholds (5.7) in (5.15) gives $\sigma^{\text{tot}} = \sigma^{\text{el}}$, as it should.

We have noted that making the partial-wave projection introduces kinematical zeros and sometimes branchpoints over and above the dynamical singularities which were discussed in the previous section. It is often desirable to remove these singularities by defining a "reduced" partial-wave amplitude

$$B_l^{ab}(s) = \frac{A_l^{ab}(s)}{(q_{s12}\,q_{s34})^l}. \tag{5.17}$$

For such a function the unitarity condition (5.12) becomes

$$B_l^{ab}(s_+) - B_l^{ab}(s_-) = 2i\sum_n \varrho_l^n(s)\,B_l^{bn}(s_+)\,B_l^{na}(s_-) + $$
$$+ \text{(many particle channels)}. \tag{5.18}$$

where

$$\varrho_l^n(s) = \frac{2q_n^{2l+1}(s)}{\sqrt{s}}. \tag{5.19}$$

Comparing this with equation (5.11) we can also write

$$B_l^{aa}(s) = \frac{e^{i\delta_l(s)}\sin\delta_l(s)}{\varrho_l^n(s)}. \tag{5.20}$$

Dispersion relations of the form (4.5) can be written for $B_l(s)$ without the kinematical problems which were noted at the end of section (II.4). This is especially important when we relax the requirement that l be integral, as we shall subsequently, for then the threshold behaviour always introduces unwanted branch points in $A_l(s)$, but not $B_l(s)$.

II.6. Asymptotic Behaviour and the Froissart-Gribov-Projection

We saw in section (I.11) that the principal difficulty in using the Mandelstam representation is the need to make subtractions before the integrals can be defined. We expect that an equation such as (3.2) (neglecting any bound-state poles for ease of writing)

$$A^{\pm}(s,t) = \frac{1}{\pi}\int_{z_s(s,t_0)}^{\infty} \frac{D_t^{\pm}(s,t')}{z'-z_s}\,\mathrm{d}z' \tag{6.1}$$

will be undefined as it stands for many values of s. But if $D_{\bar{l}}^{\pm}(s, t)$ is power bounded, i.e.

$$D_{\bar{l}}^{\pm}(s, t) \underset{z_s \to \infty}{\sim} z_s^{N(s) - \varepsilon(s)}, \quad 0 < \varepsilon(s) < 1, N(s) \text{ integer}, \quad (6.2)$$

we can make N subtractions at $z = 0$ [c.f. (I.10.8)] giving

$$A^{\pm}(s, t) = F^{N-1}(s, z_s) + \frac{z_s^N}{\pi} \int_{z_s(s, t_0)}^{\infty} \frac{D_{\bar{l}}^{\pm}(s, t')}{(z' - z_s) z'^N} \, dz' \quad (6.3)$$

where $F^{N-1}(s, z_s)$ is a polynomial fo degree $N - 1$ in z_s, and the integral now converges. Then from (1.1)

$$A_{\bar{l}}^{\pm}(s) = \frac{1}{32\pi} \int_{-1}^{1} dz_s \, P_l(z_s) \left[F^{N-1}(s, z_s) + \frac{z_s^N}{\pi} \int_{z_s(s, t_0)}^{\infty} \frac{D_{\bar{l}}^{\pm}(s, t')}{(z' - z_s) z'^N} \, dz' \right], \quad (6.4)$$

$$\text{for } l = 0, 1, 2, \ldots \ldots$$

Now $\frac{1}{2} \int_{-1}^{1} dz_s \, P_l(z_s) z_s^M = 0$, for $M < l$, by the orthogonality relation (1.2), and putting

$$\left(\frac{z_s}{z'} \right)^N = \left[1 + \left(\frac{z_s - z'}{z'} \right) \right]^N,$$

and expanding in powers of $(z_s - z')/z'$, we get

$$\frac{1}{2} \int_{-1}^{1} dz_s \, P_l(z_s) \left(\frac{z_s}{z'} \right)^N \frac{1}{(z' - z_s)} = \frac{1}{2} \int_{-1}^{1} dz_s \frac{P_l(z_s)}{z' - z_s} +$$

$$+ \frac{1}{2} \int_{-1}^{1} dz_s \, P_l(z_s) \left[-\frac{z_s^{N-1}}{z'^N} + \cdots \right]$$

$$= Q_l(z') \quad \text{for} \quad l \geqq N(s).$$

So

$$A_{\bar{l}}^{\pm}(s) = \frac{1}{16\pi} \cdot \frac{1}{\pi} \int_{z_s(s, t_0)}^{\infty} Q_l(z') \, D_{\bar{l}}^{\pm}(s, t') \, dz' \quad \text{for} \quad l \geqq N(s). \quad (6.5)$$

This integral exists because of the asymptotic behaviour of $Q_s(z)$; see equation (4.8).

So, providing the Mandelstam representation is power bounded, the higher partial waves are always completely determined by $D_{\bar{l}}^{\pm}$ (or the double spectral functions). Only the lower ones contain arbitrary subtractions.

II.7. The Froissart Bound

In fact unitarity puts limits on $N(s)$ which are highly restrictive for some regions of s, (*Froissart*, 1961c). (Our presentation here follows that of *Chew*, 1965a).

From [B 1, 3.9(1)] (see also *Squires*, 1962)

$$Q_l(z) \underset{l \to \infty}{\sim} \frac{1}{\sqrt{l}} e^{-[(l+\frac{1}{2})\,\xi\,(z)]}, \tag{7.1}$$

where

$$\xi(z) = \log(z + \sqrt{z^2 - 1}), \tag{7.2}$$

so if the integral (6.5) converges it is only the lowest values of z' in the integrand which dominate the higher partial-waves. In other words the

Fig. II.6. Peripheral interactions. (i) Projectile particle passing at range R from the target has angular momentum $q_s R$. (ii) Peripheral N-N interaction produced by π exchange.

higher partial waves are controlled by the nearest t (or u) singularities, the longest range forces: the interaction is "peripheral". Also, given that the nearest t (or u) singularity is at z_n (remember $z_n > 1$, see Fig. II.1), the higher partial-waves, those with

$$l > l_{\text{eff}} \equiv \frac{1}{\xi(z_n)},$$

will be much smaller than those with $l < l_{\text{eff}}$. In fact we can define the range of the force, R, such that (see Fig. II.6)

$$R \cdot q_s = l_{\text{eff}} \quad \text{as} \quad s \to \infty,$$

so that

$$R = \lim_{s \to \infty} \left[\frac{1}{q_s\,\xi(z_n)} \right]. \tag{7.3}$$

For example in $N - N$ scattering the longest range force is π exchange, so $t_n = m_\pi^2$, and, (I.6.16), $z_n = (1 + m_\pi^2/2q_s^2)$, giving $R = 1/m_\pi$; or, remembering that in our units $\hbar = c = 1$, the range of the force is the pion Compton wavelength, $R = \hbar/m_\pi c$.

So, provided the amplitude is power bounded, we find, combining (6.5) and (7.1),

$$A_t^{\mp}(s) \underset{l \to \infty}{\longrightarrow} f(s)\, e^{-l\xi(z_n)}, \tag{7.4}$$

where $f(s)$ is some function of s. (Note that we require $t_n > 1$. This will be true for all $s > s_0$, which is all that we need here.) Then with (7.3) we have

$$A_t^{\mp}(s) \underset{l,s \to \infty}{\longrightarrow} f(s)\, e^{-l/Rq_s} \to f(s)\, e^{-2l/R\sqrt{s}}, \tag{7.5}$$

since $q_s \underset{s \to \infty}{\longrightarrow} \sqrt{s}/2$, or

$$A_t^{\mp}(s) \underset{l,s \to \infty}{\longrightarrow} \exp\left\{ \frac{2l}{R\sqrt{s}} + \log[f(s)] \right\}. \tag{7.6}$$

Thus for large s the partial-wave series (1.3) can be truncated at

$$l \simeq \sqrt{s}\, R \log\left[f(s) \right] = c \sqrt{s} \log s$$

where c is some constant. We have seen [equation (5.8)] that, for all l and s, $|A_l(s)| < 1$, so in the forward direction, using (1.3) and the fact that $P_l(1) = 1$ for all l,

$$\left| A^{\pm}\left(s, t(z_s = 1, s) \right) \right| \underset{l \to \infty}{<} \text{const.} \sum_{l=0}^{c \sqrt{s}\, \log s} (2l + 1) \,,$$

which, on summing the arithmetic progression, gives

$$\left| A^{\pm}\left(s, t(z_s = 1, s) \right) \right| < \text{const.} \; s \; \log^2 s \,. \tag{7.7}$$

In non-forward directions [B 1, 3.9 (2), *Squires*, 1962)]

$$P_l(z_s) \xrightarrow[l \to \infty]{} \frac{g(z_s)}{\sqrt{l}} \,, \quad \text{for} \quad -1 < z_s < 1 \,,$$

with $g(z_s) < 1$, and so

$$\left| A^{\pm}(s, t) \right| \underset{s \to \infty}{<} \text{const.} \sum_{l=0}^{c \sqrt{s}\, \log s} \frac{(2l + 1)}{\sqrt{l}} < \text{const.} \; s^{\frac{3}{4}} \log^{\frac{3}{2}} s \,. \tag{7.8}$$

The presence of the logarithmic factors is not understood, and in the range of energies available to us for experiments it is very difficult to distinguish a logarithmic dependence on s. On the assumption that we can neglect such factors we obtain the result that, (7.7), (7.8),

$$A^{\pm}(s, t) = O\left(s^{N(t)}\right) \,, \tag{7.9}$$

with $N(t) \le 1$, for $t \le 0$.

Then using crossing we can make the same statement about the large t (i.e. z_s) behaviour at negative s, namely

$$A^{\pm}(s, t) = O\left(z_s^{N(s)}\right) \,, \quad \text{with} \quad N(s) \le 1 \quad \text{for} \quad s \le 0 \,. \tag{7.10}$$

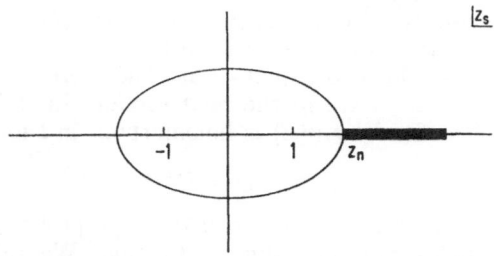

Fig. II.7. The Lehman ellipse. The region of convergence of the partial-wave series is bounded by z_n, the nearest singularity.

Thus we see that, at least for negative s, all the partial-waves except possibly the S- and P-waves are determined by $D_t^{\pm}(s, t)$ in equation (6.1). Only two arbitrary subtractions are possible in (6.1).

We may also note that (7.9), with the optical theorem (5.13), gives

$$\sigma^{\text{tot}}(s) \sim s^{N(0)-1} \le \text{const.}$$

In Section (II.1) we mentioned that the partial-wave series (1.3) must break down at the t (or u) singularity nearest to the s channel physical region. We can now appreciate better why this is so. The asymptotic form of $P_l(z_s)$ for $\mathrm{Re}\{l\} \to \infty$ satisfies [B 1, 3.9(2), *Squires*, 1962].

$$|P_l(z_s)| < \frac{1}{\sqrt{l}} \exp\{|\mathrm{Im}\, l \cdot \mathrm{Re}\,\vartheta + \mathrm{Re}\, l \cdot \mathrm{Im}\,\vartheta|\} f(z_s),$$

$$0 < \mathrm{Re}\{\vartheta\} < \pi.$$

(7.11)

Combining (7.11) with (7.4) we see that the partial-wave series will converge if

$$|\mathrm{Im}\,\vartheta| \le \xi(z_n) = \cosh^{-1}(z_n).$$

This defines an ellipse in the complex z_s plane, with foci at $z = \pm 1$, and semi-major radius z_n (*Lehmann*, 1958); see Fig. (II.7).

II.8. Analytic Continuation in Angular Momentum

The equation (6.5) can be used to define $A_{\bar{l}}^{\pm}(s)$ for all values of l, even if they are not integers as we have been assuming so far. The only singularities in l of $Q_l(z)$ are poles at the negative integers [B 1, 3.3(3)], so the function $A_{\bar{l}}^{\pm}(s)$ defined in this way is holomorphic in l for $\mathrm{Re}\{l\} > l_M(s)$, where $l_M(s)$ is the power of the divergence of $D_{\bar{l}}^{\pm}(s, z_s)$ in (6.2). And we know, (7.10), that for $s \le 0$ we have $l_M(s) \le 1$. Hence (6.5) defines, for all l such that $\mathrm{Re}\{l\} \ge l_M(s)$, an analytic continuation in l, of a function which, in order to emphasize the functional dependence, we shall denote by $A^{\pm}(s, l)$. The reader is asked to beware confusing this with $A^{\pm}(s, t)$. Subsequently we shall revert to the $A_{\bar{l}}^{\pm}(s)$ notation, even for non-integer l.

It is not immediately apparent that there is any merit in this particular continuation since only $A^{\pm}(s, l)$ for integer l has any physical significance, and it might seem that any interpolation between the integers that we cared to think of would be of equal value. The usefulness of this continuation lies, as we shall see in the next section, in the fact that the function $A^{\pm}(s, l)$ defined by (6.5) is holomorphic in l for $\mathrm{Re}\{l\} > l_M(s)$ and, (7.4),

$$A^{\pm}(s, l) = O\left[e^{-l\xi(z_n)}\right].$$

(8.1)

That it is the unique continuation with these properties follows from a theorem on complex variables due to Carlson. We refer the reader to *Titchmarsh*, 1939, p. 186, for the proof of this theorem, and simply quote:

Carlson's Theorem

If $f(z)$ is regular and of the form $O(e^{k|z|})$, where $k < \pi$, for $\mathrm{Re}\{z\} \ge 0$, and $f(z) = 0$ for $z = 0, 1, 2, 3, \ldots$, then $f(z) = 0$ identically.

So Carlson's Theorem tells us that $A^{\pm}(s, l)$ for all $\mathrm{Re}\{l\} > l_M(s)$ is completely determined by its values at the integers. For example if we try to add to values of $A^{\pm}(s, l)$ obtained from (6.5) $[\equiv A^{FG}(s, l)$ say] some function which vanishes at the integers, such as

$$A^{\pm}(s, l) = A^{FG}(s, l) + f(s) \sin \pi l \tag{8.2}$$

this new function oscillates as $l \to \infty$, and does not satisfy the requirement

$$A^{\pm}(s, l) = O[e^{|l|k}], \quad k < \pi. \tag{8.3}$$

If we insist on the function being well behaved as $l \to \infty$, then our interpolation is unique.

It was the desire to have a suitable behaviour as $l \to \infty$, which led us to introduce the amplitudes of definite signature, $A_l^{\pm}(s)$ in section (II.2) in place of $A_l(s)$. Were we to substitute (2.3) in (1.1) we should find, unlike (3.5), a term involving the integral of $Q_l(z')$ for negative z'. Now [B 1, 3.3(12)],

$$Q_l(-z) = e^{-i\pi l} Q_l(z) \tag{8.4}$$

so

$$A(s, l) = e^{-i\pi l} O[e^{-l\xi(z_n)}] \tag{8.5}$$

and does not satisfy (8.3). This can equally well be seen by including a term with $z_n < -1$ in (7.1) above.

Also we can see that the alternative definition of partial-wave amplitudes, (1.1), is no good for interpolation in l because $P_l(z)$ is not suitably behaved as $|l| \to \infty$, (7.11).

There is thus a close connection between the divergences of the Mandelstam representation and our ability to continue the scattering amplitudes of definite signature, $A^{\pm}(s, l)$, in angular momentum. But in order to examine this connection more closely we need a representation of $A^{\pm}(s, t)$ in terms of $A^{\pm}(s, l)$ and z_s which is valid outside the region of convergence of the partial wave series, i.e. outside the Lehmann ellipse. Such a representation can be obtained by using the Sommerfeld-Watson transform.

II.9. The Sommerfeld-Watson Transform

Let us suppose that $A^{\pm}(s, l)$ is an analytic function of l throughout the right-half l-plane with only isolated singularities, so that we can continue it even below $\text{Re}\{l\} = l_M(s)$ and still obtain the physical values at integer l. This is evidently a drastic assumption which we shall have to try to justify later. If we adopt it, we can replace the partial wave series (1.3) by a contour integration in the complex l-plane, as shown in Fig. (II.8) (a method used by *Sommerfeld*, 1949, following a technique of *Watson*, 1918).

$$A^{\pm}(s, t) = -\frac{16\pi}{2i} \int_{C_0} (2l + 1) \, A^{\pm}(s, l) \frac{P_l(-z_s)}{\sin \pi l} \, dl. \tag{9.1}$$

The contour C_0 is chosen to include the positive integers and zero, but to avoid any singularities of $A^{\pm}(s, l)$. The integrand has a pole at each integer, n, when $\sin(\pi l) \xrightarrow[l \to n]{} (-1)^n (l - n) \pi$. The residue of the pole is thus

$$\frac{2\pi i \, P_n(-z_s) \, A^\pm(s, n) \, (2n + 1)}{(-1)^n \, \pi} = 2i \, P_n(z_s) \, A^\pm(s, n) \, (2n + 1) \,, \quad (9.2)$$

using the fact [B 1, 3.3(10)] that $P_n(-z_s) = (-1)^n P_n(z_s)$. (This was of course our reason for taking $-z_s$ as the argument of P_l in (9.1)].

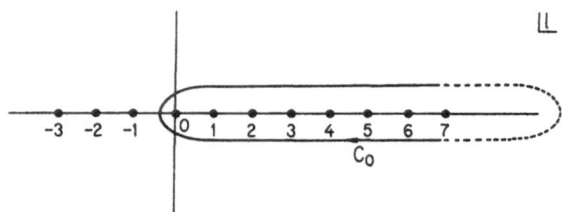

Fig. II.8. A contour, C_0, in the complex angular momentum plane.

So Cauchy's Theorem gives (note that the clockwise path of integration gives a negative sign)

$$A^\pm(s, t) = 16\pi \sum_{\substack{l \text{ positive} \\ \text{integer}}} (2l + 1) \, A^\pm(s, l) \, P_l(z_s)$$

as required.

Suppose now we displace the contour C_0 to C_1, as shown in Fig. (II.9), with a line parallel to the imaginary axis at $\mathrm{Re}\{l\} = L_1$, and a semi-circle at infinity. Providing $L_1 > l_M(s)$ we know that no singularities of

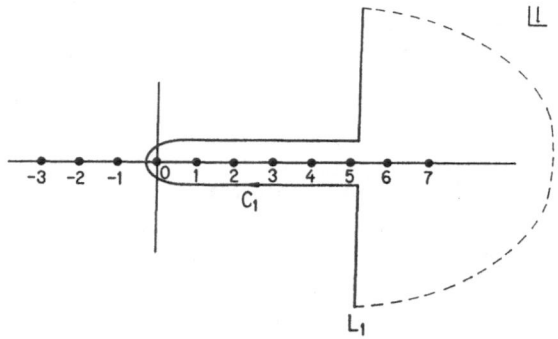

Fig. II.9. A second contour, C_1, with a semi-circle at ∞.

$A^\pm(s, l)$ will be encountered in this displacement so $\int_{C_0} = \int_{C_1}$. The contribution from the semi-circle will vanish due to (8.1), and the fact that [from (7.10)]

$$\left| \frac{P(-z_l)}{\sin \pi l} \right| < \frac{1}{\sqrt{l}} \, \frac{\exp\{|\mathrm{Im}\,\vartheta \, \mathrm{Re}\,l + (\pi - \mathrm{Re}\,\vartheta) \, \mathrm{Im}\,l|\}}{\exp\{\pi \, |\mathrm{Im}\,l|\}} \, f(z) \xrightarrow[l \to \infty]{} 0 \,. \quad (9.3)$$

This means that the representation will be valid in a larger region than
the Lehmann ellipse; in fact the region of convergence is independent
of $\mathrm{Im}\vartheta$, and includes all $z \leqq 1$. [Remember that (8.1) is strictly valid
only for $s > s_0$; see note following (7.4). We shall indicate the continuation to other values of s in Chapter III.]

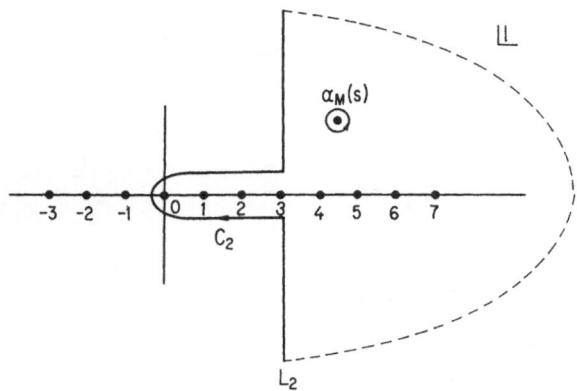

Fig. II.10. A third contour, C_2, exposing a pole at $\alpha_M(s)$.

If we now lower L we shall encounter singularities of $A^\pm(s, l)$.
Suppose that the "leading singularity" encountered (i.e. the right-most
in the complex l-plane) is a pole at $l = \alpha_M(s)$, with residue $\beta_M(s)$, so
that

$$A^\pm(s, l) \approx \frac{\beta_M(s)}{l - \alpha_M(s)} \quad \text{as} \quad l \to \alpha_M(s) . \tag{9.4}$$

Then in exposing this pole, as shown in Fig. (II.10), we get

$$A^\pm(s, t) = -\frac{16\pi}{2i} \int_{C_2} (2l + 1)\, A^\pm(s, l)\, \frac{P_l(-z_s)}{\sin\pi l}\, \mathrm{d}l - $$
$$- 16\pi^2\, [2\alpha_M(s) + 1]\, \beta_M(s)\, \frac{P_{\alpha_M(s)}(-z_s)}{|\sin\pi\,\alpha_M(s)} . \tag{9.5}$$

Let us look immediately at the asymptotic z_s behaviour of this expression. We first express $P_\alpha(z)$ in terms of hypergeometric functions
[B 1, 3.2 (23)]

$$P_\alpha(z) = \frac{\Gamma\left(-\alpha - \frac{1}{2}\right)}{\Gamma(-\alpha)}\, \frac{(2z)^{-\alpha-1}}{\pi^{\frac{1}{2}}}\, F\left(\frac{\alpha}{2} + \frac{1}{2}, \frac{\alpha}{2} + 1, \alpha + \frac{3}{2}, \frac{1}{z^2}\right) + $$
$$+ \frac{\Gamma\left(\alpha + \frac{1}{2}\right)}{\Gamma(\alpha + 1)}\, \frac{(2z)^\alpha}{\pi^{\frac{1}{2}}}\, F\left(-\frac{\alpha}{2}, -\frac{\alpha}{2} + \frac{1}{2}, -\alpha + \frac{1}{2}, \frac{1}{z^2}\right). \tag{9.6}$$

As $z \to \infty$ the hypergeometric functions tend to 1, and

$$P_\alpha(z) \underset{z \to \infty}{\sim} z^{|\alpha + \frac{1}{2}| - \frac{1}{2}}, \quad \alpha \neq \text{negative integer}. \tag{9.7}$$

So, since $\text{Re}\{\alpha_M(s)\} > L_2$ in Fig. (II.10), the second term in (9.5) will dominate asymptotically if $\text{Re}\{\alpha_M(s)\} > -\dfrac{1}{2}$, and we have

$$A^\pm(s, t) \underset{z_s \to \infty}{\sim} (z_s)^{\text{Re}\{\alpha_M(s)\}}. \tag{9.8}$$

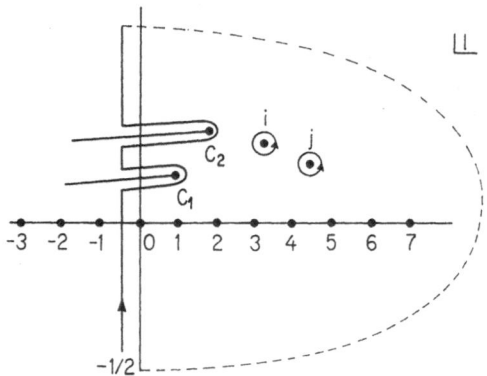

Fig. II.11. The contour for equation (9.10), showing two poles, i and j, and two branch cuts, C_1 and C_2.

Since $\alpha_M(s)$ is the right-most singularity in the l-plane, we can identify this as the asymptotic behaviour which prevents the convergence of (6.5), i.e.

$$\text{Re}\{\alpha_M(s)\} = l_M(s).$$

If on the other hand the singularity which we meet in deforming the contour is a branch point at $\alpha_c(s)$, we can draw a branch cut in the l-plane running back towards negative $\text{Re}\{l\}$, as shown in Fig. (II.11). The amplitude is then given by

$$A^\pm(s, t) = -\frac{16\pi}{2i} \int_{C_2} (2l + 1) A^\pm(s, l) \frac{P_l(-z_s)}{\sin \pi l} \, dl, \tag{9.9}$$

and its asymptotic behaviour will be

$$A^\pm(s, t) \underset{z_s \to \infty}{\sim} (z_s)^{\text{Re}\{\alpha_c(s)\}}$$

apart from logarithmic factors.

   As far as we know, the only singularities in the l-plane are likely to be poles and branch points. Because of (9.7) we can minimize the contribution of the vertical path of integration by moving the contour back to $\mathrm{Re}\{l\} = -\frac{1}{2}$ [Fig. (II.11)], and obtain

$$
A^{\pm}(s, t) = -\frac{16\pi}{2i} \int\limits_{-\frac{1}{2}-i\infty}^{-\frac{1}{2}+i\infty} (2l + 1)\, A^{\pm}(s, l)\, \frac{P_l(-z_s)}{\sin\pi l}\, \mathrm{d}l -
$$

$$
- \sum_{i\text{ poles}} 16\pi^2\, [2\alpha_i(s) + 1]\, \beta_i(s)\, \frac{P_{\alpha_i(s)}(-z_s)}{\sin\pi\,\alpha_i(s)} - \qquad (9.10)
$$

$$
- \sum_{j\text{ cuts}} \frac{16\pi}{2i} \int\limits_{C_j} (2l + 1)\, A^{\pm}(s, l)\, \frac{P_l(-z_s)}{\sin\pi l}\, \mathrm{d}l \ .
$$

The first term, the so called "background integral", vanishes as $z_s \to \infty$ because of (9.7), leaving a sum of "Regge poles" and "Regge cuts" representing the divergent part of the amplitude (in z_s or t). Potential scattering has no cuts, only poles, and it is these which were discovered by *Regge* in his pioneering work (*Regge*, 1959, 1960; *Bottino* et al., 1962). Cuts are however to be expected in relativistic scattering, and we shall discuss the reasons for this and their properties in Chapter V. For the remainder of this chapter we shall refer mainly to the poles. The poles in the l-plane will also give rise to poles of $A^{\pm}(s, t)$ in the s-plane, at values of $s\ (= s_R)$ such that $\alpha_i(s_R) = $ integer, since $\sin[\pi\alpha_i(s)] \to 0$ as $s \to s_R$. The function $\alpha_i(s)$ is referred to as a "Regge trajectory". The Froissart bound requires that $\alpha_i(s) \leqq 1$ for $s \leqq 0$, for all trajectories.

If the assumptions used in obtaining (9.10) are correct, that is the partial-wave amplitude is analytic in the right half l-plane with only isolated singularities, we can continue the Froissart-Gribov projection even below $l_M(s)$. If we put the first term of (9.10) into (6.5) the integral will exist for $\mathrm{Re}\{l\} > - 1/2$. The problem lies in the divergent pole and cut contributions. Concentrating on the poles for simplicity, we note that a single Regge trajectory, $\alpha(s)$, contributes to the amplitude

$$
A^{\pm}_R(s, t) = - 16\pi^2\, [2\alpha(s) + 1]\, \beta(s)\, \frac{P_{\alpha(s)}(-z_s)}{\sin\pi\,\alpha(s)} \qquad (9.11)
$$

The corresponding t-discontinuity can be obtained by remembering that [see B 1, 3.2 (10)]

$$
\mathrm{Im}[P_\alpha(z)] = - P_\alpha(-z)\, \sin\pi\alpha; \quad z < -1\,,
$$
$$
= 0 ; \qquad\qquad\quad z \geqq -1\,, \qquad (9.12)
$$

so

$$
D^{\pm}_{Ri}(s, t) = 16\pi^2\, [2\alpha(s) + 1]\, \beta(s)\, P_{\alpha(s)}(z_s)\, ; \quad z_s > 1\,. \qquad (9.13)
$$

Of course really $D^{\pm}(s, t)$ should begin at $z_s = z_s(s, t_0)$ not $z_s = 1$ (see Fig. II.1), but, as we shall discuss in Chapter III, there is a cancellation of this discontinuity with that of the background integral in the region $1 < z_s < z_s(s, t_0)$. Here we are only concerned with the large z_s

behaviour which gives rise to a singularity. If we put (9.13) in (6.5) and use the fact that [B 1, 3.12 (4)]

$$\int_1^\infty P_\alpha(z)\, Q_l(z)\, \mathrm{d}z = \frac{1}{(l-\alpha)\,(l+\alpha+1)} \tag{9.14}$$

we get

$$A_R^\pm(s, l) = \frac{[2\alpha(s)+1]\,\beta(s)}{[l-\alpha(s)]\,[l+\alpha(s)+1]}\,. \tag{9.15}$$

So, if we use (6.5) to continue $A^\pm(s, l)$ below $l_M(s)$, we obtain poles in s at $s = s_R$ such that $\alpha(s_R) = l$, an integer, for each trajectory. Since there are no s poles in $D_i^\pm(s, t)$ it is only the divergence of (6.5) which can generate them in $A^\pm(s, l)$.

The hypothesis that the scattering amplitude can be continued in angular momentum, which is postulate (vi) of Chapter I, is thus seen to determine the subtractions which are required in the Mandelstam representation. If we find that [c.f. (9.13)]

$$D_i^\pm(s, t) \xrightarrow[t \to \infty]{} \Gamma_i(s)\, t^{\alpha_i(s)}\,, \tag{9.16}$$

we know that the subtraction will correspond to a Regge pole such as (9.11). The Regge pole term corresponds to the hitherto arbitrary subtraction term $F^{N-1}(s, z_s)$ in (6.3). In general of course there will be several poles and cuts in the right-half angular-momentum plane, and the asymptotic behaviour will involve a collection of terms like (9.16), but, at least in principle, if we have sufficiently accurate information as to the asymptotic behaviour of the amplitude, we can determine each of the singularities. The implications of this will be discussed more fully in Chapter VI, but it is immediately evident that arbitrary subtractions are no longer allowed. For suppose we let $N = 1$ in (6.3), permitting the arbitrary function $F^0(s)$ to appear. This function, being independent of t, contributes only to the S-wave, $A_0^\pm(s)$, and could be written $F^0(s)\, \delta_{l,0}$. Similarly for $N = 2$ we get a polynomial of the first degree in t which contributes only to the S- and P-waves; and so on. So arbitrary subtractions correspond to Kronecker deltas in the l-plane, and of course our analyticity postulate specifically excludes such terms. The low partial waves must be attainable by analytic continuation from the higher ones, which are already known from (6.5), and the asymptotic behaviour of the amplitude must be the same as that of the discontinuity.

The Sommerfeld-Watson transform also gives us another way of seeing why Carlson's theorem gives a unique interpolation between the integers. For suppose "per imposibile" that there were two functions, $A^\pm(s, l)$ and $A_1^\mp(s, l)$, which were equal at all the integers $> l_M(s)$ and bounded as in (8.3). Then we could construct another function

$$A_2^\mp(s, l) = A^\pm(s, l) + \frac{A^\pm(s, l) - A_1^\mp(s, l)}{l - l_1}$$

with some arbitrary l_1, which equals $A^\pm(s, l)$ at the integers and is also bounded as in (8.3). However, if we make a Sommerfeld-Watson transform of $A^\pm(s, t)$ with $A^\pm_{\frac{1}{2}}(s, l)$ as its partial wave amplitudes, the extra pole at l_1, would give a term in $A^\pm(s, t)$ which behaved like z^{l_1}, as $z \to \infty$. Since l_1, is arbitrary this situation is impossible, and the only way of avoiding it is for $A^\pm(s, l_1) = A^\pm_{\frac{1}{1}}(s, l_1)$ for all $l_1 > l_M(s)$. So $A^\pm(s, l)$ is unique.

After this rather mathematical discussion we shall begin, in the next section, to explore the physical meaning of Maximal Analyticity of the second kind.

II.10. Regge Poles

The sort of trajectories which are found in potential scattering, and which we can anticipate in particle physics (see *Ahmadzadeh* et al., 1963; *Lovelace* and *Masson*, 1962, for some examples), look typically as in

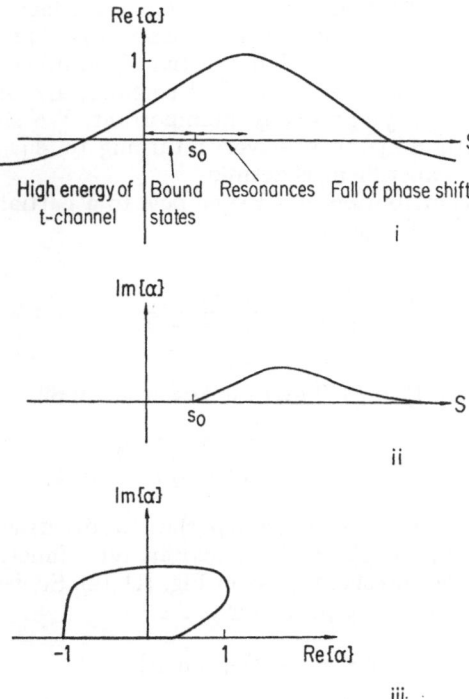

Fig. II.12. The behaviour of a typical trajectory: (i) shows the physical significance of the various regions of Re$\{\alpha\}$. The end point, $\alpha(\pm \infty)$, of the highest trajectories is -1 in potential scattering.

Figs. (II.12), and each gives a contribution to $A^\pm(s, t)$ of the form (9.11). Remembering that the even part of $A(s, t)$ is contained in A^+, and the odd part in A^-, we find that the contribution to the physical amplitude is [see (2.4)].

$$A(s, t) = -16\pi^2 [2\alpha(s) + 1]\, \beta(s) \frac{P_\alpha(-z_s) \pm P_\alpha(z_s)}{\sin \pi \alpha(s)}, \qquad (10.1)$$

the $\pm$ sign depending on whether the trajectory is in A^+ or A^-, that is whether the trajectory has positive or negative "signature". Since [B 1, 3.2(10)]

$$P_\alpha(-z) = e^{-i\pi\alpha}\, P_\alpha(z) - \frac{2}{\pi} \sin \pi\alpha\, Q_\alpha(z) \qquad (10.2)$$

it is convenient to re-write (10.1) as

$$A(s, t) = -16\pi^2 [2\alpha(s) + 1]\, \beta(s) \times$$

$$\times \left\{ [1 \pm e^{-i\pi\alpha(s)}] \frac{\sin \pi\alpha(s)}{P_\alpha(-z_s)} \mp \frac{2}{\pi} Q_\alpha(-z_s) \right\}. \qquad (10.3)$$

The factor $[1 \pm e^{-i\pi\alpha(s)}]$ is called the "signature factor". Its presence means that a given trajectory contributes a pole to the physical amplitude only at alternate integers; even l for positive signature, odd l for negative signature. The Q_α term is evidently not singular, and since [see (12.8)] $Q_\alpha(z) \sim z^{-\alpha-1}$ it is asymptotically unimportant. We shall continue to work mainly with $A^\pm(s, t)$, however, returning to $A(s, t)$ only when a comparison with experiment is desired.

To find the contribution of a Regge pole to a partial-wave we make use of the formula [B 1, 3.12(7)]

$$\frac{1}{2} \int_{-1}^{1} P_\alpha(-z)\, P_l(z)\, dz = \frac{1}{\pi} \frac{\sin \pi\alpha}{(\alpha - l)\,(\alpha + l + 1)}, \qquad l \text{ integer, } \alpha \text{ anything,}$$

$$(10.4)$$

so that putting (9.11) in (3.7) gives (*Chew* et al., 1962).

$$A_l^\pm(s) = \frac{[2\alpha(s) + 1]\, \beta(s)}{[l - \alpha(s)]\,[\alpha(s) + l + 1]}$$

as in (9.15). The properties of $\alpha(s)$ and $\beta(s)$ will be discussed in Chapter III, but usually we expect $\alpha(s)$ to be a real analytic function of s, with a branch point at the threshold s_0, as in Fig. (II.12). So we can separate α into its real and imaginary parts, for real s,

$$\alpha(s) = \alpha_R(s) + i\, \alpha_I(s),$$

$$\alpha_I(s) = 0 \quad \text{for} \quad s < s_0, \quad s \text{ real.} \qquad (10.5)$$

Then if, as in the previous section, we choose s_R such that $\alpha_R(s_R) = l$, an integer, and expand α_R in the region of s_R,

$$\alpha(s) = l + \alpha_R'(s_R)\,(s - s_R) + \cdots + i\, \alpha_I(s_R) \qquad (10.6)$$

we get from (9.15)

$$A_{\vec{r}}^{\pm}(s) \approx \frac{\beta(s_R)}{\alpha'(s_R)(s_R - s) - i\,\alpha_I(s_R)}, \quad s \approx s_R. \tag{10.7}$$

Putting $s = E^2$, E the total centre of mass energy, and $s_R = M^2$ gives

$$A_{\vec{r}}^{\pm}(s) \approx \frac{\dfrac{\beta(M^2)}{2M\,\alpha_R'(M^2)}}{(M - E) - i\,\dfrac{\alpha_I(M^2)}{\alpha_R'(M^2)\,2M}}.$$

This corresponds to a Breit-Wigner resonance of mass M and width $\Gamma = \frac{\alpha_I(M^2)}{\alpha_R'(M^2)\,M}$, [c.f. (I.8.1)]. For $s < s_0$, with $\alpha_I = 0$, we have a bound-state pole.

So we see that bound-state and resonance poles are of exactly the same type, the only difference being that the former occur if the trajectory passes through an appropriate (i.e. even/odd for $\pm$ signature) integer with s below threshold, while the latter occur if this happens above threshold. A single trajectory might contain, say, a bound state of spin zero and a resonance of spin two. We have already noted that for $s \leqq 0$ the Froissart bound requires that all trajectories have $\alpha(s) \leqq 1$, but for $s > 0$ they can rise to larger values of angular momentum, with particles at each alternate integer. The phase shift correspondingly rises through $\pi/2, 3\pi/2 \ldots$. The return of α through the various integers at higher energies [Fig. (II.12)] corresponds to the phase shift returning through $\ldots 3\pi/2, \pi/2$, and of course these points do not correspond to particles. Maximal Analyticity of the second kind puts all the particles on a similar footing, all are composites, the stable nucleon just as much as the unstable N*, or the stable deuteron, which we have always regarded as composite (see *Chew* and *Frautschi*, 1961b; *Blankenbecler* and *Goldberger*, 1962; *Frautschi, Gell-Mann*, and *Zachariasen*, 1962; *Chew*, 1962b).

The existence or non-existence of elementary (non-composite) particles is thus seen to be intimately connected with the validity of postulate (vi). This connection will receive further discussion in Chapters VI and VII, but it would seem that if we can demonstrate that all the particles do lie on Regge trajectories, and postulate (vi) is valid, then no particles can be elementary. In Chapter VIII we shall present the experimental evidence for this.

The behaviour of the trajectories for $s < 0$ is also of considerable physical importance, as this region [or some similar region for unequal-mass kinematics, see section (I.6)] corresponds to the t-channel physical region. We noted in the last section that a trajectory $\alpha(s)$ gives a contribution to the asymptotic behaviour of the amplitude, (9.11),

$$A^{\pm}(s, t) \sim t^{\alpha(s)}.$$

If α is the right-most singularity in the l-plane, this will be the asymptotic behaviour of the t-channel scattering amplitude. We shall see in Chapter VIII that such a correlation between the s-channel Regge poles and the t-channel asymptotic behaviour is well verified in several processes. This constitutes the most crucial test, and the principal success, of the Regge theory. Conversely, observation of the asymptotic behaviour of the amplitude in the t-channel permits a determination of the l-plane singularities for $s < 0$.

The physical significance of the various regions of a trajectory is summarized in Fig. (II.12a).

II.11. Singularities of Partial-Wave Amplitudes for Non-integer l

In Section (II.4) we discussed the singularities in s of the amplitudes $A_l^{\pm}(s)$ for integer l, and found that, in addition to the right-hand singularities, corresponding to the s-channel poles and thresholds, there were also left-hand singularities stemming from the branch points of $Q_l(z)$ at $z = \pm 1$. For non-integer l the left-hand singularities are more complicated, because now $Q_l(z)$ has four branch-points, and is cut between $z = -\infty$ and -1, as well as -1 to $+1$.

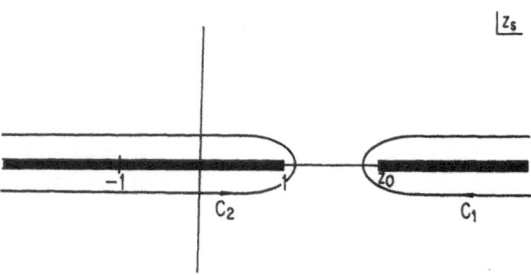

Fig. II.13. Integration contours in the complex z_s plane for equation (3.9) with complex l (c.f. Fig. II.2).

Both the equations (3.9) are still valid for the partial wave projection, but now the contour C_2 has to enclose the whole real z axis from $z = -\infty$ to 1 (Fig. II.13). The imaginary parts of $Q_l(z)$ are [from B 1, 3.3 (11) and 3.3 (12)] [c.f. (3.8)]

$$\operatorname{Im}\{Q_l(z)\} = -\frac{\pi}{2} P_l(z) , \quad -1 < z < 1 , \tag{11.1}$$
$$= \sin \pi l \, Q_l(-z) , \quad -\infty < z < -1 ,$$

so our equations for $A_l^{\pm}(s)$ become [c.f. (3.6), (3.7)].

$$A_{\bar{l}}^{\mp}(s) = \frac{1}{16\pi^2} \int\limits_{z_s(s, t_0)}^{\infty} D_{\bar{l}}^{\mp}(s, t') Q_l(z') \, dz' , \tag{11.2}$$

and

$$A_{l}^{\pm}(s) = \frac{1}{32\pi} \int\limits_{-1}^{1} P_l(z') A^{\pm}(s, t') \, dz' - \\ - \frac{\sin \pi l}{16\pi^2} \int\limits_{-\infty}^{-1} Q_l(-z') A^{\pm}(s, t') \, dz' . \tag{11.3}$$

However the extra s cut for $z < -1$ is eliminated if instead of $A_{\bar{l}}^{\mp}(s)$ we consider $B_{\bar{l}}^{\mp}(s)$ defined by (5.17). We have [B 1, 3.2 (12)].

$$Q_l(-z \pm i\varepsilon) = -e^{\pm i\pi l} Q_l(z) , \quad z > 1 , \tag{11.4}$$

and

$$(q_{s12} q_{s34})^{-l} = e^{\mp i\pi l}(-q_{s12} q_{s34})^{-l} , \tag{11.5}$$

so for $t > 0$, where from (I.6.14) $s \pm i\varepsilon$ corresponds to $z \pm i\varepsilon$,

$$\frac{Q_l(z \pm i\varepsilon)}{(q_{s12} q_{s34} \pm i\varepsilon)^l} = -\frac{Q_l(-z)}{(-q_{s12} q_{s34})^l} \tag{11.6}$$

has no cut for $z < -1$.

On the other hand for $-1 < z < 1$ we have [B 1, 3.2 (10)]

$$\sin(\pi l) Q_l(z \pm i\varepsilon) = \frac{\pi}{2} [e^{\mp i\pi l} P_l(z) - P_l(-z)] ,$$

and $P_l(z)$ has no cut for $-1 < z < 1$, so

$$\frac{Q_l(z + i\varepsilon)}{(q_{s12} q_{s34} + i\varepsilon)^l} - \frac{Q_l(z - i\varepsilon)}{(q_{s12} q_{s34} - i\varepsilon)^l} = -\frac{i\pi P_l(-z)}{(-q_{s12} q_{s34})^l} .$$

Corresponding to (11.2) we have

$$B_{\bar{l}}^{\mp}(s) = \frac{1}{16\pi^2} \int\limits_{z_s(s, t_0)}^{\infty} D_{\bar{l}}^{\mp}(s, t') Q_l(z') \, dz' \frac{1}{(q_{s12} q_{s34})^l} . \tag{11.7}$$

This integral only involves $t > 0$. The left-hand cut discontinuity has contributions both from the discontinuity of $Q_l(z_s)$, and from that of $D_{\bar{l}}^{\mp}(s, t)$ in the negative s region. From (11.1) and (2.8) we get

$$\text{Im}\{B_{\bar{l}}^{\mp}(s)\}_{LH} = \frac{1}{32\pi} \int\limits_{-1}^{z_s(s, t_0)} P_l(-z') D_{\bar{l}}^{\mp}(s, t') \, dz' \frac{1}{(-q_{s12} q_{s34})^l} + \\ + \frac{1}{16\pi^2} \int\limits_{a(s)}^{b(s)} Q_l(z') [\varrho_{tu}(t', u') \pm \varrho_{tu}(u', t')] \, dz' \frac{1}{(q_{s12} q_{s34})^l} , \tag{11.8}$$

where $u' + t' + s = \Sigma$, and dz' is related to dt' by (I.6.14). The region of integration in the second term is the region in which the double spectral functions are non-zero at a given s. If we use (8.4), and the fact that interchanging t' and u' is equivalent to reversing the sign of z, we get

$$\operatorname{Im}\{B_l^{\mp}(s)\}_{LH} = \frac{1}{32\pi} \int_{-1}^{z_s(s,\,t_0)} P_l - (z') \, D_l^{\mp}(s,t') \, dz' \, \frac{1}{(-q_{s12}\,q_{s34})^l} +$$

$$+ \frac{1}{16\pi^2} \int_{a(s)}^{b(s)} Q_l(z') \, [1 \mp e^{-i\pi l}] \, \varrho_{tu}(t',u') \, dz' \, \frac{1}{(q_{s12}\,q_{s34})^l} \, . \tag{11.9}$$

For physical l the last term does not contribute and we revert to (4.6). The right-hand cut discontinuity comes simply from the cuts of $D_l^{\mp}(s,t)$ so, from (2.8),

$$\operatorname{Im}\{B_l^{\mp}(s)\}_{RH} = \frac{1}{16\pi^2} \int_{z_s(s,\,t_0)}^{\infty} [\varrho_{st}(s,t') \pm \varrho_{su}(s,t')] \, Q(z') \, dz' \, \frac{1}{(q_{s12}\,q_{s34})^l} \, . \tag{11.10}$$

A more thorough discussion of the relation between z-plane singularities and those in s and t is left for Chapter III.

II.12. The Mandelstam-Sommerfeld-Watson Transform

Because of the asymptotic behaviour of $P_l(z)$ given by equation (9.1), we cannot move the contour of integration to $\operatorname{Re}\{l\} < -1/2$ without the background integral coming to dominate the singularities in the right-half l-plane, at large z_s. At the same time, because [B 1, 3.3(10)]

$$P_\alpha(z) = P_{-\alpha-1}(z)$$

we get, for each trajectory α, an associated trajectory at $-\alpha - 1$ which is unwanted. *Mandelstam* (1962) has shown how to remove these difficulties, and demonstrate the dominance of the right-hand singularities at large z_s even if the background integral is taken along a line $\operatorname{Re}\{l\} < -1/2$. It is especially important that this should be possible in situations when the external particles have spin (see Chapter IV).

The trick (*Mandelstam*, 1962) is first to re-write (1.3) in the form

$$A^{\pm}(s,t) = 16\pi \sum_{l=0}^{\infty} \left\{ (2l+1) \, A^{\pm}(s,l) \, P_l(z_s) + \right.$$

$$+ \frac{1}{\pi} (-1)^{l-1} (2l) \, A^{\pm}\left(s, l - \frac{1}{2}\right) Q_{l-1/2}(z_s) \bigg\} - \tag{12.1}$$

$$- 16\pi \sum_{l=0}^{\infty} \frac{1}{\pi} (-1)^{l-1} (2l) \, A^{\pm}\left(s, l - \frac{1}{2}\right) Q_{l-1/2}(z_s) \, .$$

The first term ($l = 0$) in these extra summations is evidently zero, but it is convenient to retain it as it simplifies the writing.

Now [B 1, 3.3(3)]

$$\frac{P_l(z)}{\sin \pi l} - \frac{1}{\pi} \frac{Q_l(z)}{\cos \pi l} = -\frac{1}{\pi} \frac{Q_{-l-1}(z)}{\cos \pi l}, \tag{12.2}$$

and of course the poles of $(\cos \pi l)^{-1}$ will occur at $l =$ half-odd-integer, so we can make a Sommerfeld-Watson transform of the term in braces, $\{\ \}$, in (12.1), corresponding to (9.10), and obtain

$$A^{\pm}(s, t) = \frac{16}{2i} \int_{-\frac{1}{2}+\varepsilon+i\infty}^{-\frac{1}{2}+\varepsilon-i\infty} (2l+1) A^{\pm}(s, l) \frac{Q_{-l-1}(-z_s)}{\cos \pi l} \, dl +$$

$$+ \sum_{i \text{ poles}} 16\pi \left[2\alpha_i(s) + 1 \right] \beta_i(s) \frac{Q_{-\alpha_i(s)-1}(-z_s)}{\cos \pi \alpha_i(s)} +$$

$$+ \sum_{j \text{ cuts}} \frac{16}{2i} \int_{C_j} (2l+1) A^{\pm}(s, l) \frac{Q_{-l-1}(-z_s)}{\cos \pi l} \, dl -$$

$$- 16\pi \sum_{l=0}^{\infty} \frac{1}{\pi} (-1)^{l-1} (2l) A^{\pm}\left(s, l - \frac{1}{2}\right) \times$$

$$\times Q_{l-1/2}(-z_s), \quad 0 \le \varepsilon \le \frac{1}{2}. \tag{12.3}$$

We have put the integration contour for the background integral just to the right of $\operatorname{Re}\{l\} = -1/2$ to avoid the pole of $(\cos \pi l)^{-1}$. However, if we now move the path of integration down to some $L < -1/2$, as in Fig. (II.14), we expose these extra poles at $l = l' \equiv -1/2, -3/2, \ldots$, so that (12.3) becomes

$$A^{\pm}(s, t) = \frac{16}{2i} \int_{L+i\infty}^{L-i\infty} (2l+1) A^{\pm}(s, l) \frac{Q_{-l-1}(-z_s)}{\cos \pi l} \, dl + \sum_{i \text{ poles}} + \sum_{j \text{ cuts}} -$$

$$- 16\pi \sum_{l'=-L'}^{-\frac{1}{2}} (2l'+1) A^{\pm}(s, l') Q_{-l'-1}(-z_s) \frac{(-1)^{l'}-1/2}{\pi} -$$

$$- 16\pi \sum_{l=0}^{\infty} \frac{(-1)^{l-1}}{\pi} (2l) A^{\pm}\left(s, l - \frac{1}{2}\right) Q_{l-1/2}(-z_s), \tag{12.4}$$

where $(-L')$ is the next-half integer above L.

If we now replace the summation index l' by $l = -l' - 1/2$ in the fourth term on the right-hand side, we get

$$16\pi \sum_{l=0}^{L'-\frac{1}{2}} \frac{(-1)^{-l-1}}{\pi} (-2l) A^{\pm}\left(s, -l - \frac{1}{2}\right) Q_{-l-1/2}(-z_s). \tag{12.5}$$

This will cancel with the first $L' - 1/2$ terms in the last summation of (12.4) if

$$A\left(s, l - \frac{1}{2}\right) = A\left(s, -l - \frac{1}{2}\right), \quad l \text{ integer}, \tag{12.6}$$

i.e. if the amplitude is symmetrical about $l = -1/2$ for half-odd-integer values of l. This property is known to hold for many potential problems

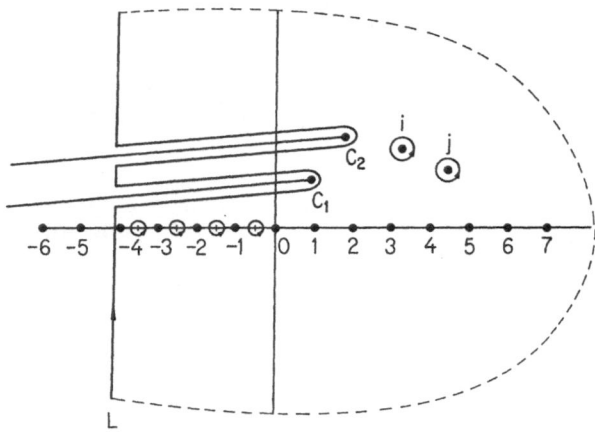

Fig. II.14. The integration contour for equation 12.4, with the *l*-plane singularities of Fig. II.11. We pick up extra poles at the negative half-odd-integers.

(*Mandelstam*, 1962) and will also be true for strong interactions as long as the Froissart-Gribov projection can be used for the required values of l; for the relation then follows from [B 1, 3.3(3)]. So we end up with

$$A^\pm(s, t) = \frac{16}{2i} \int\limits_{L-i\infty}^{L+i\infty} (2l + 1)\, A^\pm(s, l) \frac{Q_{-l-1}(-z_s)}{\cos \pi l}\, dl + \sum_{i \text{ poles}} + \sum_{j \text{ cuts}} -$$

$$- \sum_{L'+\frac{1}{2}}^{\infty} \frac{(-1)^{l-1}}{\pi} (2l)\, A\left(s, -l - \frac{1}{2}\right) Q_{l-\frac{1}{2}}(-z_s)\ . \tag{12.7}$$

The advantage of this expression in terms of $Q_{-l-1}(-z_s)$ over (9.10) is that, since [B 1, 3.2(41)]

$$Q_l(z) = \pi^{\frac{1}{2}} \frac{\Gamma(l+1)}{\Gamma\left(l + \frac{3}{2}\right)} (2z)^{-l-1} F\left(\frac{l}{2} + 1, \frac{l}{2} + \frac{1}{2}, l + \frac{3}{2}, \frac{1}{z^2}\right), \tag{12.8}$$

we have

$$Q_l(z) \xrightarrow[z \to \infty]{} \pi^{\frac{1}{2}} \frac{\Gamma(l+1)}{\Gamma\left(l + \frac{3}{2}\right)} (2z)^{-l-1} \tag{12.9}$$

so that the first and last terms in (12.7) certainly decrease like z^L for $L < -1/2$, while a Regge pole or cut term at $l = \alpha_i(s)$ still gives a contribution which goes like $(z_s)^{\text{Re}\{\alpha_i(s)\}}$ at large z_s. The s-plane poles in this representation correspond to the poles of $Q_{-\alpha-1}(-z_s)$ where $(-\alpha-1)$ = a negative integer. The apparent poles at $\alpha = 1/2$-odd-integer from the vanishing of $\cos \pi \alpha$ are of course spurious, as is evident from the fact that we have simply replaced $P_\alpha(-z_s)/\sin \pi \alpha$ by

$$\left[\frac{Q_\alpha(-z_s)}{\pi \cos \pi \alpha} - \frac{1}{\pi} \frac{Q_{-\alpha-1}(-z_s)}{\cos \pi \alpha}\right]$$

in which the poles cancel. For positive values of α these poles cancel with the poles in $\Gamma(-\alpha + 1/2)$ of (12.9). For negative values the symmetry (12.6) ensures that β vanishes.

In the next chapter we shall discuss the relation between these methods of representing an amplitude in terms of its l-plane singularities, and the Mandelstam representation.

Chapter III. Regge Poles and the Mandelstam Representation

III.1. Some Properties of the Regge Functions

In this chapter we shall show how to combine the Regge decomposition of the amplitude described in Chapter II with the Mandelstam representation of Chapter I. An immediate consequence of doing this is that we can deduce the analytic properties of the Regge pole functions, that is the trajectory, $\alpha(s)$, and the residue, $\beta(s)$. The results are very similar to those obtained in potential scattering by *Bottino* et al., 1962. (See also *Squires*, 1963a, and *Newton*, 1964).

From (II.6.5) we can write (*Oehme* and *Tiktopoulos*, 1962; *Barut* and *Zwanziger*, 1962)

$$A_l^{\pm}(s) = E_l^{\pm}(s) + F_l^{\pm}(s) , \qquad (1.1)$$

where

$$E_l^{\pm}(s) = \frac{1}{16\pi} \cdot \frac{1}{\pi} \int_T^{\infty} Q_l(z') \, D_l^{\pm}(s, t') \, \mathrm{d}t' , \qquad (1.2)$$

and $F_l^{\pm}(s)$ is given by the same integral from t_0 to T. Since $F_l^{\pm}(s)$ is defined by a finite integral it must be holomorphic in $\mathrm{Re}\{l\} > -1$, with just poles at the negative integers, since this is the behaviour of $Q_l(z)$. The other l plane singularies of $A_l^{\pm}(s)$ stem from the asymptotic t behaviour of the integrand in (1.2), as we explained in Section (II.9), and are therefore contained in $E_l^{\pm}(s)$. In fact if

$$D_l^{\pm}(s, t) \sim t^{\alpha(s)} ,$$

then, since (II.12.9)

$$Q_l(z) \sim t^{-l-1} ,$$

we have

$$E_l^{\pm}(s) \sim \int_T^{\infty} t^{-l-1+\alpha(s)} \, \mathrm{d}t = -\frac{e^{[\alpha(s)-l]\log T}}{\alpha(s)-l} , \qquad (1.3)$$

giving a pole at $l = \alpha(s)$. On the other hand, if $D_l^{\pm}(s, t)$ is not simply power bounded, but contains logarithmic factors, these will give rise to branch points in l; but in this chapter we shall concentrate on the poles.

The position of a pole is given by

$$[E_l^{\pm}(s)]^{-1} = 0 , \quad \text{at} \quad l = \alpha(s) , \qquad (1.4)$$

but, because of the singularity of $E_l^{\pm}(s)$ at the threshold, it is better to use the "reduced" amplitude (II.5.17) and write

$$[q_{s12} \, q_{s34}]^l \, [E_l^{\pm}(s)]^{-1} = 0 , \quad \text{at} \quad l = \alpha(s) . \qquad (1.5)$$

Correspondingly the residue of the pole, defined in (II.9.4), is, by Cauchy's Theorem,

$$\beta(s) = \frac{1}{2\pi i} \oint \mathrm{d}l \, E_l^{\pm}(s) \qquad (1.6)$$

or

$$\gamma(s) \equiv \beta(s) \, [q_{s12} \, q_{s34}]^{-l} \tag{1.7}$$

$$= \frac{1}{2\pi i} \oint \mathrm{d}l \, [q_{s12} \, q_{s34}]^{-l} \, E_l^{\pm}(s) \,, \tag{1.8}$$

where the integral is taken in a path around the point $l = \alpha(s)$. Equations (1.5) and (1.8) enable us to find the analytic properties in s of $\alpha(s)$, and of the "reduced residue", $\gamma(s)$.

If $\mathrm{Re}\{l\} > \mathrm{Re}\{\alpha_M(s)\}$, where $\alpha_M(s)$ is the highest lying l-plane singularity [see Section (II.9)], the integral (1.2) converges, and so $E_l^{\pm}(s) \, [q_{s12} \, q_{s34}]^{-l}$ has s-plane singularities of the type which we discussed for the full partial-wave amplitude in Section (II.11). These are the right-hand cut for $s > s_0$ given by (II.11.10), and the left-hand cut given by (II.11.9). But, since we are integrating from $t = T$ rather than $= t_0$ in (1.2), the left-hand branch point is at $s \approx -T$ for large T. The actual value is given by substituting $T = t_n$ (II.4.3) and taking the -1 alternative. We can take T as large as we like, and so cause the left-hand cut of $E_l^{\pm}(s)$ to receed as far as we wish towards $-\infty$. This means that the singularities of $\alpha(s)$ and $\gamma(s)$, which stem from those of $E_l^{\pm}(s)$, do not include the left-hand branch points of the partial wave amplitude. Another way of seeing this is to note that the left-hand cut discontinuity (II.11.9) is holomorphic in $\mathrm{Re}\{l\} > -1$, and hence it can not contribute to the l-plane singularities of $E_l^{\pm}(s)$ in $\mathrm{Re}\{l\} > -1$.

So the only relevant singularities of $E_l^{\pm}(s)$ are for $s > s_0$. If we now continue in l to $\mathrm{Re}\{l\} < \mathrm{Re}\{\alpha_M(s)\}$, using the analytic continuation of (1.2) described in Section (II.9), we shall encounter new singularities of $E_l^{\pm}(s)$. But there will not be any l independent singularities in the s-plane, for if there were they would have suddenly to appear at $\mathrm{Re}\{l\} = \mathrm{Re}\{\alpha_M(s)\}$, in contradiction to the continuity theorems for functions of two complex variables (see *Oehme* and *Tiktopoulos*, 1962). The only new singularities of $E_l^{\pm}(s)$ are the Regge singularities given by, for example, (1.5).

The implicit function theorem then tells us that if $[E_l^{\pm}(s)]^{-1}$ is regular in the neighbourhood of the point $s = s_p$ with $l = \alpha(s_p) \equiv \alpha_p$, and if

$$\frac{\partial}{\partial l} \, [E_l^{\pm}(s_p)]^{-1} \Big|_{l=\alpha_p} \neq 0 \,, \tag{1.9}$$

then $\alpha(s)$ is also a regular function in the neighbourhood of s_p. For we can expand $[E_l^{\pm}(s)]^{-1}$ in a Taylor series about s_p (*Cheng*, 1963),

$$[E_l^{\pm}(s)]^{-1} = a \, [\alpha(s) - \alpha_p] + b(s - s_p) + c \, [\alpha(s) - \alpha_p]^2 + d(s - s_p)^2 + \cdots,$$

and, if $a \neq 0$, we have from (1.5).
$$\tag{1.10}$$

$$\alpha(s) = \alpha_p - \left(\frac{b}{a}\right)(s - s_p) + \cdots,$$

and $\alpha(s)$ is analytic near s_p. But if $a = 0$ [i.e. (1.9) is not satisfied] and $c \neq 0$, then

$$\alpha(s) = \alpha_p \pm \left(-\frac{b}{c}\right)^{\frac{1}{2}} (s - s_p)^{\frac{1}{2}} + \cdots,$$

and so there are two Regge trajectories which cross at $s = s_p$, and s_p is a branch point for both. But if $b = 0$ also there need not be a branch point, so the fact that two trajectories collide may, but need not, result in a branch point in $\alpha(s)$.

In fact we shall see that such collisions must occur for Fermions in order to satisfy the MacDowell symmetry (see Section IV.6). They have been found also to occur in the solution of various potential scattering problems (see *Lovelace* and *Masson*, 1962; *Ahmadzadeh* et al., 1963; *Ahmadzadeh*, 1963; *Warburton*, 1964a, 1965) and discussed by *Desai* and *Newton*, 1963b, and *Azimov* et al., 1963. Since, as we shall show in Chapter VI, the "potentials" which generate trajectories in strong-interaction physics are expected to include combinations of attraction and repulsion, it is likely that such branch-points will occur (see *Bali* et al., 1967). In potential scattering a can only be equal to zero for $l < -1/2$ (see *Newton*, 1964, p. 50) so we can expect such cuts of $\alpha(s)$ only for fairly large negative s for the higher trajectories, but of course if the external particles have spin the branch-points could be higher in J, and hence appear at smaller $|s|$.

Apart from such cuts the only singularities of $\alpha(s)$ will occur where $[E_l^{\neq}(s)]^{-1}$ is singular, i.e. for $s > s_0$. Because of (1.8), identical remarks apply to $\gamma(s)$, but not of course to $\beta(s)$ which has the branch points of $[q_{s12}\, q_{s34}]^l$.

In conclusion, we expect that except where trajectories cross $\alpha(s)$ and $\gamma(s)$ will be real analytic functions cut from s_0 to ∞.

We can also make some general remarks about the threshold behaviour of $\alpha(s)$, similar to the potential scattering results of *Bottino* et al., 1962 (*Barut* and *Zwanziger*, 1962, *Taylor*, 1962). Let us consider an amplitude in which the lowest threshold, s_0, is the two particle elastic threshold. The reduced amplitude $B_l(s)$ defined in (II.5.17), satisfies, from (II.5.20),

$$\Delta_s\{[B_l(s)]^{-1}\} = -i\,\varrho(s)\,(q_{s12}^2)^l \, ,$$

with $\varrho(s)$ defined in (II.5.7), when the discontinuity is taken across the elastic unitary cut.

The function

$$-\frac{i\,\varrho(s)\,(-q_{s12}^2)^l}{\cos\pi l} = -\frac{i\,\varrho(s)\,(q_{s12}^2)^l\,e^{\pm\,i\pi l}}{\cos\pi l}$$

has the same discontinuity as $[B_l(s)]^{-1}$ for $s_0 < s < s_I$, where s_I is the first inelastic threshold. Thus the function

$$Y(s, l) \equiv \cos\pi l\,[B_l(s)]^{-1} + i\,\varrho(s)\,(-q_{s12}^2)^l \tag{1.11}$$

has no branch cuts for $s < s_I$. The position of a trajectory $\alpha(s)$ is given by

$$[B_l(s)]^{-1} = 0 \quad \text{at} \quad l = \alpha(s) \tag{1.12}$$

or

$$Y(s, l) = i\,\varrho(s)\,(-q_{s12}^2)^l \quad \text{at} \quad l = \alpha(s) \tag{1.13}$$

and if we expand $Y(s, l)$ in s around the threshold s_0, and l around $l = \alpha_0 \equiv \alpha(s_0)$, and define

$$Y_s \equiv \frac{\partial Y(s, l)}{\partial s}\bigg|_{\substack{l = \alpha_0 \\ s = s_0}} \tag{1.14}$$

and

$$Y_l \equiv \frac{\partial Y(s, l)}{\partial l}\bigg|_{\substack{l = \alpha_0 \\ s = s_0}} \tag{1.15}$$

we get

$$Y[s, \alpha(s)] = Y(s_0, \alpha_0) + [\alpha(s) - \alpha_0] Y_l + (s - s_0) Y_s + \cdots. \tag{1.16}$$

And, from (1.13),

$$Y[s, \alpha(s_0)] = i \varrho(s) (-q_{s12}^2)^{\alpha_0} \quad \text{for} \quad s \text{ near } s_0,$$

which, because of the relation given in (II.5.7) for $\varrho(s)$, gives

$$Y[s, \alpha_0] = -\frac{2}{\sqrt{s_0}} (-q_{s12}^2)^{\alpha_0 + \frac{1}{2}} \quad \text{for} \quad s \text{ near } s_0. \tag{1.17}$$

So

$$Y[s_0, \alpha_0] = 0, \quad \text{if} \quad \alpha_0 > -\frac{1}{2}, \tag{1.18}$$

and

$$Y[s, \alpha_0] \xrightarrow[s \to s_0]{} \infty, \quad \text{if} \quad \alpha_0 < -\frac{1}{2}. \tag{1.19}$$

Combining (1.16) and (1.17) we get

$$\alpha(s) = \alpha_0 - \frac{2}{\sqrt{s_0}} \frac{1}{Y_l} \left(\frac{s - s_0}{4}\right)^{\alpha_0 + \frac{1}{2}} e^{-i\pi\left[\alpha(s) + \frac{1}{2}\right]} -$$
$$- (s - s_0) \left(\frac{Y_s}{Y_l}\right) + \cdots, \quad \alpha_0 > -\frac{1}{2}. \tag{1.20}$$

For $\alpha_0 < -1/2$ we expand $\overline{Y}(s, l) \equiv [Y(s, l)]^{-1}$, and find instead of (1.20)

$$\alpha(s) = \alpha_0 - \frac{2}{\sqrt{s_0}} \overline{Y}_l^{-1} \left(\frac{s - s_0}{4}\right)^{-\left[\alpha(s) + \frac{1}{2}\right]} e^{i\pi\left[\alpha(s) + \frac{1}{2}\right]} -$$
$$- \left(\frac{\overline{Y}_s}{\overline{Y}_l}\right)(s - s_0) + \cdots. \tag{1.21}$$

So from (1.20) we find, if we approach the threshold from above,

$$\text{Im}\{\alpha(s)\} \approx \frac{2}{\sqrt{s_0}} Y_l^{-1} \sin\left[\pi\left(\alpha_0 + \frac{1}{2}\right)\right] \left(\frac{s - s_0}{4}\right)^{\alpha_0 + \frac{1}{2}}, \tag{1.22}$$

and

$$\text{Re}\{\alpha(s)\} \approx \alpha_0 - \frac{2}{\sqrt{s_0}} Y_l^{-1} \cos\left[\pi\left(\alpha_0 + \frac{1}{2}\right)\right] \left(\frac{s - s_0}{4}\right)^{\alpha_0 + \frac{1}{2}} -$$
$$- (s - s_0) \left(\frac{Y_s}{Y_l}\right), \tag{1.23}$$

while from below

$$\alpha(s) = \text{Re}\{\alpha(s)\}.$$

The slope of the trajectory at threshold,

$$\frac{d \, \text{Re}\{\alpha(s)\}}{ds}\bigg|_{\substack{s \text{ near } s_0 \\ \text{but} > s_0}} = \text{const.} \cos\left[\pi\left(\alpha_0 + \frac{1}{2}\right)\right]\left(\alpha_0 + \frac{1}{2}\right)\left(\frac{s - s_0}{4}\right)^{\alpha_0 - \frac{1}{2}}, \tag{1.24}$$

is infinite for $-1/2 < \alpha_0 < 1/2$, but not for $\alpha_0 > 1/2$. For $-1/2 < \alpha_0 < 0$, if we approach the threshold from below, where $\alpha(s)$ is real, we find that the sign of the slope is opposite to (1.24) so there is a cusp, but there is no cusp if $\alpha_0 > 0$. Some examples of the behaviour of $\mathrm{Re}\{\alpha(s)\}$ as the

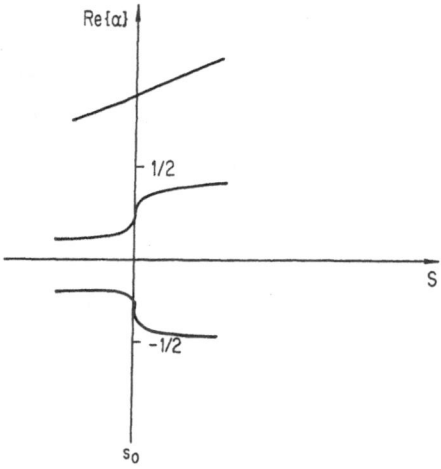

Fig. III.1. Examples of the behaviour of trajectories passing though threshold (after *Newton*, 1964).

trajectory passes through threshold are shown in Fig. (III.1). A more complete discussion will be found in *Newton*, 1964, chapter 9. At multiparticle thresholds the phase-space factor is different, and leads us to expect a more regular behaviour of $\mathrm{Re}\{\alpha(s)\}$ as we cross the threshold (*Freund*, 1962). So far there does not appear to be any experimental verification of this sort of threshold behaviour, and we shall see in Chapter VIII that a rather smooth behaviour is suggested by such evidence as we possess, though of course direct observation of the trajectories in the threshold region is not possible. This is not really surprising because we know (*Warburton*, 1964 b) that in potential scattering these cusp effects, etc., are so small as to be completely negligible, and we can anticipate that this will be true also for strong interaction physics.

Another peculiar threshold property can be deduced from the behaviour of (1.17) as $l \to -1/2$. We have

$$Y(s, l) \xrightarrow[l \to -\frac{1}{2}]{} -\frac{2}{\sqrt{s_0}} \tag{1.25}$$

and so (1.13) becomes

$$\frac{2}{\sqrt{s_0}} = \frac{2}{\sqrt{s_0}} (q^2_{s12})^{l+\frac{1}{2}} e^{-i\pi\left(l+\frac{1}{2}\right)} \tag{1.26}$$

or

$$1 = e^{[\log(q^2_{s12}) - i\pi]\left(l+\frac{1}{2}\right)}. \tag{1.27}$$

This equation is satisfied by $l = \alpha_n$ for any α_n such that

$$[\log(q_{s_{12}}^2) - i\pi]\left(\alpha_n + \frac{1}{2}\right) = \pm 2\pi n i \qquad (1.28)$$

or

$$\alpha_n = \frac{\pm 2\pi n}{\pi + i\log(q_{s_{12}}^2)} - \frac{1}{2}. \qquad (1.29)$$

So there are an infinite number of trajectories converging on $\alpha = -1/2$. as $q_{s_{12}}^2 \to$ zero. This accumulation of trajectories at threshold was noted by *Desai* and *Newton*, 1963b; and by *Gribov* and *Pomeranchuk*, 1962c (see also *Desai* and *Sakita*, 1964).

The main conclusion that we wish to draw from this section is the real analyticity of $\alpha(s)$ and $\gamma(s)$, unless the trajectories cross. This means that we can hypothesize dispersion relations such as

$$\alpha(s) = \alpha(\infty) + \frac{1}{\pi}\int_{s_0}^{\infty} \frac{\text{Im}\{\alpha(s')\}}{s' - s}\,ds' \qquad (1.30)$$

and

$$\gamma(s) = \frac{1}{\pi}\int_{s_0}^{\infty} \frac{\text{Im}\{\gamma(s')\}}{s' - s}\,ds'. \qquad (1.31)$$

An example of (1.30) is displayed in Fig. (II.12). We have supposed that $\alpha(s)$ requires a single subtraction. In potential scattering $\alpha(\infty)$ is always a negative integer, some trajectories beginning at -1, others at -2, and so on. In relativistic scattering the situation is more complicated, as we shall discuss in Chapter VIII, but (1.30) may well be acceptable. No subtraction is given for $\gamma(s)$ on the assumption that it vanishes as $s \to \infty$.

In Chapter V we shall show that unitarity demands that $\text{Im}\{\alpha(s)\}$ should be positive, so that $\alpha(s)$ is a Herglotz function (*Herglotz*, 1911). This means that, from (1.30),

$$\frac{d}{ds}[\alpha(s)] = \frac{1}{\pi}\int_{s_0}^{\infty} \frac{\text{Im}\{\alpha(s')\}}{(s' - s)^2}\,ds' \qquad (1.32)$$

and all higher derivatives will be positive for all $s < s_0$. In potential scattering we have the result (*Regge*, 1960; see also *Squires*, 1963a) that

$$\frac{d\alpha(E)}{dE}\bigg|_{E \approx 0} \approx \frac{r^2}{(2\alpha + 1)}, \qquad (1.33)$$

where $E = \sqrt{s}$, and r is the radius of the state in question. If we identify this radius with the range of the force, R, we can expect that the slope of a given trajectory will be related, roughly, to the range of the force generating that trajectory (see Section II.7) by

$$\frac{d\alpha(s)}{ds} = \frac{R^2}{2M[2\alpha(s) + 1]} \qquad (1.34)$$

where M is the mass of the particles involved. In as much as the ranges of the forces in many strongly interacting systems are fairly similar we expect the various trajectories to have comparable slopes. We shall see in Chapter VIII that this is indeed the case.

In the next section we go on to consider the implications of the analyticity properties of $\alpha(s)$ and $\gamma(s)$ for the analyticity of the Regge pole terms as a whole.

III.2. Regge Poles and the Mandelstam Representation

In Chapter I we explained that the four-line connected part is expected to satisfy the Mandelstam representation, with double spectral functions which are non-zero only within certain boundaries $[s = b_1(t)$ etc., see (I.11.3) et seq.], whose positions depend on the masses of the external particles. The detailed calculation of the positions of these boundaries was discussed in Section (I.12). For amplitudes of definite signature we end up with a representation such as (II.2.6). In what follows it will be convenient to think of the amplitude as having only one double spectral function (say ϱ_{st}), the contributions of the others being readily obtained by permuting the variables s, t and u. So from (III.2.6), interchanging the order of integration, the Mandelstam representation becomes

$$A^{\pm}(s, t) = \frac{1}{\pi^2} \int_{s_0}^{\infty} ds' \int_{b(s')}^{\infty} dt'' \frac{\varrho_{st}(s', t'')}{(s' - s)(t'' - t)} . \tag{2.1}$$

On the other hand in Chapter II we sought to represent the amplitude in terms of a sum of Regge poles and cuts, plus a background which was chosen to vanish at least as fast as an inverse square-root as the variables approach infinity [see (II.9.10) or (II.12.7)]. There will of course be poles and cuts in each of the channels s, t and u, but again in this Chapter we shall find it convenient to work with just s-channel poles, and to neglect cuts altogether, so we have from (II.9.10)

$$A^{\pm}(s, t) = -\sum_i 16\pi^2 [2\alpha_i(s) + 1] \beta_i(s) \frac{P_{\alpha_i(s)}(-z_s)}{\sin \pi \alpha_i(s)} -$$

$$- \frac{16\pi}{2i} \int_{-\frac{1}{2}-i\infty}^{-\frac{1}{2}+i\infty} (2l + 1) A^{\pm}(s, l) \frac{P_l(-z_s)}{\sin \pi l} dl , \tag{2.2}$$

or from (II.12.7)

$$A^{\pm}(s, t) = \sum_i 16\pi [2\alpha_i(s) + 1] \beta_i(s) \frac{Q_{-\alpha_i(s)-1}(-z_s)}{\cos \pi \alpha_i(s)} +$$

$$+ \frac{16}{2i} \int_{L-i\infty}^{L+i\infty} (2l + 1) A^{\pm}(s, l) \frac{Q_{-l-1}(-z_s)}{\cos \pi l} dl -$$

$$- \sum_{L'+\frac{1}{2}}^{\infty} \frac{(-1)^{l-1}}{\pi} (2l) A\left(s, l - \frac{1}{2}\right) Q_{l-\frac{1}{2}}(-z_s), \tag{2.3}$$

where $(-L')$ is the next half-odd-integer above L. The advantage of (2.3) over (2.2) is that it allows us to include all the poles with $\mathrm{Re}\{\alpha_i(s)\} > L$ explicitly (note that L is in general negative), and make the background term vanish as fast as we like $(\sim t^L)$. But (2.2) is sometimes simpler to use.

The equivalence of (2.1) and (2.2) or (2.3) is by no means obvious, because the Regge pole terms themselves have discontinuities and double spectral functions which are non-zero outside the region prescribed by Mandelstam. To show this let us first consider a simple case, such as $\pi - \pi$ scattering, where all the external particles have the same mass. Then we have seen [Section (I.12)] that the ϱ_{st} double spectral function exists only in the region shown in Fig. (I.16). However, if we take a single Regge pole from (2.3)

$$A^{\pm}_{R}(s, t) = 16\pi [2\alpha(s) + 1]\, \beta(s)\, \frac{Q_{-\alpha(s)-1}(-z_s)}{\cos\pi\alpha(s)} \qquad (2.4)$$

this expression certainly has singularities in undesirable regions. The trajectory $\alpha(s)$ has been shown, in the previous section, to be a real analytic function of s, with a dynamical cut from the s channel threshold s_0 ($=4m_\pi^2$ in this case) to ∞ (unless two trajectories cross, in which case we should have to include the contributions of both in our discussion).

Also

$$z_s = 1 + t/2q_s^2 \qquad (2.5)$$

has a pole at $q_s^2 = 0$ which introduces an unwanted kinematical cut into $Q_{-\alpha(s)-1}(-z_s)$, but this is avoided if we take [see (1.7)]

$$\beta(s) = \gamma(s)\, (q_s^2)^{\alpha(s)} \qquad (2.6)$$

so that the cut of $Q_{-\alpha(s)-1}(-z_s)$ is cancelled by that of $(q_s^2)^{\alpha(s)}$ by (II.11.6).

We have seen that this so called "reduced" residue function is also expected to be a real analytic function with a dynamical cut beginning at the threshold. It was noted in Chapter II that the Sommerfeld-Watson transform (Section II.9) depended on the partial-wave amplitudes being bounded as in (II.8.1), but this is only true for $q_s^2 > 0$ [see (II.7.1)]. The partial wave amplitude $A^{\pm}(s, l)$ defined by (II.3.6), has a branch-point at threshold which we noted in (II.4.9), and has the asymptotic behaviour

$$A^{\pm}(s, l) \underset{l \to \infty}{\sim} e^{-i\pi l}\, e^{-l\xi(z_n)}, \quad s < s_0,$$

instead of (II.7.4), because the range of integration in (II.6.5) is then over negative z_s. Hence the Sommerfeld-Watson transform cannot be performed for $s < s_0$. But

$$\hat{A}^{\pm}(s, l) \equiv e^{i\pi l}\, A^{\pm}(s, l)$$

is well behaved as $l \to \infty$, and so we can write (see *Challifour* and *Eden*, 1963a, 1963b)

$$A^{\pm}(s, z) = \frac{1}{16\pi} \sum_l (2l + 1)\, \hat{A}(s, l)\, P_l(z)$$

$$= \frac{1}{16\pi}\, \frac{1}{2i} \int_c \frac{(2l+1)\, \hat{A}^{\pm}(s, l)\, P_l(-z)}{\sin\pi l}\, \mathrm{d}l + (\text{Regge poles and cuts})$$

instead of (2.2) for $s < s_0$. Of course this representation breaks down for $s > s_0$. The ambiguity in defining the amplitudes $A^\pm(s, l)$ or $\hat{A}^\pm(s, l)$ depends on the choice of different locations for the kinematical cut from s_0, either along the negative or positive real axis.

Correspondingly, the Regge pole terms must include the kinematical factor (2.6) to kelp the t cut fixed as we move to negative q_s^2. Otherwise the relation between cuts in z_s and cuts in t would be ambiguous, since $t \pm i\varepsilon$ may correspond to $z \mp i\varepsilon$ or $z \pm i\varepsilon$ depending on whether q_s^2 in (2.5) is positive or negative. The best way of keeping this clear is to write

$$A_R^\pm(s, t) = \Gamma(s) \left(-\frac{q_s^2}{t}\right)^{\alpha(s)} (-t)^{\alpha(s)} Q_{-\alpha(s)-1}\left(-1 - \frac{t}{2q_s^2}\right), \quad (2.7)$$

where

$$\Gamma(s) = \frac{16\pi[2\alpha(s) + 1]\,\gamma(s)}{\cos\pi\alpha(s)}. \quad (2.8)$$

Then the cut in $Q_{-\alpha(s)-1}(-1 - t/2q_s^2)$ for $(-1 - t/2q_s^2) < -1$ is always cancelled by that of $(-q_s^2/t)^{\alpha(s)}$ for $(q_s^2/t) > 0$, since [(II.11.4), (II.11.5)]

$$\left(-\frac{q_s^2}{t} \pm i\varepsilon\right) Q_{-\alpha(s)-1}(-z \pm i\varepsilon) = \left(\frac{q_s^2}{t}\right)^\alpha Q_{-\alpha(s)-1}(z). \quad (2.9)$$

So the cuts in $A_R^\pm(s, t)$, as defined in (2.7), are:
at fixed s:

    a cut in t from $t = 0$ to ∞, due to the $(-t)^{\alpha(s)}$ term,
    and one from $t = 0$ to $-4q_s^2$, from (I.11.1), $-1 < z < 1$.

at fixed t:

    a cut in s from $s = s_0$ to ∞, from the cuts of $\Gamma(s)$ and $\alpha(s)$,
    and one from $q_s^2 = -\infty$ to $-t/4$, from (I.11.1), $-1 < z < 1$.

But the amplitude (2.3) must have a t discontinuity from t_0 to ∞ and an s discontinuity from s_0 to ∞ only, with the double spectral function limited to the even smaller region of Fig. (I.16). Evidently (2.3) will be compatible with the Mandelstam representation only if the background has singularities which cancel those of $A_R^\pm(s, t)$. This may seem to be asking for a miracle, but it should be remembered that what we have done in (2.3) is to make a rather odd decomposition of (2.1), the matching of the singularities should be automatic. In potential scattering, which is known both to satisfy the Mandelstam representation and exhibit Regge behaviour, it is automatic.

Similar difficulties apply to (2.2), where a single Regge term gives

$$A_R^\pm(s, t) = -G(s)\,(q_s^2)^{\alpha(s)} \frac{P_{\alpha(s)}\left(-1 - \frac{t}{2q_s^2}\right)}{\sin\pi\alpha(s)}, \quad (2.10)$$

with

$$G(s) = 16\pi^2[2\alpha(s) + 1]\,\gamma(s). \quad (2.11)$$

The singularity at $q_s^2 = 0$ is less easily dealt with in this case, because, under inversion of its argument, $P_\alpha(z)$ satisfies the relation (II.10.2) rather than the simpler (II.11.4) of $Q_\alpha(z)$. However, if we make use of (II.11.4) we can write

$$A_{\overline{R}}^{\pm}(s, t) = -G(s) \, (-q_s^2)^{\alpha(s)} \left[\frac{P_\alpha\left(1 + \frac{t}{2q_s^2}\right)}{\sin \pi \alpha(s)} + \frac{2}{\pi} Q_\alpha\left(-1 - \frac{t}{2q_s^2}\right) \right] \quad (2.12)$$

for $q_s^2 < 0$.

At fixed s this is cut in t from $t = -4q_s^2$ to ∞ ,

while at fixed t it is cut from $q_s^2 = 0$ to ∞ ,

and from $q_s^2 = -t/2$ to 0 .

Again a cancellation of the singularities outside the regions permitted by the Mandelstam representation will be needed.

In strong interactions we would often like to represent the amplitude simply in terms of Regge poles, neglecting the background because we know that it is asymptotically unimportant, but it will now be evident that we must include at least enough of the background to cancel the unwanted singularities of the Regge pole terms. And of course the possibility of making these terms satisfy the Mandelstam representation has

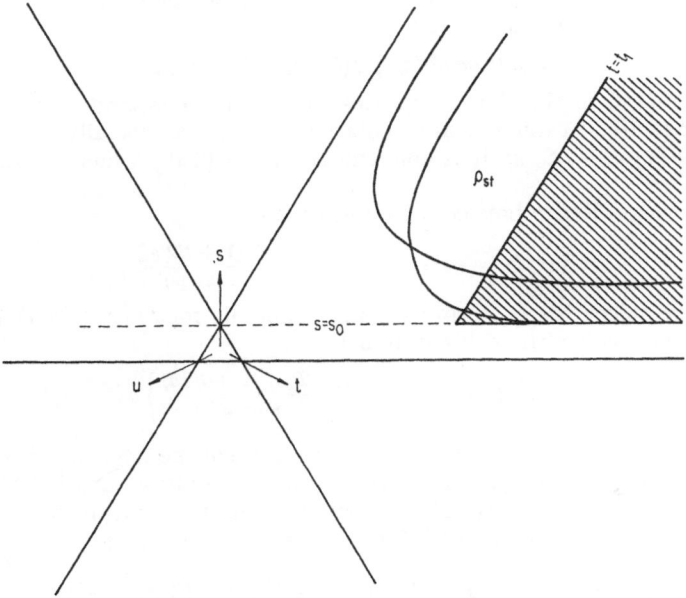

Fig. III.2. The double spectral function region for the Chew-Jones representation (shaded) compared with the true boundaries of Fig. I.16, for an s-channel Regge pole. If t_1 is sufficiently large, taking the boundary at s_0 introduces only a small discrepancy.

so far only been assumed. We have still to show that compatibility can be achieved, especially in cases where the kinematics are more complicated than the equal-mass case which has been discussed so far.

In the next few sections we shall review some of the Regge pole representations which have been proposed, firstly [Sections (III. 3–5)] with equal-mass kinematics, and then more generally.

III.3. The Chew-Jones Representation

A very simple Regge pole representation, which has proved suitable for dynamical calculations with equal mass kinematics, was proposed by *Chew* and *Jones*, 1964. They noted that an s-channel Regge pole term would be expected to provide a significant contribution to the double spectral functions only outside the crossed-channel resonance region. Thus, if we examine Fig. (III.2), we expect that for an s-channel Regge pole the important double spectral function will be in the region $t > t_1$, with t_1 sufficiently large to be above all the t-channel poles. The value of t_1 will be chosen to correspond to the energy at which Regge behaviour in the t-channel becomes evident, i.e. 2 or 3 Gev typically. For such values of t the boundary of the double spectral function will lie almost along the line $s = s_0$ [see (I.12.10)]. Hence for a single s-channel Regge pole it seems reasonable to take

$$\varrho(s, t) = \varrho^R(s, t)\, \theta(t - t_1)\, \theta(s - s_0)\,, \tag{3.1}$$

where $\varrho^R(s, t)$ is the double spectral function corresponding to either (2.7) or (2.10). In dynamical calculations we are not usually concerned with $\mathrm{Re}\{\alpha\} < -1/2$, so it is convenient to use (2.10) which we discuss first.

The single Regge pole contributes, (2.12),

$$A_R^{\mp}(s, t) = -\, G(s)\,(-q_s^2)^{\alpha(s)}\, \frac{P_\alpha(1 + t/2\, q_s^2)}{\sin \pi \alpha(s)}\,, \tag{3.2}$$

where we reject the Q_α term as being negligible for $t > t_1$ (see II.12.9). For $s > s_0$ this has the s-discontinuity

$$D_{sR}^{\pm}(s, t) = \varDelta_s\left\{-G(s)\,(-q_s^2)^{\alpha(s)}\, \frac{P_{\alpha(s)}(1 + t/2q_s^2)}{\sin \pi \alpha(s)}\right\} \theta(s - s_0) \tag{3.3}$$

due to the cuts of $\alpha(s)$ and $\gamma(s)$, as well as those of the Legendre function [see following (2.12)]. Of course this is not the only s cut of (3.2) but it is only one we require. If we now take the t-discontinuity of (3.3) in the region corresponding to (3.1) we get, using (II.9.12),

$$\varrho(s, t) = \varDelta_s\left\{G(s)\,(-q_s^2)^{\alpha(s)}\, P_{\alpha(s)}\left(-1 - \frac{t}{2q_s^2}\right)\right\} \theta(t - t_1)\, \theta(s - s_0)\,. \tag{3.4}$$

If this expression is substituted into (2.1), the s' integration, being an integration over all the s cuts of the expression in braces, $\{\ \}$, in (3.4), simply corresponds to removing the $\varDelta_s$ sign, and we are left with

$$A^{\pm}(s, t) = G(s)\,(-q_s^2)^{\alpha(s)}\, \frac{1}{\pi} \int\limits_{t_1}^{\infty} \frac{P_{\alpha(s)}(1 - t''/2q_s^2)}{t'' - t}\, dt''\,. \tag{3.5}$$

If $\alpha(s) \geqq 0$, this integral is undefined, and we have to use the dispersion relation for the Legendre function (*Chew* et al., 1962),

$$P_\alpha(-z) = -\frac{\sin \pi \alpha}{\pi} \int_1^\infty \frac{P_\alpha(z')}{z'-z}\, dz' , \tag{3.6}$$

to define the analytic continuation in α. This leads to a determined subtraction of the type discussed in Section (II.9). We thus obtain from (3.5)

$$A^\pm(s, t) = G(s)\, (-q_s^2)^{\alpha(s)} \times$$

$$\times \left[-\frac{P_{\alpha(s)}(1 + t/2q_s^2)}{\sin \pi \alpha(s)} - \frac{1}{\pi} \int_{-4q_s^2}^{t_1} \frac{P_{\alpha(s)}(-1 - t''/2q_s^2)}{t'' - t}\, dt'' \right]. \tag{3.7}$$

The difference between (3.2) and (3.7) is simply that we have removed the part of A_R corresponding to the inadmissible piece of double spectral function from $t = -4q_s^2$ to t_1. At fixed s this correction term goes asymptotically like $\sim t^{-1}$, and so is easily contained within the background. The use of such a representation for dynamical calculations will be discussed in Chapter VI.

On the other hand, we may prefer to use the representation (2.3) for an amplitude with a single Regge pole, writing

$$A^\pm(s, t) = A_{\bar{R}}^\pm(s, t) + B^\pm(s, t) , \tag{3.8}$$

and taking $A_{\bar{R}}^\pm(s, t)$ from (2.7). We have already noted the cuts of $A_{\bar{R}}^\pm(s,t)$ in the paragraph following (2.9), and if $A^\pm(s, t)$ is to satisfy the Mandelstam representation several of them must be cancelled by those of $B^\pm(s, t)$, while still retaining the property that $B^\pm(s, t)$ vanishes at least as fast as $t^{-\frac{1}{2}}$ as $t \to \infty$.

Now, (2.7),

$$A_{\bar{R}}^\pm(s, t) = \Gamma(s) \left(-\frac{q_s^2}{t} \right)^{\alpha(s)} (-t)^{\alpha(s)} Q_{-\alpha(s)-1} \left(-1 - \frac{t}{2q_s^2} \right)$$

has an s-discontinuity, for $s > s_0$, given by

$$D_{s\bar{R}}^\pm(s, t) = \Delta_s \left[\Gamma(s) \left(-\frac{q_s^2}{t} \right)^{\alpha(s)} (-t)^{\alpha(s)} Q_{-\alpha(s)-1} \left(-1 - \frac{t}{2q_s^2} \right) \right] \theta(s - s_0), \tag{3.9}$$

though this is not the only s cut of $A_{\bar{R}}^\pm$.

The t-discontinuity of this in the region $t > 0$ is then

$$\varrho(s, t) = \Delta_s \left[-\Gamma(s) \left(-\frac{q_s^2}{t} \right)^{\alpha(s)} (t)^{\alpha(s)} \sin \pi \alpha(s)\, Q_{-\alpha(s)-1} \left(-1 - \frac{t}{2q_s^2} \right) \right] \times$$

$$\times\, \theta(s - s_0)\, \theta(t)$$

$$= \Delta_s \left[-\Gamma(s) \left(\frac{q_s^2}{t} \right)^{\alpha(s)} (t)^{\alpha(s)} \sin \pi \alpha(s)\, Q_{-\alpha(s)-1} \left(1 + \frac{t}{2q_s^2} \right) \right] \times \tag{3.10}$$

$$\times\, \theta(s - s_0)\, \theta(t) ,$$

from (2.9).

This of course is still not the correct boundary for the double spectral function, but the part [compare Fig. (III.2)]

$$\bar{B}(s,t) \equiv \frac{1}{\pi^2} \int\limits_{s_0}^{\infty} \frac{ds'}{(s'-s)} \int\limits_{0}^{b_1(s')} \frac{dt''}{(t''-t)} \Delta_s \times$$

$$\times \left[-\Gamma(s') \left(\frac{q_{s'}^2}{t''} \right)^{\alpha} (t'')^{\alpha} \sin\pi\alpha \; Q_{-\alpha-1} \left(1 + \frac{t''}{2q_{s'}^2} \right) \right] \tag{3.11}$$

can easily be subtracted later. Use of the boundaries (3.10) enables us to write simpler forms for the correction terms. [In what follows we shall often drop the argument of $\alpha(s)$].

If (3.10) is used with

$$D_s^{\pm}(s,t) = \frac{1}{\pi} \int\limits_{0}^{\infty} dt'' \frac{\varrho(s,t'')}{t''-t} \tag{3.12}$$

we obtain

$$D_s^{\pm}(s,t) = \Delta_s \left[\Gamma(s) \left(-\frac{q_s^2}{t} \right)^{\alpha} (-t)^{\alpha} Q_{-\alpha-1} \left(-1 - \frac{t}{2q_s^2} \right) \right] +$$

$$+ \Delta_s \left[\frac{1}{2} \int\limits_{0}^{-4q_s^2} \frac{dt''}{t''-t} \Gamma(s) \left(-\frac{q_s^2}{t} \right)^{\alpha} (-t'')^{\alpha} P_{\alpha} \left(-1 - \frac{t''}{2q_s^2} \right) \right] \tag{3.13}$$

which differs from (3.9) by the correction terms from the extra t cut of (3.10), which is evaluated with the aid of (II.11.1).

Similarly

$$D_{\bar{t}}^{\pm}(s,t) = \frac{1}{\pi} \int\limits_{s_0}^{\infty} ds' \frac{\varrho(s',t)}{s'-s} \tag{3.14}$$

gives

$$D_{\bar{t}}^{\pm}(s,t) = -\Gamma(s) \left(-\frac{q_s^2}{t} \right)^{\alpha} (t) \sin\pi\alpha \; Q_{-\alpha-1} \left(-1 - \frac{t}{2q_s^2} \right) -$$

$$- \frac{1}{2} \int\limits_{-\infty}^{-t-s_0} \frac{ds'}{s'-s} \Gamma(s') \left(-\frac{q_s'^2}{t} \right)^{\alpha} (t)^{\alpha} \sin\pi\alpha \; P_{\alpha} \left(-1 - \frac{t}{2q_s'^2} \right), \tag{3.15}$$

and either (3.13) or (3.15) gives [remembering (3.11)],

$$A^{\pm}(s,t) = \Gamma(s) \left(-\frac{q_s^2}{t} \right)^{\alpha} (-t)^{\alpha} Q_{-\alpha-1} \left(-1 - \frac{t}{2q_s^2} \right) +$$

$$+ \frac{1}{2} \int\limits_{-\infty}^{-t-s_0} \frac{ds'}{s'-s} \Gamma(s') (-t)^{\alpha} P_{\alpha} \left(-1 - \frac{t}{2q_s'^2} \right) + \tag{3.16}$$

$$+ \frac{1}{2} \int\limits_{0}^{-4q_s^2} \frac{dt''}{t''-t} \Gamma(s) \left(-\frac{q_s^2}{t''} \right)^{\alpha} (-t'')^{\alpha} P_{\alpha} \left(-1 - \frac{t''}{2q_s^2} \right) - \bar{B}(s,t).$$

The difference between (3.16) and (2.7) is simply the removal of the contribution of the unwanted cuts of the Regge pole term. It only remains to check that the last three terms of (3.16) have an asymptotic behaviour compatible with our needs. Evidently the second term on the right-hand side of (3.16) $\sim t^{\alpha(-\infty)}$, while the third and fourth terms go like $\sim t^{-1}$, so we only require that $\alpha(-\infty) < -1/2$.

Either (3.7) or (3.16) may be used as the contribution to the amplitude of a single Regge trajectory, which satisfies the Mandelstam representation for equal-mass kinematics.

III.4. The Khuri-Jones Representation

A disadvantage of the Chew-Jones representation is that the correction terms can only be written as integrals, and these are not readily evaluated. One cannot in fact discuss the form of the contribution of a given Regge trajectory to an amplitude without being involved in explicit numerical evaluation. An alternative and earlier method was proposed independently by *Khuri*, 1963a, and *Jones*, 1962, which, though less satisfactory in some ways than the Chew-Jones method, has the advantage of being readily evaluated as a partial-wave series.

The basic idea is exactly the same. We remove from the Regge term (3.2) the part corresponding to the cut between $t = -4q_s^2$ and $t = t_1$. To this end we invoke an integral representation for the Legendre function [B 1, 3.7 (II)]

$$P_\alpha(z) = -\frac{\sqrt{2}}{\pi} \sin \pi \alpha \int\limits_0^\infty \frac{\cosh\left[\left(\alpha + \frac{1}{2}\right) x\right]}{(\cosh x + z)^{\frac{1}{2}}} \, dx , \qquad -1 < \mathrm{Re}\{\alpha\} < 0 \tag{4.1}$$

which may be re-written

$$P_\alpha(z) = -\frac{\sin \pi \alpha}{\pi \sqrt{2}} \int\limits_{-\infty}^\infty \frac{e^{\left(\alpha + \frac{1}{2}\right)}}{(\cosh x + z)^{\frac{1}{2}}} \, dx . \tag{4.2}$$

On integrating by parts, this gives

$$P_\alpha(z) = -\frac{\sin \pi \alpha}{\pi \, 2 \sqrt{2} \left(\alpha + \frac{1}{2}\right)} \int\limits_{-\infty}^\infty \frac{e^{\left(\alpha + \frac{1}{2}\right) x} \sinh x}{(\cosh x + z)^{\frac{3}{2}}} \, dx . \tag{4.3}$$

We can thus produce a representation of the Regge term (3.2) with a cut beginning at $t = t_1$, viz.

$$A^\pm(s, t) = 16\pi \, \gamma(s) \, (-q_s^2)^{\alpha(s)} \frac{1}{\sqrt{2}} \int\limits_\xi^\infty \frac{e^{\left(\alpha + \frac{1}{2}\right) x} \sinh x}{(\cosh x + z_s)^{\frac{3}{2}}} \, dx , \tag{4.4}$$

where

$$\xi(s) = \cosh^{-1}\left(1 + \frac{t_1}{2q_s^2}\right) = \log\left\{\left(1 + \frac{t_1}{2q_s^2}\right) + \left[\left(1 + \frac{t_1}{2q_s^2}\right)^2 - 1\right]^{\frac{1}{2}}\right\}. \quad (4.5)$$

Equation (4.4) is only defined for $\mathrm{Re}\{\alpha\} < 0$, but we can use (4.3) to define the analytic continuation to $\mathrm{Re}\{\alpha\} > 0$ analogous to (3.7), and find

$$A^{\pm}(s, t) = 16\pi\,\gamma(s)\,(-q_s^2)^{\alpha(s)} \times$$

$$\times \left[\frac{\pi[2\alpha(s) + 1]\,P_\alpha(1 + t/2q_s^2)}{\sin\pi\alpha(s)} - \frac{1}{\sqrt{2}}\int_{-\infty}^{\xi} \frac{e^{\left(\alpha + \frac{1}{2}\right)x}\sinh x}{(\cosh x + z_s)^{\frac{3}{2}}}\,dx\right].$$

$$(4.6)$$

In fact both (4.4) and (4.6) are valid in the interval $-1 < \mathrm{Re}\{\alpha\} < 0$, and can be shown to be equal there (see *Khuri*, 1963a).

The advantage of this representation over the Chew-Jones form is that the partial-wave projection of (4.4) or (4.6) takes on a particularly simple form. If we substitute whichever is valid for the given value of α into (II.1.1), and use the inverse relation to (4.3), i.e.

$$\frac{\sinh x}{(\cosh x + z)^{\frac{3}{2}}} = -\frac{i}{\sqrt{2}}\int_{-\frac{1}{2}-i\infty}^{-\frac{1}{2}+i\infty} (2l + 1)\,P_l(z)\,\frac{e^{-\left(l + \frac{1}{2}\right)x}}{\cos\left[\pi\left(l + \frac{1}{2}\right)\right]}\,dl\,,$$

$$(4.7)$$

we find

$$A_l^{\pm}(s) = -\,\gamma(s)\,(-q_s^2)^{\alpha(s)}\,\frac{e^{-[l - \alpha(s)]\,\xi(s)}}{\alpha(s) - l}\,. \quad (4.8)$$

It will be evident that because of the asymptotic behaviour of $P_l(z)$ as $l \to \infty$ given by (II.7.11), the partial wave series (II.1.3) with (4.8) will converge for $t < t_1$. This is a consequence of our having removed the double spectral function for $t < t_1$, and the Chew-Jones formula of the previous section has a similar domain of convergence.

The main disadvantage of the Khuri-Jones representation is that in addition to the right-hand cut in s, from $s = 4m^2 \to \infty$, due to the cuts in $\alpha(s)$ and $\gamma(s)$, there is also a left-hand cut, from $4q_s^2 = -t_1$ (i.e. $s = -t_1 + 4m^2$) to $-\infty$, arising out of the behaviour of $\xi(s)$, (4.5). Such a cut is to be expected in the partial wave amplitude (see Section II.4), but not in the total amplitude. If $A^{\pm}(s, t)$ given by (4.4) or (4.6) were the exact amplitude this cut would cancel in the sum of the partial-wave series, but, as it is only an approximation, an exact cancellation cannot be expected. This defect has made the Khuri-Jones representation unsuitable for dynamical calculations, but the partial wave projection (4.8) is very instructive in showing how the force from the exchange of a Regge trajectory differs from that due to the exchange of a fixed spin (elementary) particle. We shall discuss this in Chapter VI.

A similar but more complicated expression, from which the unwanted left-hand cut has been removed, has been given by *Kretzschmar*, 1964.

III.5. The Khuri Power-Series Expansion

So far the hypothesis of Regge poles has been tied to the Legendre function partial-wave expansion, and this we have seen results in some difficulties, in that the scattering angle, z_s, is a function of both s and t. In the end, however, we are usually interested in the asymptotic properties of the amplitude in s or t, rather than in z_s, and it would seem quite reasonable to begin with a power series in these invariants rather than the Legendre series. *Khuri*, 1963b and 1963c, has shown how this can be done. Two points need to be established, firstly that a suitable interpolation of the power series coefficients exists, and secondly that we can expand in t rather than z_s.

Instead of the partial-wave series (II.1.3) we write the power series expansion

$$A^{\pm}(s, t) = \sum_{\nu=0}^{\infty} c(s, \nu)\, t^{\nu}, \quad \nu = \text{integer}. \tag{5.1}$$

From the dispersion relation (II.2.7),

$$A^{\pm}(s, t) = \frac{1}{\pi} \int_{t_0}^{\infty} \frac{D_t^{\mp}(s, t'')}{t'' - t}\, dt''. \tag{5.2}$$

we get

$$A^{\pm}(s, t) = \frac{1}{\pi} \int_{t_0}^{\infty} \frac{D_t^{\mp}(s, t'')}{t''} \left[1 + \frac{t}{t''} + \left(\frac{t}{t''}\right)^2 + \cdots \right] dt'', \tag{5.3}$$

which, when compared with (5.1), gives

$$c(s, \nu) = \frac{1}{\pi} \int_{t_0}^{\infty} D_t^{\mp}(s, t'')\, t''^{-(\nu+1)}\, dt''. \tag{5.4}$$

This expression allows us to define an interpolation of $c(s, \nu)$ in ν for all ν such that the integral (5.4) exists, and by comparing it with (II.6.5) we see that it will converge for all ν such that $\text{Re}\{\nu\} \geq l_M(s)$. So $c(s, \nu)$ is holomorphic in ν in the same region that $A(s, l)$ is holomorphic in l. In order to continue (5.4) below this bound we have to subtract from $D_t^{\mp}(s, t)$ the parts corresponding to the various right-half l-plane singularities, just as we did in Section (II.9). Let us for convenience suppose that there are no cuts, but only poles at $l = \alpha_i(s)$. Using (2.4) for the contribution of a Regge pole, together with (II.11.1), we can write

$$D_t^{\pm}(s, t) = D_t^{B}(s, t) + \sum_i 16\pi \left[2\alpha_i(s) + 1\right] \times$$

$$\times \gamma_i(s)\,(q_s^2)^{\alpha_i(s)}\,\tan\pi\,\alpha_i(s)Q_{-\alpha_i(s)-1}\,(-z_s)\,,$$
(5.5)

where $D_t^{B}(s, t)$ is the contribution of the background, and can be chosen to vanish at least as fast as $t^{-\frac{1}{2}}$ as $t \to \infty$. We can then use (II.12.8) to make an expansion of $Q_{-\alpha-1}(-z)$ in powers of z^{-2},

$$Q_{-\alpha-1}(z) = g_0(\alpha)\,z^{\alpha} + g_1(\alpha)\,z^{\alpha-2} + \cdots + g_n(\alpha)\,z^{\alpha-2n} + \cdots,\quad (5.6)$$

and perform the integral (5.4) to obtain

$$c(s, \nu) = -\sum_i 8\left[2\alpha_i(s) + 1\right]\gamma_i(s)\,\frac{\tan\pi\,\alpha_i(s)}{(4)^{\alpha(s)}}\sum_{n=0}^{\infty}\frac{\Gamma(-\alpha_i + n)}{\Gamma(-2\alpha_i + n)} \times$$

$$\times \frac{(-1)^n}{n!}\,(4q_s^2)^n\,,\,\frac{1}{\nu + n - \alpha_i}\,,$$
(5.7)

where $c^B(s, \nu)$ is the contribution of $D_t^{B}(s, t)$. Thus to each Regge pole there corresponds a sequence of poles of $c(s, \nu)$ at $\nu = \alpha_i(s),\ \alpha_i(s) - 1,$ $\alpha_i(s) - 2, \ldots$.

We can then make a Sommerfeld-Watson transform of the series (5.1) in a completely similar way to our treatment of the partial wave series in Section (II.9), and we find

$$A^{\pm}(s, t) = -\frac{1}{2i}\int_{-\frac{1}{2}-i\infty}^{-\frac{1}{2}+i\infty}\frac{c(s, \nu)}{\sin\pi\nu}\,(-t)^{\nu}\,d\nu - \sum_i \frac{8\pi\gamma_i(s)\,[2\alpha_i(s) + 1]}{(4)^{\alpha_i(s)}\cos\pi\alpha_i(s)} \times$$

$$\times \sum_{n=0}^{\infty}\frac{\Gamma(-\alpha_i + n)^2}{\Gamma(-2\alpha_i + n)}\,\frac{1}{n!}\,(4q_s^2)^n\,(-t)^{\alpha_i-n}.$$
(5.8)

It is now evident that the Khuri series corresponds simply to making a power series expansion in t of the Regge pole terms. Of course the background can be pushed back to the left of $\mathrm{Re}\{\nu\} = -1/2$ if we wish.

So corresponding to each Regge pole $\alpha_i(s)$ there is a principal Khuri pole at $\nu = \alpha_i(s)$, giving an asymptotic behaviour $t^{\alpha_i(s)}$, and an infinite sequence of "satellite" poles at $\nu = \alpha_i(s) - n$, $n = 1, 2, \ldots$, giving contributions to the asymptotic behaviour of the form $t^{\alpha_i(s)-1}\,t^{\alpha_i(s)-2}\ldots$ etc. The argument we have given could be reversed, and we should find that corresponding to each Khuri pole at $\nu = \alpha_i(s)$ there was an infinite sequence of Regge poles at $l = \alpha_i(s),\ \alpha_i(s) - 1, \ldots$. But of course, if we take the full sequence of satellite poles in (5.8), they must give contributions which cancel each other in such a way as to leave just one Regge pole at $l = \alpha_i(s)$.

Each of the terms in the sequence (5.8) is cut in t from $t = 0$ to ∞, but naturally these cuts can be made to cancel with the background for $t < t_1$, just as in Section (III.3). The way in which this is achieved has been demonstrated explicitly by *Khuri*, 1963c. In using (5.8) one would obviously want to truncate the series in n at some value $n = N_i$

(say) for each trajectory, such that $\alpha_i(s) - N_i < -1/2$ for all s, i.e. such that all further terms vanish like the background asymptotically. The principal difference between (5.8) and the usual Regge representation then lies in the asymptotic s behaviour. The Regge pole term (2.4) has its asymptotic s behaviour controlled by the behaviour of

$$\frac{[2\alpha(s) + 1]\,\gamma(s)}{\cos\pi\alpha(s)}$$

whereas (5.8) depends on

$$\frac{[2\alpha(s) + 1]\,\gamma(s)}{\cos\pi\alpha(s)}\,(4q_s^2)^N\,.$$

The whole value of Regge poles depends on the fact that the asymptotic s behaviour is given by the t- and u-channel poles, while the s-channel pole contributions vanish at large s. There are more stringent requirements on the behaviour of $\gamma(s)$ for the truncated series (5.8) to satisfy this condition than for the Chew-Jones representation, and the latter is usually to be preferred (in this connection see *Jones*, 1964).

It should now be clear that, instead of using the partial-wave expansion in s and obtaining Regge poles, we could equally well use many other types of polynomial expansion of the amplitude, and use many other arguments of these polynomials (provided they depend on s and t), and make similar Sommerfeld-Watson transforms. The complication of the singularities in the plane of the order of the polynomial (corresponding to the l-plane or the ν-plane), and the difficult cancellations required of the singularities in the argument plane (corresponding to z_s or t), make this procedure in general rather pointless. It is obviously most convenient to use that plane (for the order of the polynomial) in which there is a minimum number of poles (preferably only one) corresponding to each physical particle. This is the case with the Regge l-plane. However, as we shall see in subsequent sections, the complications of unequal mass kinematics have led some writers to doubt the primacy of the Regge plane in particle physics.

III.6. The Problem of Unequal Mass Kinematics

So far in this chapter we have been assuming that the external particles all have the same masses, so that q_s^2 and z_s are simply related to s and t via (I.6.15) and (I.6.16). There is then a very simple correspondence between the asymptotic t behaviour and the asymptotic z_s behaviour, and singularities in z_s are readily related to those in s and t, as we described in Section (III.2). The only difficulty is the branchpoint at $q_s^2 = 0$, which we discussed in Section (III.2). However, if the masses are different, so that the more general kinematical relations (I.6.12) and (I.6.14) obtain, the discussion becomes much more complicated. Singularities in z_s become multiple singularities in s, and the correspondence between the asymptotic t and z_s behaviour becomes more dependent on s. There are thus two related problems. Can we construct a suitably corrected Regge representation which satisfies the Mandelstam representation, similar to that of Section (III.3), and, if we can, is Regge asymptotic behaviour still found at all values of s?

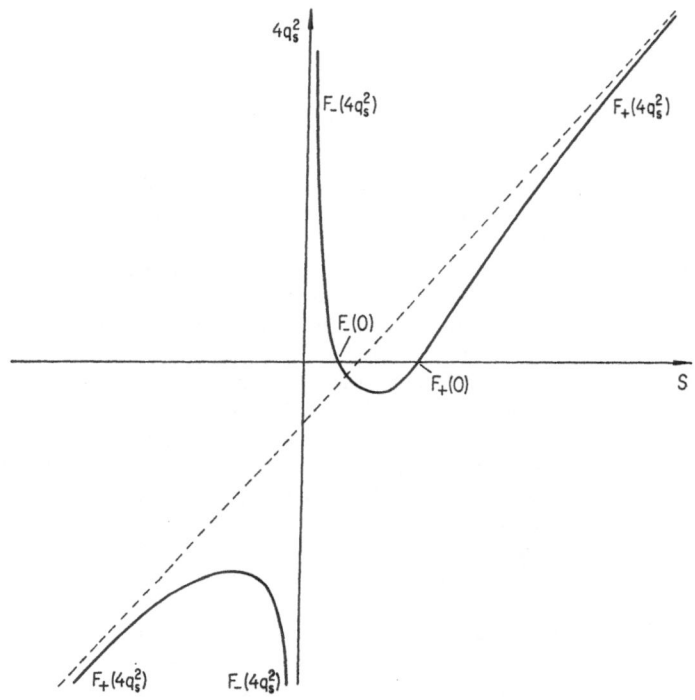

Fig. III.3. The variation of $4q_s^2$ with s from equation (6.4), showing the two branches of the relation. The dotted line corresponds to the equal mass case.

For simplicity we shall discuss in detail the situation when the s channel corresponds to the elastic scattering of two particles of masses m_1 and m_2, so that $m_3 = m_1$ and $m_4 = m_2$. The generalization to the case when all the masses are unequal is then similar but a good deal more tedious.

From (I.6.12) we have

$$q_{s\,12}^2 = q_{s\,34}^2 \equiv q_s^2 = \frac{s^2 - s\,\Sigma + \Delta^2}{4s}, \tag{6.1}$$

where

$$\Sigma = 2m_1^2 + 2m_2^2, \tag{6.2}$$

and

$$\Delta = m_1^2 - m_2^2,$$

and from (I.6.14)

$$z_s = 1 + \frac{t}{2q_s^2} = \left[1 + \frac{2st}{s^2 - s\,\Sigma + \Delta^2}\right]. \tag{6.3}$$

It is the multiple relationship between q_s^2 and s in (6.1) which causes the chief difficulty, since we have, solving for s,

$$s = \frac{1}{2}\left\{\Sigma + 4q_s^2 \pm [(\Sigma + 4q_s^2)^2 - 4\,\Delta^2]^{\frac{1}{2}}\right\} \qquad (6.4)$$

$$\equiv F_-(4q_s^2) \quad \text{or} \quad F_+(4q_s^2), \qquad (6.5)$$

where F_- or F_+ is used according to the $\mp$ sign in (6.4). A plot of $4q_s^2$ vs. s is shown in Fig. (III.3). The s-channel threshold is at

$$s_0 = (m_1 + m_2)^2 = F_+(0). \qquad (6.6)$$

We note in particular that $q_s^2 \to \infty$ as $s \to 0$ as well as $s \to \infty$, so that the simple Regge pole term,

$$A_{\overline{R}}^{\pm}(s, t) = \Gamma(s) \left(\frac{-q_s^2}{t}\right)^{\alpha(s)} (-t)^{\alpha(s)} Q_{-\alpha(s)-1}\left(-1 - \frac{t}{2q_s^2}\right), \qquad (2.7)$$

is undefined at $s = 0$. Also from (6.3) we see that there is a region of s

$$0 \geq s \geq F_-(-t) \qquad (6.7)$$

for which $|z_s| \leq 1$, so that, as $t \to \infty$, z_s is bounded by 1 at $s = 0$. This is important because it means that it is not evident that we shall find Regge asymptotic behaviour in the t-channel forward direction (i.e. at $s = 0$).

Let us then try to modify (2.7), in the spirit of Section (III.3), so that it does satisfy the Mandelstam representation. In so doing we shall find that these anomalies at $s = 0$ disappear automatically.

The cuts of (2.7) for the equal mass case were given following equation (2.9). With unequal masses the cuts in t at fixed s are evidently unaltered, as is the cut in s from $s_0 \to \infty$ which comes from $\alpha(s)$ and $\Gamma(s)$. The difference lies in the cut from $q_s^2 = -\infty$ to $-t/4$, which now becomes two cuts in the s plane:

from $s = -\infty$ to $F_+(-t)$,
and from $s = 0$ to $F_-(-t)$.

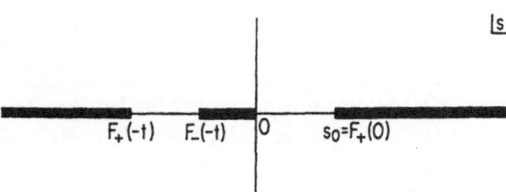

Fig. III.4. The cuts in the s-plane, at fixed (positive) t, of the Regge term (2.7), with unequal mass kinematics.

These s cuts are shown, for positive t, in Fig. (III.4). We again start from the double spectral function

$$\varrho(s, t) = \Delta_s \left[-\Gamma(s)\left(\frac{q_s^2}{t}\right)^{\alpha}(t)^{\alpha} \sin\pi\alpha\, Q_{-\alpha-1}\left(1 + \frac{t}{2q_s^2}\right)\right]\theta(s - s_0)\,\theta(t), \qquad (3.10)$$

which, substituted in (3.12), again gives $D_{\bar{s}}^{\pm}(s, t)$ as in (3.13). However if it is substituted in (3.14), because of the extra s cut we get

$$D_{\bar{t}}^{\pm}(s, t) = -\Gamma(s)\left(-\frac{q_s^2}{t}\right)^{\alpha}(t)^{\alpha}\sin\pi\alpha\; Q_{-\alpha-1}\left(-1-\frac{t}{2q_s^2}\right) - \tag{6.8}$$

$$-\frac{1}{2}\int_{-\infty}^{F_+(-t)}\frac{\Gamma(s')}{s'-s}\left(-\frac{q_s'^2}{t}\right)^{\alpha}(t)^{\alpha}\sin\pi\alpha\; P_{\alpha}\left(-1-\frac{t}{2q_s'^2}\right)ds'$$

$$-\frac{1}{2}\int_0^{F_-(-t)}\frac{\Gamma(s')}{s'-s}\left(-\frac{q_s'^2}{t}\right)^{\alpha}(t)^{\alpha}\sin\pi\alpha\; P_{\alpha}\left(-1-\frac{t}{2q_s'^2}\right)ds'\;,$$

so that finally we obtain [remembering (3.11)]

$$A^{\pm}(s, t) = \Gamma(s)\left(\frac{-q_s'}{t}\right)^{\alpha}(-t)^{\alpha}Q_{-\alpha-1}\left(-1-\frac{t}{2q_s^2}\right) +$$

$$+\frac{1}{2}\int_0^{-4q_s^2}\frac{dt''}{t''-t}\;\Gamma(s)\left(-\frac{t''}{q_s^2}\right)^{\alpha}(-t'')^{\alpha}P_{\alpha}\left(-1-\frac{t''}{2q_s^2}\right) +$$

$$+\frac{1}{2}\int_{-\infty}^{F_+(-t)}\frac{\Gamma(s')}{s'-s}\left(-\frac{q_s'^2}{t}\right)^{\alpha}(-t)^{\alpha}P_{\alpha}\left(-1-\frac{t}{2q_s'^2}\right)ds' + \tag{6.9}$$

$$+\frac{1}{2}\int_0^{F_-(-t)}\frac{\Gamma(s')}{s'-s}\left(-\frac{q_s'^2}{t}\right)^{\alpha}(-t)^{\alpha}P_{\alpha}\left(-1-\frac{t}{2q_s'^2}\right)ds' -$$

$$-\bar{B}(s, t) + \frac{1}{\pi}\int_{s_0}^{\infty}\frac{\varDelta_s\{B_M(s', t)\}}{s'-s}ds'\;.$$

The last term in (6.9) was omitted in (3.16). It is the contribution from that part of the background [see (3.8)] which satisfies the Mandelstam representation, i.e. from that part of $B^{\pm}(s, t)$ whose cuts lie only within the double spectral function boundaries. The complete function $B^{\pm}(s, t)$ is given by all the terms in (6.9) except the first. The reason for writing B_M as an integral over the s discontinuity is simply to remind ourselves that it can not have any s singularities for $s < s_0$. This representation, (6.9), satisfies the Mandelstam representation by construction, but we have still to check its asymptotic behaviour to ensure that the Regge asymptotic behaviour holds everywhere, and to convince ourselves that truly $B^{\pm}(s, t) \sim t^{-\frac{1}{2}}$, as it must by construction. As $s \to 0$ we find, from (6.1) and (6.3), that

$$q_s^2 \to \varDelta^2/4s\;, \tag{6.10}$$

and

$$z_s \to 1 + 2st/\varDelta^2$$

so the first term in (6.9) is certainly singular at $s = 0$. But since $A^{\pm}(s, t)$ satisfies the Mandelstam representation it can not possibly be singular at $s = 0$ (unless of course there is a bound-state pole, $\alpha(0) =$ an integer, but we ignore this quite different type of singularity) and some other term must have a corresponding singularity to cancel it. In fact, we can see at once that the fourth term on the right-hand side of (6.9) is also singular, and we shall make the cancellation explicit below. But let us first look at the asymptotic behaviour for $s \neq 0$.

We get

$$A^{\pm}(s,t) \xrightarrow[s \to \infty]{} \to c_1 t^{\alpha(s)} + c_2 t^{-1} + c_3 t^{\alpha(-\infty)} + c_4 t^{\alpha(0)-1} + c_5 t^{-1} + ?, \quad (6.11)$$

corresponding to the various terms in (6.9). These are similar to our discussion in Section (III.3) except that we have not been able to specify the behaviour of $B_M(s, t)$, and the behaviour of the fourth term is a good deal less than obvious. To show that it does indeed go like $t^{\alpha(0)-1}$ we write

$$\frac{1}{2} \int\limits_0^{F_-(-t)} \frac{\Gamma(s') \, (q_s'^2)^\alpha \, P_\alpha(-1-t/2q_s^2)}{s'-s} \, ds'$$

$$= \int\limits_0^{-\Delta^2/t} + \int\limits_{-\Delta^2/t}^{F_-(-t)} \frac{1}{2} \frac{\Gamma(s') \, (q_s'^2)^\alpha \, P_\alpha(-1-t/2q_s'^2)}{s'-s} \, ds'. \quad (6.12)$$

As $t \to \infty$,

$$F_-(-t) \to -\frac{\Delta^2}{t} + O\left(\frac{1}{t^2}\right)$$

so the second integral will be one power lower in t than the first, and we ignore it. We write the first integral

$$= \frac{1}{2} \int\limits_0^{-\Delta^2/t} \left\{ \Gamma(s') \, (q_s'^2)^{\alpha(s')} \, P_{\alpha(s')}\left(-1-\frac{t}{2q_s'^2}\right) - \right.$$

$$\left. - \Gamma(0) \left(\frac{\Delta^2}{4s'}\right)^{\alpha(0)} P_{\alpha(0)}\left(-1-\frac{2s't}{\Delta^2}\right) \right\} \frac{ds'}{s'-s} +$$

$$+ \frac{1}{2} \int\limits_0^{-\Delta^2/t} \Gamma(0) \left(\frac{\Delta^2}{4s'}\right)^{\alpha(0)} P_{\alpha(0)}\left(-1-\frac{2s't}{\Delta^2}\right) \frac{ds'}{s'-s}, \quad (6.13)$$

where we have subtracted from the integrand its behaviour at $s' = 0$, where $q_s'^2$ is singular, and treat it separately in the second integral. The asymptotic t behaviour of the first integral is again one power lower than the second, and we ignore it. The second integral in (6.13) is the dangerous term which we have been trying to isolate. It contains a "pinch" of the pole $s' = 0$ with the end points of integration. If we examine the expression

$$\left(\frac{\Delta^2}{4s}\right)^{\alpha(0)} Q_{-\alpha(0)-1}\left(-1-\frac{2st}{\Delta^2}\right) \quad (6.14)$$

in the light of our arguments about cuts of the Legendre functions in Section (III.2), we see that it has only one cut in s (at fixed t), from $s = 0$ to $-\Delta^2/t$; i.e. $z = -1$ to 1. The other cut for $z < -1$, cancels by the argument of Eq. (2.9). Because of (II.11.1), the second integral in (6.13) would be the dispersion integral for (6.14) over its s cuts, were it not for the fact that [see (II.12.9)], at fixed t,

$$\lim_{s \to \infty} \left(\frac{\Delta^2}{4s}\right)^{\alpha(0)} Q_{-\alpha(0)-1}\left(-1 - \frac{2st}{\Delta^2}\right) = \pi^{\frac{1}{2}} \frac{\Gamma[-\alpha(0)]}{\Gamma\left[-\alpha(0) + \frac{1}{2}\right]} (-t)^{\alpha(0)},$$

which is finite. We must thus make a subtraction at $s = \infty$ and obtain

$$\left(\frac{\Delta^2}{4s}\right)^{\alpha(0)} Q_{-\alpha(0)-1}\left(-1 - \frac{2st}{\Delta^2}\right) - \pi^{\frac{1}{2}} \frac{\Gamma[-\alpha(0)]}{\Gamma\left[-\alpha(0) + \frac{1}{2}\right]} (-t)^{\alpha(0)}$$

$$= \lim_{s_1 \to \infty} - \frac{s - s_1}{2} \int_0^{-\Delta^2/t} \frac{ds'}{(s'-s)(s'-s_1)} \left(\frac{\Delta^2}{4s'}\right)^{\alpha(0)} P_{\alpha(0)}\left(-1 - \frac{2s't}{\Delta^2}\right) \qquad (6.15)$$

$$= -\frac{1}{2} \int_0^{-\Delta^2/t} \frac{ds'}{s'-s} \left(\frac{\Delta^2}{4s'}\right)^{\alpha(0)} P_{\alpha(0)}\left(-1 - \frac{2st'}{\Delta^2}\right).$$

Hence the asymptotic t behaviour of (6.12) is the asymptotic t behaviour of the left hand side of (6.15), which is $c_4 \, t^{\alpha(0)-1}$, as we stated above. All the terms which we have dropped on the way are at worst of the form $\sim t^{\alpha(0)-2}$.

A seeming difficulty with (6.11) is that the terms which we have written explicitly give to $B^{\pm}(s,t)$ an asymptotic behaviour $\sim t^{\alpha(0)-1}$ Let us remind ourselves however that, comparing (2.7) with (6.9) we have

$$B^{\pm}(s,t) = \frac{1}{2} \int_0^{-4q_s^2} \frac{dt''}{t''-t} \Gamma(s) \left(\frac{-q_s^2}{t''}\right)^{\alpha} (-t'')^{\alpha} P_{\alpha}\left(-1 - \frac{t''}{2q_s^2}\right) +$$

$$+ \frac{1}{2} \int_{-\infty}^{F_+(-t)} \frac{\Gamma(s')}{s'-s} \left(\frac{-q_s'^2}{t}\right)^{\alpha} (-t)^{\alpha} P_{\alpha}\left(-1 - \frac{t}{2q_s'^2}\right) ds' +$$

$$+ \frac{1}{2} \int_0^{F_-(-t)} \frac{\Gamma(s')}{s'-s} \left(\frac{-q_s'^2}{t}\right)^{\alpha} (-t)^{\alpha} P_{\alpha}\left(-1 - \frac{t}{2q_s'^2}\right) ds' - \qquad (6.16)$$

$$- \bar{B}(s,t) + \frac{1}{\pi} \int_{s_0}^{\infty} \frac{\Delta_s\{B_M(s',t)\}}{s'-s} ds'.$$

We know that the left-hand side behaves like $t^{-\frac{1}{2}}$ so any worse behaviour coming from the various terms on the right must be cancelled by the final term, whose behaviour we did not specify in (6.11). We have already seen that for instance the Pomeranchon is expected to have $\alpha(0) = 1$ so the cancellation must certainly occur in any amplitude to which this trajectory contributes. For such a case the final term of (6.16)

$$\frac{1}{\pi} \int_{s_0}^{\infty} \frac{\Delta_s\{B_M(s',t)\}}{s'-s} ds'$$

must behave asymptotically like $t^{\alpha(0)-1}$ despite the fact that the integrand is, for all s', like $\sim t^{-\frac{1}{2}}$. This fact led *Goldberger* and *Jones*, 1966a, b, to doubt the possibility of reconciling the Regge and Mandelstam representations for those amplitudes with a pole such that $\alpha(0) > \frac{1}{2}$. In fact, however, there is no reason to believe that the required cancellation can not occur.

But let us first note what happens at $s = 0$. The first term in (6.9) cancels with the first term on the left hand side of (6.15), and the leading behaviour is given by the second term of (6.15) which contains no s singularity, and goes like $t^{\alpha(0)}$. Hence the Regge asymptotic behaviour of the amplitude at $s = 0$ is maintained, but it does not come from $A_{\overline{R}}^{\mp}(s, t)$, and the singularity of $A_{\overline{R}}^{\mp}(s, t)$ at $s = 0$ is indeed cancelled by $B^{\pm}(s, t)$, as we knew it must.

The left hand side of (6.15) gives us, if we expand $Q_{-\alpha(0)-1}$ in powers of z^{-2},

$$\frac{c_4'}{s} t^{\alpha(0)-1} + \frac{c_4''}{s^2} t^{\alpha(0)-2} + \cdots , \qquad (6.17)$$

where $c_4'/s = c_4$. We have noted that there will also be other terms in $t^{\alpha(0)-2}$. Our problem is thus to construct $B_M(s, t)$ in (6.16) so that it cancels this behaviour, thus making $B(s, t) \sim t^{-\frac{1}{2}}$. It has to do this despite the fact that clearly we also must have

$$\Delta_s\{B_M(s, t)\} \sim t^{-\frac{1}{2}}, \qquad (6.18)$$

if $B(s, t)$ is to be of background size, and the fact that $B_M(s, t)$ cannot have the singularity of (6.17) at $s = 0$. But of course it does not require this singularity at $s = 0$, since we know that this singularity will cancel with that of $A_{\overline{R}}^{\mp}(s, t)$. We thus need

$$B_M(s, t) \sim \frac{1}{s} t^{\alpha(0)-1} , \qquad s \neq 0 ,$$

$$\sim t^{-\frac{1}{2}}, \qquad s = 0 .$$

This requirement of non-uniform boundedness of $B_M(s, t)$ at first looks rather odd. But of course it is no odder than the non-uniformity of the other terms in (6.9). Indeed it is their non-uniformity which has driven us to such desperate lengths.

A form for $B_M(s, t)$ which appears to satisfy our requirements would be

$$B_M(s, t) = \frac{1}{\pi} \int_{s_0}^{\infty} \frac{\Delta_s\{\Gamma(s')\,(q_s'^2)^{\beta(s')}\,Q_{-\beta(s')-1}(-1-t/2q_s'^2)\}\left[\dfrac{1}{1+s't}\right]}{s'-s}\,\mathrm{d}s' , \qquad (6.19)$$

with $\beta(0) = \alpha(0) - 1$. Provided $\beta(s) < 1/2$ for all s, (6.18) is satisfied. The integral may be evaluated by "unwrapping" the contour, just as we did to get (6.9) from $D_{\overline{s}}^{\pm}(s, t)$ except that there is now an extra pole of the integrand at $s' = -1/t$. As $t \to \infty$ this pole will give a contribution $t^{\beta(0)}/s$, so, provided $\beta(0) = \alpha(0) - 1$, this can be made to cancel with (6.17). At $s = 0$, however, there is no singularity of $B_M(s, t)$. These points were emphasized by *Omnès* and *Leader*, 1967, and *Leader* (private communication). Of course there is no special merit in (6.19) except to show that a term with suitable properties can be found. Whether this cancellation with the background is the way which nature has chosen is an experimental question which we shall discuss in Section (III.8). Another possible mechanism whereby the cancellation is achieved by introducing further trajectories, known as "daughters", will be discussed in the next section.

If $\alpha(-\infty)$ were greater than $-1/2$ there would appear to be some difficulty in carrying out the programme we have outlined, as it would be necessary to find some region of s (not necessarily on the physical sheet) for which $\alpha(s) < -1/2$, make the Sommerfeld-Watson transform, and then analytically continue in s to the desired region. To the best of our knowledge this problem has never been tackled. There are also formal problems in moving the background integral to the left of $\text{Re}\{l\} = -1/2$ because of the accummulation of poles which occurs there at threshold.

III.7. "Daughter" Trajectories

In the last section we noted that Regge asymptotic behaviour applies even at $s = 0$, where the Regge pole term itself becomes undefined, but there is a cancellation of the Regge term with $B^\pm(s, t)$. Because of these complications there would seem to be some advantage in working instead with the Khuri power series representation of Section (III.5), in which a uniform asymptotic t behaviour for all s is guaranteed. However, a glance at (5.8) indicates that difficulties are also to be expected here with unequal mass kinematics, because of the behaviour, (6.10), of q_s^2 at $s = 0$. Each of the terms in the sum over n in (5.8) has an nth order pole in s at $s = 0$. How are these singularities to be cancelled? There are two possibilities. We could go through similar arguments to those of the previous section to arrange a cancellation with the background. This has been discussed by *Fearing*, 1967, but, since, if such a cancellation occurs, the Chew-Jones representation of the last section is almost certainly to be preferred, we will not repeat the arguments here. We have mentioned the Khuri power series because it makes most clear the other type of cancellation, first discussed by *Freedman* and *Wang*, 1966, 1967a, a cancellation between different trajectories.

If we call the highest lying trajectory in (5.8) $\alpha_0(s)$, we have, writing out the first terms of the sum over n

$$-8\pi \frac{\gamma_0(s) \, [2\alpha_0(s) + 1]}{(4)^{\alpha_0(s)} \cos \pi \alpha_0(s)} \frac{\Gamma[-\alpha_0(s)]}{\Gamma[-2\alpha_0(s)]} \left\{ (-t)^{\alpha_0(s)} - \frac{\alpha_0(s)}{2} 4q_s^2 (-t)^{\alpha_0(s)-1} + \cdots \right\}.$$

$$(7.1)$$

The second term which has a simple pole in s at $s = 0$, could be cancelled by the leading term of a second Regge pole $\alpha_1(s)$, provided that the residue of this pole, $\gamma_1(s)$, has a pole at $s = 0$, and $\alpha_1(0) = \alpha_0(0) - 1$. Explicitly, we require that

$$\gamma_1(s) = -\frac{\gamma_0(0)\,[2\alpha_0(0) + 1]\,\Delta^2}{4s} + \text{(regular part)} \qquad (7.2)$$

near $s = 0$. We should also require a third trajectory $\alpha_2(s)$ such that

$$\alpha_2(0) = \alpha_0(0) - 2\,,$$

whose residue has a double pole at $s = 0$, so that its leading term will cancel the third term of α_0 and the second term of α_1, and so on. An infinite set of trajectories with ever more singular residues is required. The trajectory $\alpha_1(s)$, is known commonly as the first "daughter" of $\alpha_0(s)$; $\alpha_2(s)$ is the second daughter, and so on. [These "daughters", which are characterized by their singular residues, should not be confused with the Khuri "satellites" of Section (III.5)].

Thus if we add the contributions of the parent and first daughter we get [absorbing the various factors in (7.1) and (7.2) into the functions $G(s)$, $a(s)$, and $b(s)$ which are regular at $s = 0$]

$$G(s)\left\{(-t)^{\alpha_0(s)} + \frac{a(s)}{s}(-t)^{\alpha_0(s)-1} + \cdots + \frac{b(s)}{s}(-t)^{\alpha_1(s)} + \cdots\right\}, \qquad (7.3)$$

and at $s = 0$ this becomes, if we expand in s and effect the cancellation of the $1/s$ singularity,

$$G(0)\left\{(-t)^{\alpha_0(s)} + c\,[\alpha_0'(0) - \alpha_1'(0)]\,\Delta^2 \log t\,(-t)^{\alpha_0(0)-1} + d\,(-t)^{\alpha_0(0)-1} + \cdots\right\}, \qquad (7.4)$$

where c and d are constants. The logarithmic term is peculiar to the unequal mass case.

If only one of the mass differences $(m_1 - m_3)$ and $(m_2 - m_4)$ is non-zero the first Khuri satellite contribution can be shown not be singular, and there will be no need for odd-order daughters; only the even ones are required.

Though this argument has been presented in the Khuri representation, it is equally possible to introduce daughters to effect the cancellations required by the Chew-Jones representation, and this was discussed by *Freedman, Jones* and *Wang*, 1967. For if we have a daughter trajectory with $\alpha_1(0) = \alpha_0(0) - 1$, as before, and

$$\Gamma_1(s) = -\frac{\Delta^2}{4s}\,[2\alpha_0(s) + 1]\,\Gamma_0(0)\,, \quad \text{near} \quad s = 0\,,$$

our calculation of the behaviour at $s = 0$ [Eq. (6.12) et seq.] must be modified, because the end points of integration now pinch on a double pole at $s = 0$. So the second integral in (6.13) gives, with $\Gamma_1(s)$ instead of $\Gamma(s)$ terms behaving like $t^{\alpha_1(0)}/s$, as is evident from an application of the residue theorem. This term is just what we need to cancel (6.16). Of course, unless we invoke cancellation with the background at some point in the sequence we shall again need an infinite set of trajectories, with increasingly singular residues, to cancel every term.

The complexity of the angular momentum plane required by the daughter hypothesis might make one doubt the compatibility of the Regge and Mandelstam representations for unequal mass kinematics, unless of course one is prepared to accept the arguments for the cancellation of the $t^{\alpha(0)-1}$ terms by the background. *Freedman* and *Wang* tried to give additional supporting evidence for daughters in their paper, based on the properties of the Bethe-Salpeter equation, which is well known to have a relation to Regge poles in perturbation theory (see Chapter VII). They showed that daughters might follow from the four dimensional symmetry of the Bethe-Salpeter equation at $s = 0$, a fact previously noted by *Domokos* and *Suranyi*, 1964. More recently these extra symmetries at $s = 0$ have been discussed from the more general point of view of invariance under the group $O(4)$ (*Durand*, 1967; *Nakanishi*, 1967; *Freedman* and *Wang*, 1967b; *Finkelstein* and *Wang*, 1967; *Domokos*, 1967), which also involves questions about "conspiracies", and which we shall discuss in the next chapter. At present these arguments appear to be explanations of why daughters may exist rather than proofs that they must, and the more economical cancellation with the background seems to us equally satisfactory.

III.8. Some Experimental Consequences

Though we are leaving the discussion of the experimental consequences of Regge poles to Chapter VIII, it is appropriate to mention here some of the properties which follow from the material of this chapter. We must remind the reader that in order to make the equations more simple we have been restricting ourselves to amplitudes of definite signature, and with only one double spectral function. For physical amplitudes one will often need to permute the variables, and introduce signature factors.

Firstly we have the prediction that the Regge asymptotic behaviour will hold even in regions where the bound (6.7) on the growth of z_s might naively seem to prevent it. A well known example of this is πN elastic scattering (taken to be the s-channel) in the backward direction ($u = 0$), which is expected to be controlled at large s by the N and $N^*(\Delta)$ trajectories in the $\pi \bar{N}$ (u-)channel, despite the fact that $u = 0$ is a singular point (corresponding to $s = 0$ above), and $|z_u| \leq 1$ for $0 \leq u \leq \Delta^2/s$. This has been discussed by *Stack*, 1966; *Chiu* and *Stack*, 1967 and *Barger* and *Cline*, 1966, 1967a, and a general discussion of Regge representations with these kinematics in mind (they differ in some details from our account because of the loss of symmetry between u and t with unequal masses in the s-channel) is contained in the paper by

Goldberger and *Jones*, 1966b. A general account of such fits must be postponed until we have discussed spin, and will be given in Chapter VIII, but here we can simply remark that the asymptotic behaviour seems well verified.

The other important question concerns the existence of the daughter trajectories. We have seen that they must cancel with the parent trajectory in an amplitude of given signature. This means that they must have the same quantum numbers as the parent. However, from our definition (II.10.3) we see that as we require the cancellation to occur, not just in $A^{\pm}(s, t)$ but also in $A(s, t)$, we shall need the first, and all the other odd daughters to have the opposite signature to that of the parent, and the even daughters. Thus the parent has, at $s = 0$, the signature factor

$$1 \pm e^{i \pi \alpha_0(0)},$$

while the first daughter will have

$$1 \pm e^{i \pi \alpha_1(0)} = 1 \mp e^{i \pi \alpha_0(0)},$$

since $\alpha_1(0) = \alpha_0(0) - 1$. The cancellation will be effective only if the first daughter is of opposite signature to the parent. The second daughter should have the same signature as the parent, and so on alternately. This has the important consequence that there is no zero mass, zero spin, particle corresponding to the first daughter of the Pomeranchon trajectory.

We only know what the positions of the daughters must be at $s = 0$. If we suppose that they are likely to run more or less parallel to the parents we can expect a large number of new particles where the daughters cross the appropriate integers. For example the nucleon trajectory which is shown in Fig. (VIII.3) should have a $3^-/2$ daughter of mass about 1700 Mev. This is not all that far from the N* (1518), and it is just possible that all of the odd parity trajectories, shown in Fig. (VIII.4) are the daughters of those shown in Fig. (VIII.3). The very sensitive interference methods of *Barger* and *Cline*, 1966, 1967a, which are discussed in Section (VIII.8), would seem to more or less preclude the existence of any other such particles even if they had been overlooked in more direct methods of search.

At the moment there is no known particle which can be identified as lying on the first daughter of any meson trajectory, though we shall mention in Section (VIII.1) the possibility that the δ (985) might be a daughter of the ϱ.

Since all daughters have $\alpha(0) \leq 0$ it could be that the trajectories are very flat, and never manifest themselves as particles. Equally it could be that for some reason they have vanishing residues at the physical integers. In either event the way to look for them would be to try and observe their effects in controlling the high energy behaviour of appropriate reactions. (These methods will be reviewed in detail in Chapter VIII.) The difficulty here is that the "wrong" signature precludes the coupling of the meson trajectories (which are much more likely to be found in this way than baryons) to most of the more accessible processes. Thus the Pomeranchon, with $B = Y = I = 0$, $G = +1$, would have a daughter which could give rise to a 1^- meson. One might think of trying to observe it in $a - b$ elastic scattering (say), if the trajectory would couple to $a - \bar{a}$ or $b - \bar{b}$. But, since in any experiment at least one of a or b would have to be a nucleon (the other being possibly π, K or N), and our daughter will not couple to $\pi - \pi$ because Bose statistics eleminates a $T = 0$, 1^- state, nor to $N - \bar{N}$ or $K - \bar{K}$ because of G-parity (see e.g. *Källén*, 1964, p. 318) this would certainly be no good. Only in a double production process such as $N + N \to N* + N*$ etc. where neither vertex has a particle — antiparticle pair, could one hope to detect the first daughter. Similar remarks apply to the first daughter of the ϱ. Most other trajectories are too low-lying for their daughters to be observable, and this is also true of the better behaved second daughters of all trajectories. As G-parity also forbids most of the more straightforward decays of the boson daughters it is not particularly surprising that they have not been directly observed as particles.

At the present time the only "evidence" for daughters is purely theoretical.

Chapter IV. Spin

IV.1. The Partial-wave Expansion for Particles with Spin

In this chapter we shall generalize some of the previous work to apply to scattering of particles of non-zero spin. Clearly such a generalization is necessary in most practical applications of the theory. As we shall see the generalization is not trivial; in addition to a considerable increase in algebraic complexity, several new features appear. Fortunately in most applications the spins concerned are small, and the formalism becomes much simpler. However, it is instructive to see the elegant way in which the theory can be developed in full generality; applications to specific problems will be dealt with in Chapter VIII.

There is an extensive literature on the extension of Regge-pole ideas to particles with spin. Within the framework of potential scattering the theory was developed by *Charap* and *Squires*, 1962b, 1963a, 1963b, and by *Desai* and *Newton*, 1963a, who considered the spin 1/2 − spin 1/2 case. The relativistic situation was studied by *Calogero* et al., 1963a, 1963b, by *Gell-Mann* et al., 1963, by *Mandelstam*, 1963a, and by *Kibble*, 1963. *Andrews* and *Gunson*, 1964, gave a more formal treatment and introduced the $e_{\lambda\lambda'}^J$ functions which we shall define and use below. All this work was based on the helicity formalism of *Jacob* and *Wick*, 1959. Further progress has been made by means of the helicity crossing matrix, relating helicity amplitudes in one channel to those in another channel, which was given explicitly by *Trueman* and *Wick*, 1964 and by *Muzinich*, 1964. More recently *Cohen-Tannoudji* et al., 1967, have given an alternative derivation of the helicity crossing matrix. Using this crossing matrix *Hara*, 1964, and *Wang*, 1966a, have shown explicitly how to isolate completely the kinematical singularities of the helicity amplitudes. *Stapp*, 1967, has produced similar results using slightly different basic assumptions. General formulations of the theory have also been given recently by *Mueller* and *Trueman*, 1967a, 1967b, and by *Drechsler*, 1967a. Discussions devoted more specifically to particular phenomenological applications are contained in articles by *Wang*, 1966b, 1967; *Thews*, 1967; *Frautschi* and *Jones*, 1967a, b, c; and *Fox*, 1967. *Leader*, 1967, has studied a particular class of kinematical constraints in great detail. Finally in this literature survey we note that there exist alternative though physically equivalent, methods of "Reggeizing", i.e. using "invariant amplitudes " rather than helicity amplitudes. Such methods are briefly discussed by *Durand*, 1967; *Taylor*, 1967, and *Jones* and *Scadron*, 1967.

In this chapter we shall use the helicity formalism of *Jacob* and *Wick*, 1959; the helicity of a particle being the projection of its spin angular momentum in the direction of its motion, as in (I.2.5). The helicity is an invariant quantity, so, for given helicities of the incoming and outgoing particles, the scattering amplitude is again just a function of the scalar invariants s and t. Thus, if we denote the helicity of the ith particle by λ_i, we can write the amplitude as $\langle \lambda_3 \lambda_4 | A(s, t)| \lambda_1 \lambda_2 \rangle$. We normalize this amplitude in an analogous way to the spin-zero case of Chapter I, so the differential cross-section, averaged over initial helicities and summed over final helicities, is given by [c.f. (I.7.10)]

$$\frac{d\sigma}{d\Omega} = \frac{q_{s34}}{q_{s12}} \frac{1}{(8\pi)^2 s} \frac{1}{(2\sigma_1 + 1)(2\sigma_2 + 1)} \sum_{\lambda_1 \lambda_2 \lambda_3 \lambda_4} |\langle \lambda_3 \lambda_4 |A(s, t)| \lambda_1 \lambda_2 \rangle|^2, \quad (1.1)$$

where σ_i is the spin of the ith particle.

The partial-wave expansion analogous to (II.1.3) has been given by *Jacob* and *Wick*, 1959, and can be written in the form

$$\langle \lambda_3 \lambda_4 |A(s, t)| \lambda_1 \lambda_2 \rangle = 16\pi \sum_{J = \text{Max}(|\lambda|, |\lambda'|)} (2J + 1) \langle \lambda_3 \lambda_4 |A_J(s)| \lambda_1 \lambda_2 \rangle d^J_{\lambda \lambda'}(\theta),$$

$$(1.2)$$

where

$$\lambda = \lambda_1 - \lambda_2, \quad (1.3a)$$

and

$$\lambda' = \lambda_3 - \lambda_4, \quad (1.3b)$$

and $d^J_{\lambda \lambda'}(\theta)$ is the rotation matrix which we shall discuss in detail in the next section. The significance of λ and λ' becomes clear if we remember that, in the centre-of-mass system, the two particles are moving in opposite directions, so that λ and λ' are the projections of the total spin in the direction of motion before and after the collision. Since the projection of the orbital angular momentum in this direction is zero, it follows that λ and λ' are, in fact, the projections of the total angular momenta in the directions of motion before and after the collision. It follows that physical values of J must be greater than the maximum of λ and λ', as indicated in (1.2). In addition, conservation of angular momentum requires that in the forward direction, for which λ and λ' are projections of total angular momentum in the same direction, the amplitude vanishes unless $\lambda = \lambda'$. This property, which of course will be an automatic consequence of our formulation, will be of use later.

For given σ_i there are $\prod_{i=1}^{4} (2\sigma_i + 1)$ helicity amplitudes. These are not independent, however, being related by parity and time reversal invariance (we shall assume that these hold for all processes which we consider). From parity we obtain (*Jacob* and *Wick*, 1959):

$$\langle -\lambda_3 - \lambda_4 |A_J(s)| -\lambda_1 - \lambda_2 \rangle = \prod_{i=1}^{4} \eta_i (-1)^{\sigma_3 + \sigma_4 - \sigma_1 - \sigma_2} \times$$

$$\times \langle \lambda_3 \lambda_4 |A_J(s)| \lambda_1 \lambda_2 \rangle, \quad (1.4)$$

where η_i is the intrinsic parity of the ith particle; and from time reversal,

$$\langle \lambda_1 \lambda_2 |A_J(s)| \lambda_3 \lambda_4 \rangle = \langle \lambda_3 \lambda_4 |A_J(s)| \lambda_1 \lambda_2 \rangle. \quad (1.5)$$

It will be convenient in much of what follows to make use of these symmetries and consider, without loss of generality, amplitudes for which

$$\lambda \geq |\lambda'|. \quad (1.6)$$

This simplifies the writing of certain equations.

We can use the orthogonality property of $d^J_{\lambda\lambda'}$ which will be given in (2.3) to invert (1.2) and obtain

$$\langle \lambda_3 \lambda_4 | A_J(s) | \lambda_1 \lambda_2 \rangle = \frac{1}{32\pi} \int\limits_{-1}^{+1} dz \, \langle \lambda_3 \lambda_4 | A(s,t) | \lambda_1 \lambda_2 \rangle \, d^J_{\lambda\lambda'}(\theta) \quad (1.7)$$

for physical values of J, i.e. $J = \lambda, \lambda + 1, \lambda + 2$, etc. This equation is analogous to (II.1.1).

IV.2. Properties of $d^J_{\lambda\mu}(\theta)$ and $e^J_{\lambda\mu}(\theta)$

Before proceeding further it is necessary to discuss the properties of the functions $d^J_{\lambda\mu}(\theta)$ which were introduced in (1.2). For ease of reference we collect together in this section all the properties we shall need, and also discuss a related set of functions $e^J_{\lambda\mu}(\theta)$ which will be used below. This section is entirely mathematical and essentially consists of a list of formulae.

The $d^J_{\lambda\mu}(\theta)$ are defined and discussed in, for example, books by *Rose*, 1957, and *Edmonds*, 1957. They satisfy the symmetry properties

$$d^J_{\lambda\mu}(\theta) = d^J_{-\lambda-\mu}(\theta) = (-1)^{\lambda-\mu} d^J_{\mu\lambda}(\theta) , \quad (2.1)$$

and

$$d^J_{\lambda\mu}(\theta) = (-1)^{J+\lambda} d^J_{\lambda-\mu}(\pi - \theta) . \quad (2.2)$$

Their orthogonality relations are

$$\int\limits_0^\pi d^J_{\lambda\mu}(\theta) \, d^{J'}_{\lambda\mu}(\theta) \sin\theta \, d\theta = \delta_{JJ'} \frac{2}{2J+1} , \quad (2.3)$$

$$\frac{1}{2} \sum_J (2J+1) \, d^J_{\lambda\mu}(\theta) \, d^J_{\lambda\mu}(\theta') = \delta(\cos\theta - \cos\theta') , \quad (2.4)$$

and

$$\sum_\lambda d^J_{\lambda\mu}(\theta) \, d^J_{\lambda\mu'}(\theta) = \delta_{\mu\mu'} . \quad (2.5)$$

The analyticity properties of $d^J_{\lambda\lambda'}(\theta)$, considered as a function of $z = \cos\theta$, can be seen by expressing them in terms of Jacobi polynomials ($P^{ab}_n(z)$):

$$d^J_{\lambda\mu}(\theta) = (-1)^{N-\mu} \left[\frac{(J+M)! \, (J-M)!}{(J+N)! \, (J-N)!} \right]^{\frac{1}{2}} \left(\frac{1-z}{2} \right)^{\frac{|\lambda-\mu|}{2}} \times$$
$$\times \left(\frac{1+z}{2} \right)^{\frac{|\lambda+\mu|}{2}} P^{|\lambda-\mu|(\lambda+\mu)}_{J-M}(z) \quad (2.6)$$

where

$$M = \max(\lambda, \mu) \quad (2.7)$$

$$N = \min(\lambda, \mu) \quad (2.8)$$

and the equation holds only if $M \geqq 0$. For other cases we can use (2.1).

The $P_n^{ab}(z)$ are polynomials for n a non-negative integer, so (2.6) shows that for such values of n the $d_{\lambda\mu}^J(\theta)$ are analytic functions of z, apart from possible singularities at $z = \pm 1$ which are given explicitly by this equation.

It is sometimes useful to be able to express the $d_{\lambda\mu}^J(\theta)$ in terms of the more familiar Legendre polynomials. This can be done by means of the so-called Clebsch-Gordan series

$$d_{\lambda\mu}^M(\theta)\, d_{\lambda\mu}^J(\theta) = \sum_{\tau=-M}^{M} P_{J+\tau}(z)\, C\,(J\,M\,J+\tau;\lambda,-\lambda)\, C\,(J\,M\,J+\tau;\mu,-\mu) \quad (2.9)$$

together with the expression

$$d_{\lambda\mu}^M(\theta) = (-1)^{N-\lambda}\frac{(2M)!}{(\lambda+\mu)!\,|\lambda-\mu|!}\left(\frac{1+z}{2}\right)^{\frac{\lambda+\mu}{2}}\left(\frac{1-z}{2}\right)^{\frac{|\lambda-\mu|}{2}} \quad (2.10)$$

valid for $M \geqq 0$. We can use (2.9) for non-integral values of J if we continue the Clebsch-Gordan coefficients by means of the explicit expressions of *Wigner* and *Racah* (see *Charap* and *Squires*, 1962b), i.e.

$$C\,(J+a,\,b,\,J+c;\,d,\,e) = (2J+2c+1)^{\frac{1}{2}} \times \quad (2.11)$$

$$\times \left[\frac{(2J+a-b+c)!\,(-a+b+c)!\,(a+b-c)!\,(J+c+d+e)!\,(J+c-d-e)!}{(2J+a+b+c+1)!\,(J+a-d)!\,(J+a+d)!\,(b-e)!\,(b+e)!}\right]^{\frac{1}{2}} \times$$

$$\times \sum_v \frac{(-1)^{v+b+e}}{v!}\,\frac{(J+b+c+d-v)!\,(J+a-d+v)!}{(-a+b+c-v)!\,(J+c+d+e-v)!\,(J+a-b-d-e+v)!}\,.$$

In order to discuss the analyticity of the $d_{\lambda\mu}^J(\theta)$ in the J-plane it is convenient to express them in terms of the hypergeometric function:

$$d_{\lambda\mu}^J(\theta) = (-1)^{N-\lambda}\left(\frac{1-z}{2}\right)^{\frac{|\lambda-\mu|}{2}}\left(\frac{1+z}{2}\right)^{\frac{\lambda+\mu}{2}}\left[\frac{(J+M)!\,(J-N)!}{(J+N)!\,(J-M)!}\right]^{\frac{1}{2}} \times$$

$$\times \frac{1}{|\lambda-\mu|!}\,F\left(-J+M,\,J+M+1,\,M-N+1;\frac{1}{2}\,(1-z)\right) \quad (2.12)$$

for $M \geqq 0$. The hypergeometric function is analytic in J, so that the positions of the J-plane singularities are explicit in (2.12), and arise from the square root factor. If we recall that the function $w!$ has a simple pole in the w-plane whenever w equals a negative integer, we see that the singularities of $d_{\lambda\mu}^J$ are at those integral values of $(J-M)$ for which either

$$N \leqq J < M\,, \quad (2.13a)$$

or

$$-M \leqq J < -N\,. \quad (2.13b)$$

The values in (2.13a) correspond to the scattering from a possible physical state to an impossible physical state, and the relevant amplitudes are called "sense-nonsense" amplitudes (*Gell-Mann*, 1962a). From (2.12) we see that $d_{\lambda\mu}^J(\theta)$ behaves as $(J-M-n)^{\frac{1}{2}}$, $n = 1, 2, 3, \ldots$, near a sense-nonsense value of J. We discuss sense-nonsense amplitudes in detail in Section (IV.5) and also Sections (V.4)–(V.6).

The form of $d^J_{\lambda\mu}(\theta)$ for large z can also be obtained from (2.12) by using the asymptotic form of the hypergeometric function. We find

$$d^J_{\lambda\mu}(\theta) = \frac{(-1)^{N-\mu}\,(i)^{M-N}}{(2J+1)}\left[\frac{(J+M)!\,(J-N)!}{(J-M)!\,(J+N)!}\right]^{\frac{1}{2}} \times$$
$$\times\left[\frac{(2J+1)!}{(J+M)!\,(J-N)!}\left(\frac{z}{2}\right)^J\left(1+O\left(\frac{1}{z^2}\right)\right)- \right. \qquad (2.14)$$
$$\left. -\frac{(-2J-1)!}{(-J-1+M)!\,(-J-1-N)!}\left(\frac{z}{2}\right)^{-J-1}\left(1+O\left(\frac{1}{z}\right)\right)\right]$$

for $M \geqq 0$. We see from this expression that in $\mathrm{Re}\,J \geqq -1/2$, we have

$$d^J_{\lambda\mu}(\theta) \sim z^J, \qquad (2.15)$$

except for sense-nonsense values of J, for which $d^J_{\lambda\mu}=0$, and nonsense-nonsense values of J, for which

$$d^J_{\lambda\mu}(\theta) \sim z^{-J-1}. \qquad (2.16)$$

An important property which follows from (2.14) is that the leading term "factorizes" in the sense that

$$d^J_{\lambda\lambda}(\theta)\,d^J_{\mu\mu}(\theta) \to (d^J_{\lambda\mu}(\theta))^2 \qquad (2.17)$$

in the limit of large z (in $\mathrm{Re}\,J > -1/2$, and away from nonsense states). This fact was noticed by *Fox* and *Leader*, 1967.

We turn now to the functions $e^J_{\lambda\mu}(\theta)$ which we shall require in Section (IV.4) and later sections. They are related to the $d^J_{\lambda\mu}(\theta)$ in a similar way to that in which the $Q_l(z)$ are related to the $P_l(z)$, and play essentially the same rôle as that of the $Q_l(z)$ in the theory with particles of zero spin. These functions were first introduced explicitly by *Andrews* and *Gunson*, 1964. It is worth noticing that the $e^J_{\lambda\mu}$ which we consider here are not in any way related to those of *Gell-Mann* et al. 1963, which are merely combinations of two $d^J_{\lambda\mu}(\theta)$ functions appropriate to states of given parity.

We can define the $e^J_{\lambda\mu}$ by an equation analogous to (2.6), i.e.

$$e^J_{\lambda\mu}(\theta) = (-1)^{N-\lambda}\left[\frac{(J+M)!\,(J-M)!}{(J+N)!\,(J-N)!}\right]^{\frac{1}{2}}\left(\frac{1-z}{2}\right)^{\frac{|\lambda-\mu|}{2}}\left(\frac{1+z}{2}\right)^{\frac{\lambda+\mu}{2}} \times$$
$$\times Q^{|\lambda-\mu|,(\lambda+\mu)}_{J-M}(z) \qquad (2.18,$$

for $M \geqq 0$. We supplement this with the reflection property analogous to (2.1), i.e.

$$e^J_{\lambda\mu}(\theta) = (-1)^{\lambda-\mu}\,e^J_{-\lambda-\mu}(\theta) = (-1)^{\lambda-\mu}\,e^J_{\mu\lambda}(\theta). \qquad (2.19)$$

The $Q^{ab}_n(z)$ in (2.18) are Jacobi functions of the second kind. They are related to Jacobi functions of the first kind by a relation analogous to Neumann's formula (II.3.4)

$$Q^{ab}_n(z) = \frac{1}{2}\,(z-1)^{-a}(z+1)^{-b}\int\limits_{-1}^{+1}\frac{dz'}{(z-z')}\,(1-z')^a\,(1+z')^b\,P^{ab}_n(z') \qquad (2.20)$$

for n a non-negative integer [B 2, 10.8 (20)].

We can combine (2.20) with (2.18) and (2.6) to obtain

$$\left(\frac{1-z}{2}\right)^{\frac{|\lambda-\mu|}{2}}\left(\frac{1+z}{2}\right)^{\frac{(\lambda+\mu)}{2}}e_{\lambda\mu}^{J}(\theta)\,(-1)^{|\lambda-\mu|}$$

$$=\frac{1}{2}\int\limits_{-1}^{+1}\frac{dz'}{(z-z')}\,d_{\lambda\mu}^{J}(\theta')\left(\frac{1-z'}{2}\right)^{\frac{|\lambda-\mu|}{2}}\left(\frac{1+z'}{2}\right)^{\frac{(\lambda+\mu)}{2}} \tag{2.21}$$

for integral $J \geqq M$. This equation plays the role for particles with spin that Neumann's formula does for particles of zero spin (c.f. Section (II.3)].

To find the behaviour of $e_{\lambda\mu}^{J}(\theta)$ for large z we use the relation between $Q_n^{ab}(z)$ and the hypergeometric function

$$Q_n^{ab}(z) = \frac{1}{2}\frac{(n+a)!\,(n+b)!}{(2n+a+b+1)!}\left(\frac{z-1}{2}\right)^{-(n+a+1)}\left(\frac{z+1}{2}\right)^{-b}\times$$

$$\times F\left(n+1, n+a+1, 2n+a+b+2, \frac{2}{1-z}\right), \tag{2.22}$$

from which

$$Q_n^{ab}(z) = \frac{1}{2}\frac{(n+a)!\,(n+b)!}{(2n+a+b+1)!}\left(\frac{z}{2}\right)^{-n-a-b-1}\left(1+O\left(\frac{1}{z}\right)\right). \tag{2.23}$$

Combining (2.23) with (2.18) we have

$$e_{\lambda\mu}^{J}(\theta) = \frac{(-1)^{N-\lambda}}{2\,(2J+1)!}\left[(J+\lambda)!\,(J-\lambda)!\,(J+\mu)!\,(J-\mu)!\right]^{\frac{1}{2}}\times$$

$$\times\,e^{\pm i\frac{\pi}{2}(\lambda-\mu)}\left(\frac{z}{2}\right)^{-J-1}\left(1+O\left(\frac{1}{z}\right)\right), \tag{2.24}$$

the $\pm$ occurring for $\mathrm{Im}\,z \gtrless 0$.

A relation between the $d_{\lambda\mu}^{J}$ and the $e_{\lambda\mu}^{J}$ which is sometimes useful is

$$\frac{\pi\,d_{\lambda\mu}^{J}(\theta)}{\sin\pi(J-\lambda)} = \frac{e_{\lambda\mu}^{J}(\theta)}{\cos\pi(J-\lambda)} - \frac{e_{-\lambda-\mu}^{J-1}(\theta)}{\cos\pi(J-\lambda)}, \tag{2.25}$$

which is analogous to (II.12.2).

In developing the theory for particles with spin it is clear that it is always possible to use (2.9) and (2.10) so that everything can be done in terms of the more familiar P_J and Q_J functions. This method was used for example by *Calogero* et al., 1963a, in generalizing the Mandelstam form of the Sommerfeld-Watson transform, and is used extensively by *Mueller* and *Trueman*, 1967a, 1967b. Since the resulting expressions tend to be rather larger we shall here use the more compact expressions made possible by the introduction of the $e_{\lambda\mu}^{J}$ functions.

IV.3. Kinematical Singularities and Single Variable Dispersion Relations

In Chapter II we defined a continuation of the partial-wave amplitude A_l to complex l by means of a t-dispersion relation. In order to make the analogous definition for the non-zero spin case we must first remove certain "kinematical singularities" from the amplitude. These singularities in t, for fixed s, arise from the singularities of $d^j_{\lambda\lambda'}(\theta)$ at $z = \pm 1$, which are exhibited explicitly in (2.6). It is one of the great values of the helicity formalism that it reveals so clearly these kinematical singularities.

Using (1.2) and (2.6) we see that the function

$$\langle \lambda_3 \lambda_4 | \hat{A}(s, t) | \lambda_1 \lambda_2 \rangle = \left(\frac{1-z}{2}\right)^{-\frac{\lambda-\lambda'}{2}} \left(\frac{1+z}{2}\right)^{-\frac{\lambda+\lambda'}{2}} \times \tag{3.1}$$
$$\times \langle \lambda_3 \lambda_4 | A(s, t) | \lambda_1 \lambda_2 \rangle$$

has no singularities in the z-plane as long as the partial-wave expansion converges; it is therefore analogous to the amplitude $A(s, t)$ for the spin zero case and has a right-hand physical cut and a left-hand cut coming from unitarity in the u-channel. We can therefore write a t-channel dispersion relation (cf. Section I.10)

$$\hat{A}_H(s, t) = \frac{1}{\pi} \int\limits_{t_0}^{\infty} \frac{dt'}{t'-t} D_{Ht}(s, t') + \frac{1}{\pi} \int\limits_{u_0}^{\infty} \frac{du'}{u'-u} D_{Hu}(s, u') \tag{3.2}$$

where we have ignored the need for subtractions or pole terms — their role is exactly the same as in Chapter II. We have also introduced the label H as a shorthand for the helicity quantum numbers $(\lambda_1, \lambda_2, \lambda_3, \lambda_4)$.

It is worth noting that this method is the simplest available for locating kinematic singularities and zeros of scattering amplitudes. Other methods, for example using invariant amplitudes (e.g. *Goldberger* et al., 1960, for the $N - N$ problem) become very cumbersome unless the spin values are small. Of course at this stage we have only considered singularities in t for fixed s; in Section (IV.6) we shall cross to the t-channel and use the same method to eliminate s-singularities.

The fact that the helicity amplitudes have kinematical singularities means that the argument of Chapter III should be applied to the amplitudes with these factors removed. This has an important consequence when we consider cases where, in the forward (or backward) direction, z does not go to infinity when the cross-channel energy goes to infinity. In Section (III.6) we showed that the Regge formula still works for the spinless case. Applying an analogous method to the general case we should find that the kinematical factors,

$$\left(\frac{1 \pm z}{2}\right)^{\frac{\lambda \pm \lambda'}{2}},$$

would remain exactly as they are, and that only for the remaining z-dependence would the substitution $z \sim t$ be valid. This point has been emphasized by *Wang*, 1967, and by *L. Jones*, 1967. We discuss cases where this is important in Section (VIII.7).

IV.4. The Generalised Froissart-Gribov Projection

To obtain the Froissart-Gribov definition of the continued amplitude we substitute (3.2) into (1.7) and invert the order of integration. We use the generalized Neumann's formula (2.21), and obtain

$$\langle \lambda_3 \lambda_4 | A_J(s) | \lambda_1 \lambda_2 \rangle = (16\pi)^{-1} \frac{1}{\pi} \int\limits_{z_s(s,t_0)}^{\infty} dz_s \Big\{ (-1)^{\lambda-\lambda'} D_{Ht}(s,t) \times$$

$$\times e_{\lambda\lambda'}^J(\theta_s) \left(\frac{1-z_s}{2}\right)^{\frac{\lambda-\lambda'}{2}} \left(\frac{1+z_s}{2}\right)^{\frac{\lambda+\lambda'}{2}} + (-1)^{J+2\lambda+\lambda'} \times \qquad (4.1)$$

$$\times D_{Hu}(s,t) \, e_{\lambda-\lambda'}^J(\theta_s) \left(\frac{1+z_s}{2}\right)^{\frac{\lambda+\lambda'}{2}} \left(\frac{1-z_s}{2}\right)^{\frac{\lambda-\lambda'}{2}} \Big\}$$

for $\lambda \geq |\lambda'|$. Here we have transformed the second integral to one over positive z by means of the reflection property of the $d_{\lambda\mu}^J(\theta)$. As in Chapter II the presence of the factor $(-1)^J$ in the second term of (4.1) makes this an unsuitable form for continuing to complex J. We therefore replace $A_J(s)$ by two amplitudes of definite signature $A_J^{\mp}(s)$ defined by

$$\langle \lambda_3 \lambda_4 | A_J^{\mp}(s) | \lambda_1 \lambda_2 \rangle = (16\pi^2)^{-1} \int\limits_{z_s(s,t_0)}^{\infty} dz_s \Big\{ (-1)^{\lambda-\lambda'} D_{Ht}(s,t) \times$$

$$\times e_{\lambda\lambda'}^J(\theta_s) \left(\frac{1-z_s}{2}\right)^{\frac{\lambda-\lambda'}{2}} \left(\frac{1+z_s}{2}\right)^{\frac{\lambda+\lambda'}{2}} \pm (\cos\pi\lambda - \sin\pi\lambda) \times \quad (4.2)$$

$$\times (-1)^{\lambda+\lambda'} D_{Hu}(s,t) \, e_{\lambda-\lambda'}^J(\theta) \left(\frac{1+z_s}{2}\right)^{\frac{\lambda+\lambda'}{2}} \left(\frac{1-z_s}{2}\right)^{\frac{\lambda-\lambda'}{2}} \Big\}$$

for $\lambda \geq |\lambda'|$. We use this equation to define the continued partial-wave amplitude in the region of the J-plane for which the integrals converge. We see by comparing (4.2) with (4.1) that $A_J^{\mp}(s)$ equals the physical amplitude when J is an even (odd) non-negative integer, if the physical values are integers, and for $(J - 1/2)$ equal to an even (odd) non-negative integer, if the physical values are half-odd-integral.

The partial-wave expansion (1.2) can be written in terms of the definite signature amplitudes as (*Charap* and *Squires*, 1963a)

$$\langle \lambda_3 \lambda_4 | A(s,t) | \lambda_1 \lambda_2 \rangle = 16\pi \sum_{J=\lambda}^{\infty} (2J+1) \{ \langle \lambda_3 \lambda_4 | A_J^+(s) | \lambda_1 \lambda_2 \rangle \times$$

$$\times d_{\lambda\lambda'}^+(J,z) + \langle \lambda_3 \lambda_4 | A_J^-(s) | \lambda_1 \lambda_2 \rangle \, d_{\lambda\lambda'}^-(J,z) \} \qquad (4.3)$$

for $\lambda \geq |\lambda'|$, where we have introduced

$$d_{\lambda\lambda'}^{\pm}(J,z) = d_{\lambda\lambda'}^J(\theta) \pm (\cos\pi\lambda - \sin\pi\lambda) \, d_{\lambda-\lambda'}^J(\pi-\theta) . \qquad (4.4)$$

In the derivation of (4.2) etc., we have ignored the question of subtractions. These can be treated exactly as in the zero-spin case (Chapter II), and the result is that (4.2) gives the physical amplitude for all physical J for which the integral converges. Suppose, for some s,

$$A(s,t) \leq O(z_s^{\alpha}) \qquad (4.5)$$

for large z, then from (3.1)

$$D_t(s, t) \leq O(z_s^{\alpha - M}) \tag{4.6}$$

with M given by (2.7), so, using the asymptotic form of $e_{\lambda\lambda'}^J$ given in (2.24), we see that the integral converges for

$$\operatorname{Re} J > \alpha . \tag{4.7}$$

From (4.2), (2.18) and (2.22) it follows that $\langle \lambda_3 \lambda_4 | A \neq (s) | \lambda_1 \lambda_2 \rangle$, for $\lambda \geq |\lambda'|$, has fixed square-root branch points at the sense-nonsense points

$$J = \lambda - 1, \lambda - 2, \lambda - 3, \ldots |\lambda'| \tag{4.8a}$$

and at the points

$$J = - |\lambda'| - 1, \ -|\lambda'| - 2, \ -|\lambda'| - 3, \ldots - \lambda . \tag{4.8b}$$

However, these fixed branch points in $A_{\neq H}^{\pm}(s)$ cancel with exactly similar branch points in the $d_{\lambda\lambda'}^{\pm}(J, z)$ functions [see Section (IV.2)] so that they do not yield branch points in the summand of (4.3). We shall leave until the next chapter the important question of whether there are any fixed poles in the summand.

Note that the argument for the existence of the branch points of $A_{\neq H}^{\pm}(s)$ arising from (4.2), depends upon the integral in (4.2) being convergent for some s at the appropriate value of J.

IV.5. The Sommerfeld-Watson Transform

As in Section (II.9) we write (4.3) as a contour integral

$$\langle \lambda_3 \lambda_4 | A (s, t) | \lambda_1 \lambda_2 \rangle = - \frac{16\pi}{2i} \int_C \frac{d J (2J + 1)}{\sin \pi (J + \lambda)} \times$$

$$\times \{A_{\neq H}^{+}(s) d_{\lambda\lambda'}^{+}(J, -z) + A_{\bar{J}H}^{-}(s) d_{\lambda\lambda'}^{-}(J, -z)\} \tag{5.1}$$

for $\lambda \geq |\lambda'|$, where we have again used H as a shorthand for the four helicity indices. The contour C encircles that part of the real axis in the J-plane with $\operatorname{Re} J \geq \lambda$. We now open out the contour, as in Section (II.9), and move it to the left, so that it runs along the line $\operatorname{Re} J = -1/2$ (or just to the right of this if physical values of J are half-odd-integral). The large semi-circle at $|J| = \infty$ gives a zero contribution as in the zero-spin case (see *Calogero* et al., 1963a, for details).

For simplicity we assume that the only J-plane singularities of the term in curly brackets in (4.1) are poles [the contribution of cuts can be treated exactly as in Section (II.9)], so we obtain

$$(16\pi)^{-1} \langle \lambda_3 \lambda_4 | A (s, t) | \lambda_1 \lambda_2 \rangle = - \frac{1}{2i} \int_{-\frac{1}{2} - i\infty}^{-\frac{1}{2} + i\infty} \frac{d J (2J + 1)}{\sin \pi (J + \lambda)} \times$$

$$\times [A_{\neq H}^{+}(s) d_{\lambda\lambda'}^{+}(J, -z) + A_{\bar{J}H}^{-}(s) d_{\lambda\lambda'}^{-}(J, -z)] -$$

$$- \pi \sum_{\substack{i \\ \mathrm{Re}\,\alpha_i > -\frac{1}{2}}} \frac{(2\alpha_i^{\pm} + 1)}{\sin \pi (\alpha_i^{\pm} + \lambda)} \, \beta_{iH}^{\pm}(s) \, d_{\lambda\lambda'}^{\pm}(J, -z) -$$

$$- \sum_{J=\lambda'}^{\lambda-1} (2J + 1) \, [A_{JH}^{+}(s) \, d_{\lambda\lambda'}^{+}(J, -z) + A_{JH}^{-}(s) \, d_{\lambda\lambda'}^{-}(J, -z)] -$$

$$- \sum_{J=0\left(\frac{1}{2}\right)}^{\lambda'-1} (2J + 1) \, [A_{JH}^{+}(s) \, d_{\lambda\lambda'}^{+}(J, -z) + A_{JH}^{-}(s) \, d_{\lambda\lambda'}^{-}(J, -z)],$$

$$\tag{5.2}$$

where $\beta_{iH}^{\pm}(s)$ is the residue of the pole in $A_{JH}^{+}(s)$ at $J = \alpha_i^{\pm}(s)$. The last two sums in (5.2), which have no analogue in the zero spin case, arise from the integral (or half-odd-integral) values of J between zero and $\lambda - 1$, which are not included in the original partial-wave expansion (1.2). We shall see below the reason for writing them as separate sums over sense-nonsense and nonsense-nonsense values.

Using the large z behaviour of $d_{\lambda\lambda'}^J$ given in (2.15) and (2.16) we see that the first two terms in (5.2) behave for large z like $z^{-1/2}$ and $z^{\alpha_M(s)}$ (α_M being the rightmost Regge trajectory), respectively, and the final term like z^{-1} or $z^{-1/2}$ according to whether physical J-values are integral or half-odd-integral. In the third term, where the sum is over sense-nonsense values, the $d_{\lambda\lambda'}(J, -z)$ have square root zeros so, provided the A_{JH}^+ are finite at the sense-nonsense value of J, this term will not occur. However, for the "wrong signature" values of J (wrong signature means positive signature for odd J, and negative signature for even J, for the integral J case, for example), it would be possible for A_{JH}^+ to behave like $(J - J_0)^{-1/2}$ near the relevant value of J (i.e. J_0) since such a contribution would be removed by the signature factor — this fact will be important later. It is of course very desirable that these terms should vanish, since, apart from the square-root zero $(J - J_0)^{1/2}$, the $d_{\lambda\lambda}^J$ would behave like z^{J_0} at the sense-nonsense values of J, and the third term of (5.2) would give rise to a non-Regge type behaviour (i.e. $t^{\lambda-1}$). We shall discuss these points further in the next chapter.

Assuming that the third term in (5.2) can be dropped, we obtain Regge behaviour, plus a background term which is $O(t^{-1/2})$ when integral J are physical, and $O(t^{-1/2+\varepsilon})$ for arbitrarily small positive ε when half-odd-integral J are physical.

The next step is to extend this behaviour to lower powers of z or t by the analogue of Mandelstam's method for zero spin. This has been done by *Calogero* et al., 1963a, and by *Drechsler*, 1967a. The details are rather complicated so we do not include them here, except to note that the result again depends on the symmetry

$$A_{HJ}^{\pm} = A_{H-J-1}^{\mp} \tag{5.3}$$

being valid for integral (or half-odd-integral) values of J [cf. (II.12.6)]. The result is to permit one to replace $d_{\lambda\mu}^J$ in (5.2), say, by the combination of $e_{\lambda\mu}^J$ and $e_{-\lambda-\mu}^{-J-1}$ given in (2.25), and to drop the $e_{\lambda\mu}^J$ term, thus obtaining the simple Regge behaviour $t^{\alpha(s)}$, for all values of α.

It is convenient here to recall the important exception to the statement that we obtain Regge behaviour, $t^{\alpha (s)}$, for high t in all cases, noted at the end of Section (IV.3). This is when the masses are such that z_s does not become large at high energy. In this case we must use the procedure of Section (III.6). However we must do this with the amplitude $\hat{A}$ (3.1) which has the kinematical singularities removed. The arguments of Section (IV.3) then show that $\hat{A} \sim t^{\alpha-M}$. The kinematical factors exhibited in (3.2) do not behave like t^M for large t at $s = 0$ (since $z_s \equiv 1$), but like constants, so $t^{\alpha-M}$ is the true behaviour in this case. We shall see the importance of this in Section (VIII.7).

IV.6. Kinematical Constraints on the Residue Functions

In Section (IV.3) we showed that the helicity amplitudes $A_H(s, t)$ have certain kinematical singularities in t, at values corresponding to $z_s = \pm 1$. We also saw that it was easy to remove these kinematical singularities. There will be similar singularities in the s-plane at values of s corresponding to $z_t = \pm 1$ (where z_t is the cosine of the t-channel centre-of-mass scattering angle). Clearly we can locate and remove these by a similar procedure, except that this must be applied to the t-channel helicity amplitudes rather than the s-channel helicity amplitudes which we have discussed so far.

The procedure is straightforward in principle. We first write the t-channel helicity amplitudes in terms of the s-channel helicity amplitudes by means of the helicity crossing matrix $M_{H_t H_s}$, i.e.

$$A^t_{H_t}(s, t) = M_{H_t H_s}(s, t) \, A^s_{H_s}(s, t) \tag{6.1}$$

where $A^t_{H_t}(s, t)$ is the helicity amplitude in the t-channel, and we are using $A^s_{H_s}$ rather than just A_H to denote the s-channel helicity amplitudes. We use H_s to represent the set $\lambda_1, \lambda_2, \lambda_3, \lambda_4$ of s-channel helicities and H_t to represent the set $\mu_1, \mu_2, \mu_3, \mu_4$ of t-channel helicities. The matrix $M_{H_t H_s}(s, t)$ has been given by *Trueman* and *Wick*, 1964, and *Muzinich*, 1964, as

$$M_{H_t H_s}(s, t) = d_{\lambda_1 \mu_1}(\chi_1) \, d_{\lambda_2 \mu_2}(\chi_2) \, d_{\lambda_3 \mu_3}(\chi_3) \, d_{\lambda_4 \mu_4}(\chi_4) \tag{6.2}$$

where the angles χ_i are given by

$$\cos \chi_1 = \frac{[-(s + m_1^2 - m_2^2)(t + m_1^2 - m_3^2) - 2m_1^2(m_3^2 - m_1^2 + m_2^2 - m_4^2)]}{4 \sqrt{st} \, q_{s12} \, q_{t13}} \tag{6.3a}$$

$$\cos \chi_2 = \frac{[(s + m_2^2 - m_1^2)(t + m_2^2 - m_4^2) - 2m_2^2(m_3^2 - m_1^2 + m_2^2 - m_4^2)]}{4 \sqrt{st} \, q_{s12} \, q_{t24}} \tag{6.3b}$$

$$\cos \chi_3 = \frac{[(s + m_3^2 - m_4^2)(t + m_3^2 - m_1^2) - 2m_3^2(m_3^2 - m_1^2 + m_2^2 - m_4^2)]}{4 \sqrt{st} \, q_{s34} \, q_{t13}} \tag{6.3c}$$

$$\cos \chi_4 = \frac{[-(s + m_4^2 - m_3^2)(t + m_4^2 - m_2^2) - 2m_4^2(m_3^2 - m_1^2 + m_2^2 - m_4^2)]}{4 \sqrt{st} \, q_{s34} \, q_{t24}}. \tag{6.3d}$$

Here $q_{t\,13}$ and $q_{t\,24}$ are center of mass momenta in the t-channel, analogous to those in the s-channel defined in Section I.6. We shall not go through the derivation of this crossing matrix, but merely note that it is a consequence of our initial postulates; in particular, Lorentz invariance plays a crucial role.

The t-plane kinematical singularities of $A_{H_t}^s$ and the s-plane kinematical singularities of $A_{H_t}^t$ can be exhibited explicitly by the procedure of Section (IV.3), so (6.1) can be used to locate and remove the kinematical singularites in the s and t planes.

The details of the procedure are somewhat complicated so we refer to the papers of *Hara*, 1964, and *Wang*, 1966 a, and here we quote their results. In the following we denote s-channel amplitudes free of kinematical singularities or zeros by $\hat{\hat{A}}(s, t)$, and we relate them to the amplitudes $\hat{A}(s, t)$ defined in (3.1). Except where indicated otherwise the helicities are always $\lambda_3, \lambda_4, \lambda_1, \lambda_2$, and we do not indicate them explicitly. Note that for the case where physical J-values are half-odd-integral the $\hat{\hat{A}}$ amplitudes may still contain a kinematical singularity at $s = 0$ in the s-plane, but this is removed by considering the w-plane where $w = \sqrt{s}$.

I. Equal mass, $m_i = m$

$$\hat{\hat{A}}(s, t) = (s - 4m^2)^{\alpha/2}\,(s)^{-\beta/2}\,\hat{A}(s, t) \tag{6.4}$$

where

$$\alpha = 1/2 \sum_{i=1}^{4} v_i + \mathrm{Max}_\eta \left\{ \sum_{i=1}^{4} (\sigma_i - v_i/2) \right\} - |\lambda - \lambda'| - |\lambda + \lambda'| \tag{6.5}$$

and

$$\beta = \mathrm{Max}_\delta \left\{ |(|\sigma_4 - \sigma_2| - |\sigma_3 - \sigma_1|)| + 1 \right\} \tag{6.6}$$

with

$$v_i = 1 \quad \text{or} \quad 0 \tag{6.7}$$

according to whether $2\sigma_i$ is odd or even respectively,

$$\eta = \frac{\eta_1\,\eta_3}{\eta_2\,\eta_4}\,(-1)^{2(\sigma_1 + \sigma_2)} \tag{6.8}$$

$$\delta = \eta\,(-1)^{|\lambda-\lambda'|} \tag{6.9}$$

and the term $\mathrm{Max}_\eta\{X\}$ means the greatest even (odd) integer smaller than or equal to X when $\eta = +1\,(-1)$. The η_i are the intrinsic parities of the indicated particles.

II. $m_1 = m_3, m_2 = m_4$

$$\hat{\hat{A}}(s, t) = \{[s - (m_1 + m_2)^2]\,[s - (m_1 - m_2)^2]\}^{\alpha/2}\,s^{1/2|\lambda - \lambda'|}\,\hat{A}(s, t)\,. \tag{6.10}$$

III. $m_1 = m_2, m_3 = m_4$

$$\hat{\hat{A}}(s, t) = (s - 4m_1^2)^{1/2\,\alpha\pm}\,(s - 4m_2^2)^{1/2\,\beta\pm}\,s^{-\gamma/2}\,[\langle\lambda_3\lambda_4\,|\hat{A}|\,\lambda_1\lambda_2\rangle \pm$$
$$\pm\,\langle-\lambda_3\lambda_4\,|\hat{A}|\,\lambda_1\lambda_2\rangle] \tag{6.11}$$

where

$$\alpha^\pm = -|\lambda - \lambda'| + \text{Max}_{\pm\,\eta_{12}}\{\sigma_1 + \sigma_2 - 1/2\,(v_1 + v_2) + 1/2\,(|\lambda - \lambda'| - |\lambda + \lambda'|)\} + \\ + 1/2\,(v_1 + v_2)\;. \tag{6.12}$$

$$\beta^\pm = -|\lambda - \lambda'| + \text{Max}_{\pm\,\eta_{34}}\{\sigma_3 + \sigma_4 - 1/2\,(v_3 + v_4) + 1/2\,(|\lambda - \lambda'| - |\lambda + \lambda'|)\} + \\ + 1/2\,(v_3 + v_4) \tag{6.13}$$

$$\gamma = \text{Max}_{\eta_s}\{|(|\sigma_4 - \sigma_2| - |\sigma_3 - \sigma_1|)|\} \tag{6.14}$$

$$\eta_s = \frac{\eta_1\,\eta_3}{\eta_2\,\eta_4}\,(-1)^{2(\sigma_1 + \sigma_2)}\,(-1)^{\lambda - \lambda'} \tag{6.15}$$

$$\eta_{12} = \frac{\eta_1\,\eta_3}{\eta_2\,\eta_4}\,(-1)^{\sigma_1 + \sigma_2 + \lambda_1 + \lambda_2} \tag{6.16}$$

$$\eta_{34} = \frac{\eta_1\,\eta_3}{\eta_2\,\eta_4}\,(-1)^{\sigma_3 + \sigma_4 + \lambda_3 + \lambda_4}\;. \tag{6.17}$$

IV. $m_1 = m_2$, $m_3 \neq m_4$ (including the possibility that $m_1 = m_3$, say)

$$\hat{\hat{A}}\,(s,t) = (s - 4m_1^2)^{\alpha^\pm/2}\,[s - (m_3 + m_4)^2]^{\beta^\pm/2}\,s^{\gamma^\pm/2}\,[s - (m_3 - m_4)^2]^{\bar\beta^\pm/2}\,\times \\ \times\,[\langle \lambda_3 \lambda_4 |\hat{A}|\,\lambda_1 \lambda_2\rangle \pm \langle -\lambda_3 - \lambda_4\,|\hat{A}|\,\lambda_1\lambda_2\rangle] \tag{6.18}$$

where

and

and

if

and

$$\bar\beta^\pm = \beta^\pm \quad \text{for} \quad v_3 = v_4 = 0 \tag{6.19a}$$

$$\bar\beta^\pm = \beta^\mp \quad \text{for} \quad v_3 = v_4 = 1 \tag{6.19b}$$

$$\gamma^\pm = \text{Max}_{\mp\,\eta_{12}}\{\sigma_1 + \sigma_2 - 1\} + 1 \tag{6.20a}$$

$$v_1 = v_2 = 1$$

$$\gamma^\pm = \text{Max}_{\mp\,\eta_{12}}\{(\sigma_1 + \sigma_2)\} \quad \text{for} \quad v_1 = v_2 = 0\;. \tag{6.20b}$$

V. All Masses Different; Physical Spin Integral

$$\hat{\hat{A}}\,(s,t) = [s - (m_1 + m_2)^2]^{\alpha^\pm/2}\,[s - (m_1 - m_2)^2]^{\bar\alpha^\pm/2}\,\times \\ \times\,[s - (m_3 + m_4)]^{\beta^\pm/2}\,[s - (m_3 - m_4)]^{\bar\beta^\pm/2}\,s^{\bar\gamma/2}\,\times \tag{6.21} \\ \times\,[\langle \lambda_3 \lambda_4\,|A\,(s,t)|\,\lambda_1 \lambda_2\rangle \pm \langle -\lambda_3 - \lambda_4\,|A\,(s,t)|\,\lambda_1\lambda_2\rangle]$$

where in addition to the quantities defined above

$$\bar\alpha^\pm = \alpha^\pm \quad \text{if} \quad v_1 = v_2 = 0 \tag{6.22a}$$

$$= \alpha^\mp \quad \text{if} \quad v_1 = v_2 = 1 \tag{6.22b}$$

and

$$\bar\gamma = \text{Max}\,(|\lambda - \lambda'|,\;|\lambda + \lambda'|)\;. \tag{6.23}$$

VI. All Masses Different; Physical J Half-odd-integral

$$\hat{\hat{A}}\,(s,t) = (4s\,q^2_{s\,12})^{a^\pm/2}\,(4s\,q^2_{s\,34})^{b^\pm/2}\,(s)^{\bar\gamma/2}\,\times \\ \times\,[\langle \lambda_3 \lambda_4\,|\hat{A}\,(s,t)|\,\lambda_1 \lambda_2\rangle \pm \langle -\lambda_3 - \lambda_4\,|\hat{A}\,(s,t)|\,\lambda_1\lambda_2\rangle] \tag{6.24}$$

where

$$a^\pm = -|\lambda - \lambda'| + \text{Max}_{\pm\,\eta_{12}}\{\sigma_1 + \sigma_2 + 1/2\,(|\lambda - \lambda'| - |\lambda + \lambda'|)\} \\ b^\pm = -|\lambda - \lambda'| + \text{Max}_{\pm\,\eta_{34}}\{\sigma_3 + \sigma_4 + 1/2\,(|\lambda - \lambda'| - |\lambda + \lambda'|)\}\;. \tag{6.25}$$

From the above equations we see that there are in general kinematical singularities and/or zeros at $s = 0$, $s = (m_1 \pm m_2)^2$ and $s = (m_3 \pm m_4)^2$. The points $s = (m_1 \pm m_2)^2$ are the physical threshold and "pseudo-threshold" for the $1 + 2$ channel, and similarly for the $3 + 4$ channel. The behaviour at these thresholds, as given by the above expressions, corresponds to the usual threshold behaviour of a partial-wave amplitude [see (II.4.9)], the orbital angular momentum being the lowest allowed for the helicity state considered.

The kinematical singularities and zeros of these amplitudes will also occur in the partial-wave amplitudes, and hence in general (see below for possible exceptions) also in the residue functions of Regge-poles. Many authors have therefore extracted them as explicit factors from the residue functions. Except possibly for the point $s = 0$ these points are outside the t-channel physical region (which is where the Regge-pole formula is used), so it is clear that, in the absence of further assumptions about the "reduced residue" (i.e. the residue with the "kinematical effects" factored out), this is a harmless but unnecessary exercise. However it is sometimes assumed that, at least over part of the physical region, the kinematical effects dominate the variation of the residue so that the reduced residue is approximately constant. Some caution is necessary here as this procedure might well be misleading. In particular it should be noted that physical quantities (which depend also upon the complex conjugate of helicity amplitudes) are not analytic functions of s and t. Thus a particular t-channel physical quantity when expressed in terms of t-channel helicity amplitudes continued to a non-physical point may remain finite, whereas if it is expressed in terms of s-channel amplitudes similarly continued it may become infinite. In such a case the suggested "peaking" towards this point may be spurious. An example is considered by *Lin*, 1967, who notes that infinities occur in some of the s-channel helicity amplitudes for $p\bar{p} \to p\bar{N}^*$ at $s = 0$. Now the boundary of the t-channel physical region for this process ($pp \to pN^*$) approaches $s = 0$ as $t \to \infty$ (see Section VIII.3), so these singularities might be expected to produce a peaking in the physical cross-section near the forward direction. However, as Lin points out, these physical infinities arise from the crossing matrix, which is in fact bounded for all physical s and t (the angles χ_i are physical angles in the physical region), so they do not dominate the cross-section and it would be misleading to include them explicitly in the residue functions.

One important role of the branch points in s is that they ensure that the contribution of a given Regge pole has the same phase for each helicity amplitude in the cross-channel physical region. This is because the phase factor i^{M-N} of (2.14), for example, cancels with similar factors from the s-cuts for negative s. This can be seen most simply by working with the t-channel helicity amplitudes, for it is clear that these do not have kinematical cuts in the t-channel physical region.

In general a zero at $s = 0$ in a t-channel amplitude, which arises from angular momentum conservation as discussed below (1.4), and which appears explicitly in the t-channel partial-wave expansion, will not show up as a zero in a particular s-channel helicity amplitude (unless it is of a square-root type), but rather as a linear relation between s-channel helicity amplitudes. In dealing with these zeros it is therefore convenient to refer directly to the t-channel helicity amplitudes which are related by crossing (of course in physical applications it is the t-channel helicity amplitudes which are in fact required). The natural way to obtain such a zero in the amplitude is to require that the residue of any contributing trajectory also has the zero. However it may be that, at least in some cases, such a procedure is incorrect when the zero does not correspond to a branch point, and that it arises from two or more trajectories whose contributions to the amplitude in question exactly cancel each other at $s = 0$, but whose contributions individually do not cancel. This possibility, which was first suggested in the context of N N scattering by *Volkov* and *Gribov*, 1963, and has since been studied in detail by *Leader*, 1967, is referred to as a "conspiracy". The motivation for the idea of such conspiracies is the fact that a kinematical zero in a residue, which would be required without any conspiracy, affects other amplitudes through the factorization theorem and leads to the contribution of a given trajectory being zero even when there is no kinematic reason for this. The fact that the residues of s-channel Regge poles, expressed in terms of *t-channel* helicities, factorize follows from the known factorization when expressed in terms of s-channel helicities, and the form of the crossing matrix (6.2). *Leader*, 1967, has analysed in detail the form of the constraints for general spins and masses, and distinguishes three possible ways in which they might be satisfied, namely "evasion", in which the residues have the appropriate zeros and no compensating trajectories are required; "conspiracy" in which a finite number of trajectories having different quantum numbers are so arranged that they satisfy the constraints provided they are all included; and finally "daughters", which are an infinite set of trajectories having the same quantum numbers. *Leader* shows that a non-trivial evasion (i.e. one which does not involve complete decoupling) is always possible. Thus there is no compelling reason for having conspiracies of any sort, although, as we shall see, the consequences of not having conspiracies are in some cases unpleasant. We discuss these matters further in Section (IV.9) and also at various places in Chapter VIII, where we discuss possible experimental tests for conspiracies, and give additional references.

The partial-wave amplitudes contain the threshold factors, i.e. powers of $s - (m_1 \pm m_2)^2$ and $s - (m_3 \pm m_4)^2$ of the full amplitude, and in addition they contain threshold factors arising from the projection (4.2). From the latter we find that $A_{HJ}(s)$ contains the factor $(q_{s12}\, q_{s34})^{(J - \lambda)}$ in addition to any threshold factors occuring in $A_H(s, t)$. In consequence it turns out that the power of (q_{s12}) (for example) is the lowest value of l (the orbital angular momentum) compatible with the given J and helicity values in the initial (i.e. 1, 2) state.

We note that there are additional constraints on the partial wave amplitudes at the physical thresholds which arise from the fact that only one l-value (i.e. $l = 0$) is contributing. These constraints have the form of relations between the various amplitudes at threshold (*H. F. Jones*, 1967). It is not clear whether they should be regarded as significant in phenomenological analyses since they hold at a point outside the range where Regge pole fits are made (i.e. $s \leqq 0$). The only case where they have been used is in the work of *Frautschi* and *Jones*, 1967b, who apparently find some effects of these constraints in the physical region [see Section (VIII.6.h)].

A further source of kinematical factors in the residue is the requirement, mentioned in Section (IV.5), that the sense-nonsense amplitude should be finite. In fact, as we shall see in Section (V.6) this only applies to the "right-signature" amplitude. The description "right-signature" applies when the value of J is even (odd) for a positive (negative) signature amplitude, when physical J-values are integral, and correspondingly when physical J-values are half-odd-integral. Later we shall also use the term "wrong-signature", which carries the obvious meaning. Since there is a square-root branch point at these sense-nonsense values of J [see (4.8)], it follows that the residue will behave like $(J - J_0)^{1/2}$ for $J \approx J_0$, where J_0 is a right signature sense-nonsense value of J. Some authors have also included the factor $(J + J_0 + 1)^{1/2}$ to extract the behaviour near $J = -J_0 - 1$, suggested by the symmetry property (5.3). However it should be noted that this symmetry property relates amplitudes of opposite signature, so this other factor will only occur if the wrong signature sense-nonsense amplitude is also zero (we shall see in Section (V.6) that this is probably not the case).

Finally we note that the factorization theorem [Section (V.1)] also makes restrictions on the residues. It allows us to write

$$\beta_{\lambda_3 \lambda_4 \lambda_1 \lambda_2} = \gamma_{\lambda_3 \lambda_4} \, \gamma_{\lambda_1 \lambda_2} \tag{6.26}$$

for all allowed values of the λ_i. If we combine this equation with the knowledge that a right-signature sense-nonsense coupling contains the factor $(J - J_0)^{1/2}$ then we see that, because the sense-sense and nonsense-nonsense amplitudes are analytic at $J = J_0$, one of them must contain the factor $(J - J_0)$. Thus a Regge trajectory passing through a right-signature value of J for which there are both nonsense and sense states, decouples either from the sense or the nonsense states, i.e. it "chooses" nonsense or sense respectively. At present it is an experimental question which of these occurs [see Section (VIII.6d)].

IV.7. Parity

Since parity is a good quantum number for strong interactions it is convenient in many applications of Regge pole theory to introduce amplitudes with specific parity. To this end we introduce the states

$$|JM\,\lambda_1\lambda_2\rangle_\pm = \frac{1}{\sqrt{2}} \left[|JM\,\lambda_1\lambda_2\rangle \pm \eta_1\eta_2(-1)^{\sigma_1+\sigma_2-v} |JM - \lambda_1 - \lambda_2\rangle \right] \tag{7.1}$$

where η_1 and η_2 are the intrinsic parities of particles 1 and 2, and $v = 0$ or $1/2$ according to whether physical J-values are integral or half-odd-integral. These states are eigenstates of parity, satisfying

$$P\,|JM\,\lambda_1\lambda_2\rangle_\pm = \pm\,(-1)^{J-v}\,|JM\,\lambda_1\lambda_2\rangle_\pm\,. \qquad (7.2)$$

Here we have used equation (41) of *Jacob* and *Wick*, 1959.

The matrix elements of a parity conserving operator, A, between these good parity states are given by

$$\begin{aligned}
{}_\pm\langle JM\,\lambda_3\lambda_4\,|A|\,JM\,\lambda_1\lambda_2\rangle_\pm &= \langle JM\,\lambda_3\lambda_4\,|A|\,JM\,\lambda_1\lambda_2\rangle \pm \\
&\pm\,\eta_3\eta_4(-1)^{\sigma_3+\sigma_4-v}\langle JM-\lambda_3-\lambda_4\,|A|\,JM\,\lambda_1\lambda_2\rangle.
\end{aligned} \qquad (7.3)$$

Here the $+$ sign corresponds to positive or negative parity according to whether $(J-v)$ is even or odd, and the $-$ sign corresponds in each case to the opposite parity. Now for amplitudes of well defined signature we recall that positive (negative) signature amplitudes are physical at even (odd) values of $J-v$ [see remarks following (4.2)]. Thus the following amplitudes of well defined signature (S) and parity (P) can be defined:

$$\langle\lambda_3\lambda_4\,|A_J^{SP}|\,\lambda_1\lambda_2\rangle = \langle\lambda_3\lambda_4\,|A_J^S|\,\lambda_1\lambda_2\rangle + PS\,(-1)^{\sigma_3+\sigma_4-v}\,\eta_3\eta_4 \times$$
$$\times\,\langle-\lambda_3-\lambda_4\,|A_J^S|\,\lambda_1\lambda_2\rangle \qquad (7.4)$$

where, for example, $S = +1$ and $P = +1$ corresponds to the positive signature, positive parity amplitude.

An important property of the good parity amplitudes for Boson-Fermion scattering is the generalisation of the MacDowell-symmetry for πN scattering (*MacDowell*, 1959). From the helicity crossing matrix (discussed in the previous section) it follows that there is a square root branch point in the helicity amplitudes at $s = 0$. For this reason it is convenient to use the variable

$$w = \sqrt{s}\,, \qquad (7.5)$$

and the form of the branch point is such that *one* of the amplitudes on the right-hand-side of (7.4) changes sign when $w \to -w$. Hence this reflection connects amplitudes of *opposite* parity. Specifically we have (*Hara*, 1964)

$$\langle\lambda_3\lambda_4\,|A_J^{SP}(w)|\,\lambda_3\lambda_4\rangle = -(-1)^{\lambda-\lambda'}\langle\lambda_3\lambda_4\,|A_J^{S-P}(-w)|\,\lambda_1\lambda_2\rangle\,. \quad (7.6)$$

This relation has interesting consequences for fermion Regge trajectories (i.e. those which are physical at half-odd-integral values of J). First, since different parities will have different trajectories it follows that $\alpha(w)$ is not an even function of w for a fermion trajectory, so that as a function of $s = w^2$ the trajectory has a branch point at $s = 0$. Secondly, if physical states of both parities exist then the trajectory will pass through positive values of J for both positive and negative values of w. Thus it will appear as shown for example in Fig. (VIII.7).

IV.8. The Pomeranchon Contribution

For elastic scattering of spin-zero particles the Froissart bound tells us that an amplitude cannot increase with t faster than t^1, ignoring logarithmic factors. The evidence that total cross-sections tend to

8*

constants led to the idea of a trajectory which passes through $s = 0$ at the maximum position allowed by this condition, i.e. $\alpha(0) = 1$; this is the so-called Pomeranchon trajectory. In order to satisfy the Pomeranchuk theorems this particle has zero quantum numbers (apart from spin).

For particles with spin the same bound has been shown to hold, under essentially the same assumptions, by *Hara*, 1964, so again the Pomeranchon trajectory (assuming it to exist) will dominate high-energy scattering for those processes for which it can be exchanged in the cross-channel.

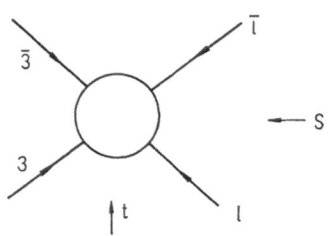

Fig. IV.1. Elastic scattering in the *t*-channel.

Hara, 1966, endeavoured to prove that the Pomeranchon contribution to the total cross-section for the scattering of two particles is independent of the helicities, i.e. is spin independent. As we shall see there is an unproved assumption in this work, but it is nevertheless worth going through the argument. We present it here in a simpler form due to *Mueller* and *Trueman*, 1967a.

We wish to consider elastic scattering in the *t*-channel so we require particles 1 and $\bar{2}$ and particles 3 and $\bar{4}$ to be the same. We use the labels 1 and 3 for the two particles [see Fig. (IV.1)] .

We want to calculate the effect of Pomeranchon exchange in the s-channel, so we only need consider s-channel amplitudes with $PC = +1$.

For such amplitudes we have

$$\langle \lambda_1 \bar{\lambda}_1 \,|A\,(s,t)|\, \lambda_3 \bar{\lambda}_3 \rangle = \langle \lambda_1 \bar{\lambda}_1 \,|A\,(s,t)|\, -\bar{\lambda}_3 - \lambda_3 \rangle \qquad (8.1)$$

(N.B.: In this section the equalities apply only to the Pomeranchon contribution, i.e. they hold for the full amplitude only in the limit $t \to \infty$).

We write the crossing relation from the s-channel to the t-channel at $s = 0$:

$$\langle \mu_1 \mu_3 \,|A^t(0,t)|\, \bar{\mu}_1 \bar{\mu}_3 \rangle = \sum_{\lambda_i} d^{\sigma_1}_{\lambda_1 \mu_1}\left(\frac{\pi}{2}\right) d^{\sigma_1}_{\lambda_1 \bar{\mu}_1}\left(\frac{\pi}{2}\right) d^{\sigma_1}_{\lambda_3 \mu_3}\left(\frac{\pi}{2}\right) \times$$
$$\times\ d^{\sigma_3}_{\lambda_3 \bar{\lambda}_3}\left(\frac{\pi}{2}\right) \langle \lambda_1 \bar{\lambda}_1 \,|A^s(0,t)|\, \lambda_3 \bar{\lambda}_3 \rangle , \qquad (8.2)$$

where we have explicitly indicated the channels by the suffices t and s, and we have used μ for the t-channel helicities [as in Section (IV.6)]. Putting (8.1) into (8.2), and using reflection properties of the $d^{\sigma}_{\lambda\mu}$, we find

$$\langle \mu_1 \mu_3 \,|A^t(0,t)|\, \bar{\mu}_1 \bar{\mu}_3 \rangle = \sum_{\lambda's} d^{\sigma_1}_{\lambda_1 \mu_1}\left(\frac{\pi}{2}\right) d^{\sigma_1}_{\lambda_1 \bar{\mu}_1}\left(\frac{\pi}{2}\right) d^{\sigma_3}_{-\bar{\lambda}_3 \mu_3}\left(\frac{\pi}{2}\right) \times$$
$$\times\ d^{\sigma_3}_{-\lambda_3 \bar{\lambda}_3}\left(\frac{\pi}{2}\right) \langle \lambda_1 \bar{\lambda}_1 \,|A^s(0,t)|\, \lambda_3 \bar{\lambda}_3 \rangle \qquad (8.3)$$
$$= \langle \mu_1 \bar{\mu}_3 \,|A^t(0,t)|\, \bar{\mu}_1 \mu_3 \rangle\, (-1)^{\bar{\mu}_3 - \mu_3} .$$

We now use conservation of angular momentum (or the analogue of (3.1)] for the t-channel amplitudes at $s = 0$. Because of (8.3) this yields *two* conditions, i.e.

$$\mu_1 - \mu_3 = \bar{\mu}_1 - \bar{\mu}_3 \tag{8.4a}$$

and

$$\mu_1 - \bar{\mu}_3 = \bar{\mu}_1 - \mu_3 \tag{8.4b}$$

which give

$$\mu_1 = \bar{\mu}_1 \tag{8.5a}$$

and

$$\mu_3 = \bar{\mu}_3 \, . \tag{8.5b}$$

These equations can now be used in the crossing relation from the t to the s-channel to obtain

$$\langle \lambda_1 \bar{\lambda}_1 \, |A^s(0, t)| \, \lambda_3 \bar{\lambda}_3 \rangle = \langle \lambda_1 \bar{\lambda}_1 \, |A^s(0, t)| \, \bar{\lambda}_3 \lambda_3 \rangle \, . \tag{8.6}$$

When we write the Pomeranchon contribution to these two amplitudes, however, it turns out that it is opposite in sign unless $|\lambda_1 - \bar{\lambda}_1|$ and $|\lambda_3 - \bar{\lambda}_3|$ are even. It follows that the Pomeranchon coupling to states for which either of these quantities is odd must be zero. The only alternative is a "conspiracy", which would here require another trajectory passing through $\alpha(s) = 1$, at $s = 0$. We reject this possibility.

Now we introduce the unproven assumption, referred to above, that only states with $|\lambda_1 - \bar{\lambda}_1|$ and $|\lambda_3 - \bar{\lambda}_3| \leq 1$ couple to the Pomeranchon trajectory at $s = 0$. Since at $s = 0$, $\alpha(s) = 1$, the other states are *nonsense* states at this value of s, and *Hara*, 1966, made this assumption on this basis. However, the Pomeranchon has *wrong signature* at $\alpha = 1$ and there is therefore no reason why it should be decoupled from nonsense states [see the discussion of Sections (IV.5) and (V.6)]. Thus this assumption is unproven and our result below may be wrong. It should be noted however that the assumption is unnecessary unless spins higher than 1/2 are involved, so, for example, the results certainly apply to the NN system.

Making the above assumption, where necessary, we immediately see that only

$$\lambda_1 = \bar{\lambda}_1 \tag{8.7a}$$

and

$$\lambda_3 = \bar{\lambda}_3 \tag{8.7b}$$

are allowed. That this leads to a high energy amplitude independent of helicities was demonstrated by *Hara*, 1966, and previously by *Peierls* and *Trueman*, 1964. To obtain this result we write (8.2) with (8.5) and (8.7),

$$\langle \mu_1 \mu_3 \, |A^t(0, t)| \, \bar{\mu}_1 \bar{\mu}_3 \rangle = \delta_{\mu_1 \bar{\mu}_1} \, \delta_{\mu_3 \bar{\mu}_3} \, \langle \mu_1 \mu_3 \, |A^t(0, t)| \, \mu_1 \mu_3 \rangle$$

$$= \sum_{\lambda_1 \lambda_3} d^{\sigma_1}_{\lambda_1 \mu_1} \left(\frac{\pi}{2} \right) d^{\sigma_1}_{\lambda_1 \bar{\mu}_1} \left(\frac{\pi}{2} \right) d^{\sigma_3}_{\lambda_3 \mu_3} \left(\frac{\pi}{2} \right) \times \tag{8.8}$$

$$\times \, d^{\sigma_3}_{\lambda_3 \bar{\mu}_3} \left(\frac{\pi}{2} \right) \langle \lambda_1 \lambda_1 \, |A^s(0, t)| \, \lambda_3 \lambda_3 \rangle \, .$$

We multiply both sides of (8.8) by $d^{q_1}_{\lambda_1 \bar{\mu}_1} \left(\frac{\pi}{2}\right) d^{q_3}_{\lambda_3 \bar{\mu}_3} \left(\frac{\pi}{2}\right)$ and sum over $\bar{\mu}_1$ and $\bar{\mu}_3$. Using the orthogonality of the d's this immediately gives

$$\langle \mu_1 \mu_3 | A^t(0, t) | \mu_1 \mu_3 \rangle = \langle \lambda_1 \lambda_1 | A^s(0, t) | \lambda_3 \lambda_3 \rangle \qquad (8.9)$$

for all values of the μ's and λ's. This proves the required result.

IV.9. The Application of Group-theoretical Methods to Regge Theory

We have noticed two apparently distinct problems concerning the behaviour of Regge trajectories at $s = 0$. There are the difficulties associated with unequal mass kinematics, discussed in Chapter III, for which we found the alternative solutions, that the background term has special properties, or that there are infinite sequences of daughter trajectories which cancel the singularities of their parents (and each other) at $s = 0$. Also we have found in Section (IV.6) that, in amplitudes for the scattering of particles with spin, the kinematical constraints require either that trajectories have vanishing residues in some amplitudes at $s = 0$ (evasion), or that they combine to give cancelling contributions (conspiracy).

Both of these phenomena may perhaps be associated with an extra symmetry possessed by the four-line connected part at $s = 0$, and there have been several attempts in recent work to express these symmetries in an appropriate group-theoretical language, and so unify the whole subject. It is also felt by some authors that from the group theory point of view the solutions to the $s = 0$ constraints which seem more "natural" are the daughter and conspirator hypotheses, rather than the ones which ignore the extra symmetry at this point, the background cancellation and evasion.

We shall see that at the present time this does not appear to be a completely compelling argument, but the group-theoretical methods certainly have great elegance, and may well be important for future developments, so it seems appropriate to say something about them here. Unfortunately the presentation of the full apparatus of representations of the complex inhomogeneous Lorentz group would require very lengthy discussion, and take us a long way from the main topics of this book, so we shall have to be content with a rather brief outline, giving references to papers where the details can be found. Introductions to the properties of the Lorentz group are to be found in *Wigner*, 1964, and *Brittin* and *Barut*, 1965.

Postulate (iii) of Chapter I requires that the S-matrix should be invariant under Lorentz transformations. This must be true first of all in the physical region of a given channel, but, with the analyticity postulate (v), the Hall-Wightman theorem (*Hall* and *Wightman*, 1957), as extended by *Stapp*, 1962, implies that the invariance must also hold in unphysical regions which can only be reached by complex Lorentz transformations. Thus the S-matrix must correspond to a representation of the complex inhomogeneous Lorentz group (or Poincaré group) $\mathfrak{P}$.

The partial-wave analysis in the s-channel physical region ($s > s_0$, $t < 0$), which we have introduced in Sections (II.1) and (IV.1), corresponds to representing the amplitude on the basis of irreducible representation of $\mathfrak{P}$ with

$$P^2 \equiv (p_1 + p_2)^2 \equiv s > 0 \,. \tag{9.1}$$

P^2 is the eigenvalue of one of the Casimir operators of the group. The rotation matrices $d^J_{\lambda\lambda'}$ are representations of the "little group" for $s > 0$, i.e. the rotation group in three dimensions $SO(3)$ [or its covering group $SU(2)$], the "little group" being defined as the group of transformations which leave the total four-momentum

$$P_\mu \equiv (p_{1\mu} + p_{2\mu}) \tag{9.2}$$

invariant (that is rotations). The Peter-Weyl theorem tells us that the matrix elements of the irreducible representations of $SO(3)$ [or strictly $SU(2)$] provide a complete basis for representing any function which is square integrable over the group manifold, so in this case the rotation functions $d^J_{\lambda\lambda'}$, with $J = 0, 1, 2, \ldots$ or $1/2, 3/2, \ldots$, used in (1.2) provide a complete basis for the expansion of any function for which the partial-wave series converges.

The great advantage of the partial-wave decomposition is that the amplitude $A_H(s, t)$ is thereby separated into its symmetry part, contained in the $d^J_{\lambda\lambda'}$ and its dynamical part, contained in $A^J_H(s)$. The J value depends upon the other Casimir operator of $\mathfrak{P}$, which has eigenvalues $= s J(J + 1)$. Of course any other complete set of functions could be used to decompose the amplitude, but only with this particular separation do we obtain the benefits resulting from the fact that $SO(3)$ is an invariance of the S-matrix.

The procedure we have used for Reggeizing the amplitude has been to start from the $SO(3)$ partial-wave series in the s-channel physical region, perform the Sommerfeld-Watson transform to give a representation in terms of complex J, and then use the analyticity of the S-matrix in s and t to transform the resulting expression [such as (II.9.10) or (5.2)] to other regions of the variables, and in particular to the t-channel physical region. It is here that we want to see if the s-channel poles control the t-channel asymptotic behaviour. But in the region $P^2 \equiv s < 0$ the little group of the Poincaré group is no longer $SO(3)$, but $SO(2, 1)$ [or its covering group $SU(1, 1)$]. In fact *Wigner*, 1939, has shown that there are four distinct classes of representations of $\mathfrak{P}$, characterized by different values of the Casimir operator, P^2, and which have different little groups. They are

    (i) Timelike, $P^2 > 0$; little group $SO(3)$
    (ii) Spacelike, $P^2 < 0$; little group $SO(2, 1)$
    (iii) Lightlike, $P^2 = 0$ and $P_\mu \neq 0$; little group $E(2)$
    (iv) Null, $P^2 = 0$ and $P_\mu = 0$; little group $SO(3, 1)$
[$E(2)$ is the Euclidean group in 2 dimensions].

The representations of $SO(2, 1)$ have been studied by *Bargmann*, 1947, and discussed from the point of view of Regge poles by *Andrews* and *Gunson*, 1964; *Sertorio* and *Toller*, 1964; *Hadjioannou*, 1966; *Joos*, 1964; *Boyce*, 1967 and *Boyce* et al., 1967. It is found that there are four families of unitary representations,

    a) Principal series: $\text{Re}\{J\} = -1/2, \ -\infty < \text{Im}\{J\} < \infty$.
    b) Supplementary series: $-1/2 < \text{Re}\{J\} < 0, \ \text{Im}\{J\} = 0$.
    c) Scalar: $J = 0$.
    d) Discrete: $J = 0, 1, 2, \ldots,$ or $-1/2, 1/2, 3/2, \ldots$.

Theorem 9 of Bargman, which is analogous to the Peter-Weyl theorem used above, tells us that any function which is square integrable over the group manifold can be expanded in terms of the principal series, and those members of the discrete series with $\text{Re}\{J\} > -1/2$. The discrete series corresponds to nonsense channel terms and is restricted to

$$J < M \equiv \max(\lambda, \lambda') . \tag{9.3}$$

These terms are cancelled by terms of the principal series for $J < M$ if the integration contour is moved from $\text{Re}\{J\} = -1/2$ to $\text{Re}\{J\} = M - 1/2$, (See e.g. *Boyce*, 1967, for details. Note also that we are neglecting questions of signature here.) and we end up with

$$A_H(s, t) = -\frac{16\pi}{2i} \int_{M-1/2-i\infty}^{M-1/2+i\infty} dJ \, \frac{(2J+1)}{\sin\pi(J-\lambda)} A_H^J(s) \, d_{\lambda\lambda'}^J(z_s) . \tag{9.4}$$

Here z_s is the cosine of the s-channel scattering angle, and is of course unphysical $(z_s > 1)$ in the region $s < 0$. The scattering amplitude is only square integrable over the group manifold if

$$A_H(s, t) = O(t^{-1/2}) \tag{9.5}$$

and evidently (9.4) is just the Sommerfeld-Watson transform of an amplitude with no singularities in $\text{Re}\{J\} > -1/2$, corresponding to (5.2). It would give back the normal partial-wave series (1.2) if we were to make the inverse Sommerfeld-Watson transformation. For sufficiently negative s (9.5) will probably hold, but as s is increased we expect dynamical Regge singularities to appear which correspond to functions which are not square integrable over the group manifold, and these have to be extracted before the group representation can be used. If we make the usual connection between asymptotic behaviour and the J plane singularities we end up with an expression just like (5.2).

So representing the amplitude in terms of an expansion in $O(2, 1)$ "partial-waves" is equivalent to making the Sommerfeld-Watson transform. This way of reproducing it is of some interest but nothing has been gained.

However, when s vanishes the correct little group to use is either (iii) or (iv) above, depending on the value of P_μ. Since

$$\sum_\mu p_{1\mu}^2 = m_1^2, \quad \text{and} \quad \sum_\mu p_{2\mu}^2 = m_2^2 \tag{9.6}$$

the fact that

$$P^2 \equiv \sum_\mu (p_{1\mu} + p_{2\mu})^2 = 0 \qquad (9.7)$$

implies

$$P_\mu \equiv (p_{1\mu} + p_{2\mu}) = 0 \qquad (9.8)$$

only if $m_1 = m_2$. In fact only if $m_1 = m_2$ and $m_3 = m_4$ can P_μ vanish for any P^2, on the mass shell. Whereas normally the fourline connected part has just two degrees of freedom, denoted by s and t, there are three degrees of freedom in satisfying the constraints (9.6) and (9.8) if $m_1^2 = m_2^2$ (and $m_3^2 = m_4^2$). This is why the little group is the larger symmetry group $SO(3, 1)$ [or its covering group $SL(2C)$] in this case. Representations of $SO(3, 1)$ have been given by *Sciarrino* and *Toller*, 1966; *Toller*, 1967, and by *Boyce* et al., 1967, who also consider the light-like case. The extra degree of freedom, when the masses are equal, exhibits itself in the fact that the principle unitary representations of $SO(3, 1)$ are labelled, not just by a single number (such as J above), but by the eigenvalues of two Casimir operators, j_0 which is discrete ($j_0 = 0, 1, 2, \ldots$ or $1/2, 3/2, \ldots$), and σ which is pure imaginary ($-i\infty < \sigma < i\infty$). The corresponding partial wave expansion can be written in the form

$$A_H(s = 0, t) = \delta_{\lambda\lambda'} \sum_{TT'} \sum_{j_0 = -T_M}^{T_M} -\frac{16\pi}{2i} \int_{-M-i\infty}^{-M+i\infty} \mathrm{d}\sigma (j_0^2 - \sigma^2) \times$$
$$\times A_{TT'}^{j_0\sigma}(s)\, d_{T\lambda T'}^{j_0\sigma} \qquad (9.9)$$

where the $d_{T\lambda T'}^{j_0\sigma}$ are $SO(3, 1)$ rotation matrices, and $A_{TT'}^{j_0\sigma}$ are the corresponding "partial-wave" amplitudes. Here

$$T_M = \min(T', T),$$

and

$$|\sigma_1 - \sigma_3| \leqq T \leqq \sigma_1 + \sigma_3, \quad |\sigma_2 - \sigma_4| \leqq T' \leqq \sigma_2 + \sigma_4$$

(these σ's being the spins of the particles, not to be confused with the integration variable). The details can be found in *Boyce* et al., 1967, and *Delbourgo* et al., 1967 a, b, whose notation we have followed except that we have used T for their J to avoid confusion with the total angular momentum. *Toller*, 1967, obtained the same results, but he used (M, λ) for our (j_0, σ).

The importance of this expression appears if we adopt the hypothesis that the σ variable is the natural variable in which to insert poles and cuts corresponding to those parts of the amplitude which are not square integrable over the group manifold at $s = 0$. The plausibility of this rests on the analogy with the way in which we inserted the Regge poles and cuts in J into (9.4) for $s < 0$. We then get, corresponding to (5.2),

$$A_H(s = 0, t) = [(9.9)] + \sum_{TT'} \sum_r g_r^{TT'}(t) [j_0^2 - a_r^2(j_0, 0)] d_{T\lambda T'}^{j_0 a_r}(z_s) \qquad (9.10)$$

from poles at $\sigma = a_r(j_0, s = 0)$, where $g_r^{TT'}(t)$ is the residue of the pole r, multiplied by appropriate factors (we neglect cuts in this discussion, but they could of course be included in the usual way). There are relations between the various T, T' terms which enable this summation to be eliminated, and the asymptotic form of the d's is independent of T and T', being of the form $d^{i p q}_{T_\lambda T'} \sim z_s^{[\sigma-1-|j_0-\lambda|]}$. These poles in the σ plane are called by *Toller* "Lorentz poles", and by *Delbourgo* et al. "Toller poles". The process of continuation in σ might thus be called "Tolleration" (but probably shouldn't). *Toller* supposes that it is these Lorentz poles which are the fundamental dynamical singularities. In order to interpret them as poles in the J-plane it is necessary to be able to expand the $d^{i p q}_{T_\lambda T'}$ in terms of the $SO(3)$ representatives $d^J_{\lambda\lambda}$. This problem [which is of course the same as decomposing representations of $SL(2C)$ in terms of those of $SU(1, 1)$] has been solved by *Sciarrino* and *Toller*, 1967, and *Akyeampong* et al., 1967a. The argument is rather complicated, but it turns out that a Lorentz pole corresponds to a family of trajectories separated by integers at $s = 0$,

$$\alpha_n(0) = a(j_0, 0) - n - 1 , \tag{9.11}$$

and with singular residues at $s = 0$ for $n \neq 0$, just like daughters and conspirators.

So given the existence of single Lorentz pole we predict the existence of daughters and conspirators. The possible types of conspiracy are more general than those introduced by *Volkov* and *Gribov* [referred to above, Section (IV.6)], however, because, whereas they considered only finite numbers of trajectories with non-singular residues, we now have infinite daughter sequences, and there is an infinite number of possible conspiring families for a given amplitude. It should be noted however that we have predicted these daughters only for $m_1 = m_2$, $m_3 = m_4$, whereas the analyticity argument for daughters given in Chapter III requires $m_1 \neq m_2$ (and $m_3 \neq m_4$ to get all orders of daughters). In this case $SO(3, 1)$ is not a symmetry of the amplitude, though we are, of course, free to postulate that the spectrum of trajectories exhibits the higher symmetry even if the amplitude does not.

Conversely, if we take a single Regge pole, this will correspond to an infinite sequence of Lorentz poles, or, looked at from the point of view of the σ-plane, an infinite set of Lorentz poles can counter-conspire to leave just a single Regge pole. This of course does nothing to remove the requirement that single (i.e. non-conspiring) Regge trajectories must have vanishing residues in some amplitudes, as has been stressed by *Finkelstein* and *Wang*, 1967, but so long as we admit the possibility of evasion (an background cancellation instead of daughters) we cannot disprove the contrary view to Toller's, that it is the Regge poles which are fundamental, and the counter-conspiracy which is natural. In fact since physical particles correspond to single Regge poles, this would probably seem the most reasonable point of view in the absence of other evidence.

With unequal masses the symmetry is only to be found off the mass-shell, since it is only there that it is possible for P_μ to vanish. This is why the Bethe-Salpeter equation, which is an off the mass-shell model, has proved so useful for generating daughters, as we noted in Chapter III. The extra symmetry at $P_\mu = 0$ was in fact remarked long ago by *Wick*,

1954, and *Cutkosky*, 1954. In working with the Bethe-Salpeter equation it is usual to make an analytic continuation to imaginary values of the energy, whereupon the $SO(3, 1)$ symmetry becomes $SO(4)$. Much recent work has been couched in this $SO(4)$ language, which has the advantage of avoiding the difficulties due to the non-compactness of $SO(3, 1)$, though the continuation in energy may result in complications.

Domokos and *Suranyi*, 1964, showed that the parameter n (say) which labels the irreducible representations of $SO(4)$ can be used to label families of trajectories (daughter sequences), the daughters corresponding to the so called "abnormal" solutions of the equation, and it is these which were rediscovered in the work of *Freedman* and *Wang*, 1967a. A good set of references to work on abnormal solutions of the Bethe-Salpeter equation can be found in *Nakanishi*, 1965. Further discussion of the connection between Lorentz poles and Regge families has been given by *Domokos*, 1967; *Freedman* and *Wang*, 1967b, and *Finkelstein* and *Wang*, 1967, and the natural way in which Lorentz poles can occur in invariant (rather than helicity) amplitudes has been noted by *Durand*, 1967; *Taylor*, 1967, and *Jones* and *Scadron*, 1967. Additional work on the Bethe-Salpeter and other types of models is reported in papers by *Chung* and *Snider*, 1967; *Chung* and *Wright*, 1967, and *Swift*, 1967.

Of course even in the equal mass case the symmetry applies only at one point $(s = 0)$, and there are two possible attitudes which can be taken as to the proper method of continuing away from this point. One, which has been proposed by *Oakes*, 1967, and *Delbourgo* et al., 1967b, is that the poles should be regarded as possessing the $SO(3, 1)$ symmetry for all s, so that a single Lorentz pole in the σ-plane at $\sigma = a(j_0, s)$ will give rise to an infinite sequence of parallel trajectories

$$\alpha_n(s) = a(j_0, s) - n - 1, \quad n = 0, 1, \ldots \tag{9.12}$$

n being the daughter number. On the other hand, since the symmetry is "broken" for $s \neq 0$ (or everywhere on the mass-shell if $m_1 \neq m_2$) one may anticipate that the daughters will depart from this prediction as we go away from the symmetry point. Perturbation methods have been used by *Domokos*, 1967, to estimate the departure from (9.12) away from the symmetry position. Calculations by *Cutkosky*, 1967, using the Bethe-Salpeter equation suggest that daughter trajectories might have an extremely different behaviour from that of the parent trajectory.

The nature of the continuation away from $s = 0$ has also been reviewed by *Bali* et al., 1967b, who point out that the sudden jump from $SO(2, 1)$ to $SO(3, 1)$ invariance at $P_\mu = 0$ occurs only in the four-line connected part, which has, in a sense, anomalously few degrees of freedom, and that with three or more particles in the incoming and outgoing states these problems do not arise, because there is then no such change in the number of degrees of freedom. In other papers *Bali* et al., 1967c, d, have also shown that Toller's methods, because they give the Sommerfeld-Watson transformed amplitude directly, avoiding the need to make an analytic continuation of crossed-channel partial-wave amplitudes, enable one to give an unambiguous meaning to Regge behaviour in multiparticle amplitudes. Hitherto there has been uncertainty as to the best choice of variables for this purpose. An ambitious scheme for using Toller's methods to "Reggeize" internal symmetries has been proposed by *Salam* and *Strathdee*, 1967.

It is rather remarkable that though these group theoretical methods reproduce the analytically continued Sommerfeld-Watson transform for relativistic scattering, there is no similar method available for non-relativistic scattering, through it is here that the necessary analytic continuations can be proved. There is of course no crossed channel in non-relativistic scattering, but one would expect the space-like region of the variables to correspond to it. In the time-like region the little group of the Galilei group is still $SO(3)$, and from this we get the usual partial-wave expansion, but in the space-like region the little group is $E(2)$ (*Inönü* and *Wigner*, 1952, see also *Ryder*, 1967) which does not have representations corresponding to the background integral. So here group theory does not appear to be able to give the results of the Sommerfeld-Watson transform. *Lévy-Leblond*, 1966, in a useful review, gives the opinion that this must cast serious doubts on the relevance of the relativistic $SO(2, 1)$ crossed-channel analysis.

It seems fairly clear that whether nature bothers to make full use of the available symmetry at $s = 0$, and contains conspirators and daughters, or whether it disregards this freedom, and contents itself with the background cancellation of singularities and evasion, is a dynamical question which can not be solved by group-theoretical methods alone. At the moment the Bethe-Salpeter model for trajectories seems to be the most convincing theoretical argument in favour of the symmetry, but this is perhaps rather weak evidence. *Domokos*, 1967, has shown that the usual S-matrix models for trajectories, such as the N/D method (which we shall examine in Chapter VI), which involve the continuation of two particle unitarity, on the mass-shell, down to $s = 0$, preclude the introduction of a sufficient number of variables for the extra symmetry to be present. We need to include coupling to many-particle channels, in which there is not the sudden change in the number of degrees of freedom, to obtain the symmetry in a mass-shell theory. The Bethe-Salpeter equation includes the coupling of such channels to some extent, and this presumably accounts for its ability to generate daughters. So perhaps the pion, which is presumably a bound state of three- (and more) particle channels, is more likely to have daughters etc. than the ϱ, which is probably mainly in two particle channels. Our present understanding of dynamics does not seem to permit a definite decision one of the other.

Given this theoretical uncertainty, we must hope that decisive evidence will be available from experiments before too long, and that we shall be able to determine whether or not daughters and conspirators exist. Unfortunately this evidence may not be very easy to come by. Some of the ways of looking for daughters were discussed in Section (III.8), and we shall mention some further tests for both daughters and conspirators in Chapter VIII.

One negative piece of evidence has been provided by *Akyeampong* et al., 1967b. We shall discuss in detail in Section (VIII.6a) how the exchange of a ϱ trajectory is able to explain the πN charge exchange differential cross-section, including a minimum at $t = -0.6$ $(\text{Gev}/c)^2$ which is interpreted as the vanishing of the ϱ's coupling to the helicity amplitude $\langle 00 |A^t| + - \rangle$ at $\alpha_\varrho(t) = 0$. *Akyeampong* et al. have shown that this decoupling does not occur with a Toller pole of the form (see 9.12)

$$a_\varrho(0, t) = \alpha_\varrho(t) + 1 \tag{9.13}$$

and that in order to obtain the dip a cancellation between two Toller poles is needed. The other pole might be either a $j_0 = 1$ [note that (9.13) has $j_0 = 0$] or a completely separate ϱ' trajectory. Since the prediction of this dip has been one of the major successes of the single Regge pole model, the failure of Toller poles to give the same results must be regarded as a serious difficulty. An essentially similar conclusion is reached by *Nath*, 1967, who analyses the process $\pi^- p \to \eta n$ in terms of the A_2 trajectory (cf. Section VIII.6c).

Chapter V. The Nature of *J*-Plane Singularities

V.1. The Analytic Continuation of Unitarity

In this chapter we shall consider the form of possible singularities of partial-wave amplitudes in the *J*-plane, and see what restrictions can be placed on them. For example, we shall see why, in certain models, the only moving singularities that can occur are poles, and why this result does not hold in general; we shall see that the presence of high spins imposes certain integral conditions on the discontinuity functions (superconvergence relations); in addition we shall be able to prove the existence of cuts in the *J*-plane and show that these allow fixed *J*-plane poles in certain amplitudes.

Many of the considerations of this chapter depend on the unitarity equation in the channel in which the partial-wave expansion is made, i.e. the *s* channel in our case. We begin by showing that the unitarity equation, which holds initially for physical values of *J*, holds also for the analytically continued amplitude. Consider, first, the case of spinless particles, and suppose there is a one-channel elastic region, $s_0 < s < s_I$, so that the unitarity equation reads [from (II.5.7)]

$$\operatorname{Im} B_l(s) = [q_{s\,12}\,q_{s\,34}]^l\,\varrho\,(s)\,|B_l(s)|^2\,, \tag{1.1}$$

for $s_0 < s < s_I$, and integral $l \geq 0$, where we have removed the threshold factor $[q_{s\,12}\,q_{s\,34}]^l$ from the amplitude by defining $B_l(s)$ as in (II.5.17), i.e.

$$B_l(s) = [q_{s\,12}\,q_{s\,34}]^{-l}\,A_l(s)\,. \tag{1.2}$$

We write the unitarity equation in the equivalent form

$$B_l^{\pm}(s) - B_{l^*}^{\mp}(s)^* = 2i\,[q_{s\,12}\,q_{s\,34}]^l\,\varrho\,(s)\,B_{l^*}^{\mp}(s)^*\,B_l^{\pm}(s) \tag{1.3}$$

for $s < s_0 < s_I$. It follows from (1.1) that this equation is true for l equal to alternate integers ≥ 0, i.e. even integers for the (+) case and odd integers for the (−) case. Also we see that both sides of the equation are analytic functions of l-apart from possible isolated singularities — and both sides satisfy the boundedness condition of Carlson's theorem (Section II.8). It then follows from Carlson's theorem that (1.3) is true for all the values of l to which the function can be continued.

It is clear that this extension of unitarity to complex *J* can be carried out in a similar manner for problems involving several coupled channels, and particles with non-zero spin. In all cases the appropriate unitarity equation — written as an analytic equation in *J* — holds for all values of *J* to which the partial-wave amplitude can be continued.

Two important results follow immediately from the unitarity equation. First, we see that a Regge trajectory cannot cross the *J* axis for *s* real and above threshold, unless, at the point where $\operatorname{Im} \alpha = 0$, the residue vanishes. This follows for example from (1.3) for the one-channel zero-spin case, since if there is a pole on the real axis, the right-hand-side of this equation would have a double pole. From the fact that Regge trajectories have $\operatorname{Im} \alpha > 0$ just above the physical threshold (essentially this is a consequence of the fact that there are no resonance poles on the physical sheet in the s-plane; see Chapter II), it is often conjectured that $\operatorname{Im} \alpha > 0$ for all s above threshold. This is in fact true in all the potential models that have been considered.

Secondly, from the unitarity equation for a many-channel problem, we see that the residue of Regge poles "factorize" (*Gell-Mann*, 1962b; *Gribov* and *Pomeranchuk*, 1962a). A general proof of this result was given by *Charap* and *Squires*, 1962a, as follows. We consider the S matrix for the N coupled channels which are open in some energy region $s_N < s < s_{N+1}$. Denoting this by $\mathbf{S}_N(J, s)$ we have the unitarity equation

$$\mathbf{S}_N(J, s)\, \mathbf{S}_N^+(J^*, s) = 1 \,, \quad s_N < s < s_{N+1} \,, \tag{1.4}$$

where $+$ means hermitian conjugate, which we rewrite in the form

$$\mathbf{S}_N(J, s) = \frac{[\text{cof } \mathbf{S}_N^+(J^*, s)]}{\det \mathbf{S}_N^+(J^*, s)} \,, \quad s_N < s < s_{N+1} \,, \tag{1.5}$$

where the numerator is the matrix of cofactors of $\mathbf{S}_N^+$.

Now Sylvester's law of nullity tells us that the rank of a product of two matrices is at least as large as the sum of the ranks of the two matrices minus the number of rows, so applying this to the product of $\mathbf{S}_N^+$ and $\text{cof}\mathbf{S}_N^+$ we have

$$\mathbf{r}[\mathbf{S}_N^+] + \mathbf{r}[\text{cof}\mathbf{S}_N^+] - N \leqq \mathbf{r}[\det\mathbf{S}_N^+ \cdot 1] \,, \tag{1.6}$$

where $\mathbf{r}[\ ]$ means the rank of the matrix. Now, if we ignore the possibility of two completely coincident Regge trajectories, the poles of $\mathbf{S}_N$ will be simple, so that, from (1.5) the zeros of $\det\mathbf{S}_N^+$ will be simple, i.e. $\mathbf{S}_N^+$ will have rank $(N - 1)$. Thus, at the pole, (1.6) yields

$$\mathbf{r}[\text{cof}\mathbf{S}_N^+] \leqq 1 \,. \tag{1.7}$$

Ignoring the trivial case of zero residues we see from (1.7) and (1.6) that the residue matrix has rank 1, i.e. all 2×2 minors are zero:

$$\beta_{ij}\,\beta_{kl} = \beta_{il}\,\beta_{jk} \tag{1.8}$$

which allows us to put

$$\beta_{ij} = \gamma_i\,\gamma_j \tag{1.9}$$

if we use time reversal invariance which makes $\mathbf{S}$ symmetrical.

Kawai, 1967, has shown that the leading high energy contributions of other types of singularity in the J-plane also factorize in the above sense, and suggests that an experimental check on the factorization theorem would not necessarily imply that the high energy behaviour was dominated by poles.

V.2. Moving Singularities in the J-Plane

In this section we shall see if anything useful can be said about the nature of singularities of partial-wave amplitudes which occur at (say) positions $J = \alpha(s)$, where $\alpha(s)$ is not a constant. From the properties of analytic functions of two variables we know that $\alpha(s)$ will be an analytic function of s, apart from possible isolated singularities.

We assume that the scattering amplitude A (s, t) is uniformly bounded by a power of t for all s in the physical sheet, say

$$|A\,(s,\,t)| = O\,(|t|^{J_{\mathbf{M}}})\,. \tag{2.1}$$

This assumption is necessary if the Mandelstam representation is to hold. It follows that the integral in the Froissart-Gribov projection converges for $\mathrm{Re}\,J > J_{\mathbf{M}}$, so, for all s on the physical sheet,

$$\mathrm{Re}\,\alpha(s) \leqq J_{\mathbf{M}}\,. \tag{2.2}$$

If we write the "solution" of the equation $J = \alpha(s)$ as

$$s = s_{\alpha}(J)\,, \tag{2.3}$$

and follow a physical sheet solution from a value J in $\mathrm{Re}\,J < J_{\mathbf{M}}$ to a value $\mathrm{Re}\,J > J_{\mathbf{M}}$, then because of (2.2) the solution must have left the physical s-sheet, i.e. it must have passed through one of the s-cuts.

The positions and origins of the s-cuts have been analysed in Sections (II.4) and (II.11). In particular we recall that the discontinuities of the left-hand cut are given by finite integrals of functions which are analytic in J (apart from *fixed* singularities). It follows that singularities cannot leave the physical s-sheet by pasing through the left-hand cuts (we ignore a possible exception to this which might arise if there was a branch point which changed its character as it moved, and could cease to be a branch point just at the point where it crossed the left-hand cut).

On that part of the right-hand cut where elastic unitarity holds we can use (1.3) to relate the amplitude on both sides of the cut, and it is easy to see that a singularity can only be lost from the physical sheet if it is a pole, otherwise the relation requires that any singularity on the second sheet is also on the first. Thus we have the extremely useful result that, where elastic unitarity holds for all energies, the only moving singularities in the J-plane are poles. Such a situation occurs for example in potential scattering.

It is clear that this result can be extended to apply when a finite number of two-body channels are coupled together (provided of course that a uniform power bound holds in each channel.) However the proof cannot be extended to remove the possibility of moving branch points crossing the right-hand cut in regions where inelastic unitarity involving 3 (or more)-particle intermediate states applies. Potential models involving three particle systems have in fact been constructed in which such moving branch points occur (see *Drummond*, 1965, and references therin), and, as we shall see below, it can be shown that they also occur in relativistic scattering processes.

V.3. Cuts in the Complex Angular Momentum Plane

In view of the simplicity of Regge pole contributions to high energy amplitudes compared with those of cuts (see Section II.9), and the fact that many simple models do not give any cuts, it was naturally hoped that cuts would be absent at least from the physically interesting regions of the J-plane. Indeed most phenomenological fitting has used only poles [for some cases where cuts have been considered see Sections (VIII.6, 7)], and there is so far no convincing experimental evidence for the existence of such cuts. However it has been demonstrated by *Mandelstam*, 1963c, that cuts must in fact be present, and indeed occur in physically interesting regions. We shall explain the reasons for this in a later section; here we shall discuss cuts which arise in a particular approximation and which are, in fact, spurious, being exactly cancelled by other terms in the exact amplitude. The reasons for our dealing with these spurious cuts are that their positions in the J-plane turn out to be identical to those of the actual cuts, and that they are involved in certain problems which arise in approximate dynamical calculations [e.g. Section (VI.7)].

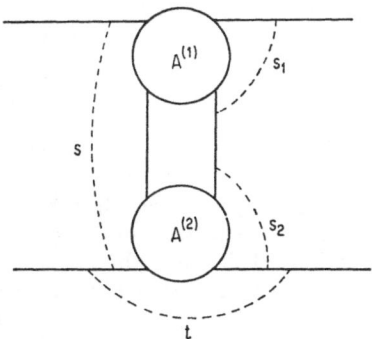

We consider the unitarity equation *in the t-channel* for the full amplitude $A(s, t)$, and keep only a two-particle intermediate state contribution [see Fig. (V.1)]. We calculate the contribution of this diagram to the double spectral function, using t-channel unitarity. This

Fig. V.1. A unitarity diagram giving Amati-Fubini-Stanghellini cuts.

calculation is analogous to that performed in Section (I.12) except that the s and t channels are interchanged. We obtain

$$\varrho_{st}^{t\,el}(s, t) = (16\pi^2\, q_t \sqrt{t})^{-1} \int \mathrm{d}s_1\, \mathrm{d}s_2 \frac{D_s^{(1)}(s_1, t)\, D_s^{(2)}(s_2, t)^*}{K^{1/2}(s, s_1, s_2, t)}, \qquad (3.1)$$

where we have just kept the contribution of the s-discontinuity for ease of writing.

We assume Regge-pole behaviour for $A^{(1)}$ and $A^{(2)}$, and hence obtain Regge-pole behaviour for the discontinuities $D_s^{(1)}$ and $D_s^{(2)}$, i.e.

$$D^{(i)}(s_i, t) \sim \Gamma(s_i)\, t^{\alpha_i(s_i)} \qquad (3.2)$$

for large t. Then, for large t we have

$$\varrho_{st}^{t\,el}(s, t) \sim (16\pi^2\, q_t \sqrt{t})^{-1} \int \mathrm{d}s_1\, \mathrm{d}s_2 \frac{\Gamma_1(s_1)\, \Gamma_2^*(s_2)\, t^{\alpha_1(s_1) + \alpha_2^*(s_2) - 1}}{K^{1/2}(s, s_1, s_2, t)}. \qquad (3.3)$$

It is clear from (3.3) that the high t behaviour of $\varrho_{st}^{t\,el}(s, t)$ corresponds to a continuous superposition of Regge poles, i.e. to a cut. The right-hand branch point is at a position $\alpha_c(s)$ given by

$$\alpha_c(s) = \mathrm{Max}\,[\alpha_1(s_1) + \alpha_2(s_2) - 1], \qquad (3.4)$$

subject to $2ss_1 + 2ss_2 + 2s_1s_2 - s^2 - s_1^2 - s_2^2 \geqq 0$.

If we assume that $d\alpha_i(s)/ds \geqq 0$ for $s < 0$ [see Section (III.1)], then we can immediately obtain some conditions on $\alpha_c(s)$. For $s < 0$ we have

$$\alpha_c(s) \geqq \alpha_1(s) + \alpha_2(0) - 1 , \qquad (3.5a)$$

and

$$\alpha_c(s) \geqq \alpha_1(0) + \alpha_2(s) - 1 . \qquad (3.5b)$$

Also

$$\alpha_c(0) = \alpha_1(0) + \alpha_2(0) - 1 . \qquad (3.6)$$

Further, if we take $A_1 \equiv A_2$ and use a linear approximation for $\alpha_1 = \alpha_2 = \alpha$, namely

$$\alpha(s) = \alpha(0) + s\alpha'(0) , \qquad (3.7)$$

where

$$\alpha'(0) > 0 , \qquad (3.8)$$

then

$$\alpha_c(s) = 2\alpha(s/4) - 1 . \qquad (3.9)$$

From (3.6) we see that if only one of the trajectories passes through $J = 1$ at $s = 0$ [it cannot be higher than this from the Froissart bound — Section (II.7)] then the cut and the other pole are at the same point at $s = 0$. Further, from (3.9) we see the tendency for cuts to have a less steep slope than poles — this leads us to expect that they are more likely to be significant in fits to high energy cross-sections at large angles than near the forward direction.

The fact that the iteration of two-particle unitarity behaves as described above has been noted by many authors, and leads to difficulties, for example in the iteration of one-particles exchange diagrams used in dynamical calculations, particularly when the exchanged particle has spin one or above [see Section (VI.7)]. That this type of iteration also produces apparent cuts in the l-plane was first noted by *Amati* et al., 1962.

It is clear that the argument we have used for obtaining these angular-momentum cuts is likely to be wrong since we used two-particle t-channel unitarity in the region of high t. In fact *Mandelstam*, 1963b, was able to show that these particular cuts cancel with similar terms coming from many-body intermediate states. We refer to the original paper, and to the chapter VII, for details of the proof (which is quite involved), and here merely note, for future use, an essential point of the

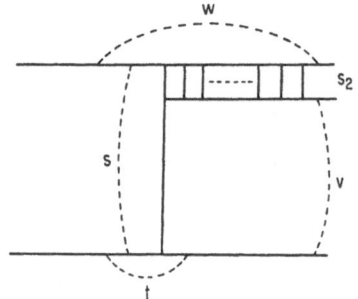

Fig. V.2. A Feynman diagram which behaves as t^{-1}.

argument. Consider the Feynman diagram shown in Fig. V.2). For large values of t, and fixed values of the other invariants, the diagram behaves like t^{-1} (see Chapter VII for general discussion of the high energy behaviour of Feynman diagrams). This is true regardless of how many "rungs" there are in the ladder and remains true even if the ladder sums to give a bound state or Regge pole, i.e. the residue of a pole in the s_2 plane also behaves like t^{-1} for large t. We now combine Fig. (V.2) with a similar diagram, using three particle s-channel unitarity, thereby obtaining Fig. (V.3). On performing the unitarity integral this diagram again behaves like t^{-1} for large t. It therefore has no moving singularities in the J-plane. However, from the previous discussion, the t-channel

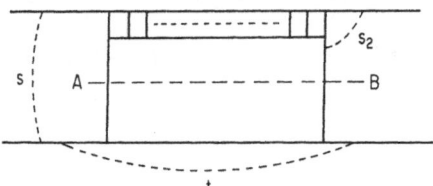

Fig. V.3. A Feynman diagram which behaves as t^{-1} although its t-channel two-particle discontinuity behaves like $t^{\sigma + \alpha - 1}$.

two-particle unitarity contribution to Fig. (V.3) (i.e. that obtained by cutting it along the line $A B$) behaves like $t^{\sigma + \alpha(\hat{s}_2) - 1}$, where σ is the spin of the particle indicated in Fig. (V.3), and $\hat{s}_2$ is the highest value of s_2 involved in the t-channel unitarity integral, and we have assumed that the ladder sums to give a Regge trajectory $\alpha(s_2)$. This therefore shows that there must be a cancellation between the two-particle t-channel unitarity integral and the contributions of many particle states to t-channel unitarity. Note that we have taken here a slightly simpler case than that discussed before [Fig. (V.1)], since we have assumed that A_1 is given just by elementary particle exchange, rather than by a Regge-pole contribution as in (3.2). However the argument works equally well for the diagram of Fig. (V.1).

The high t behaviour of Fig. (V.2) can be generalized to apply also to more complicated diagrams, e.g. Fig. (V.4), which also behaves like t^{-1} for large t. However when two of these diagrams are combined by s-channel unitarity to form a single diagram, e.g. Fig. (V.5), the proof that this behaves like t^{-1} no longer holds and there are convincing reasons for believing that such diagrams do give rise to cuts in the J-plane

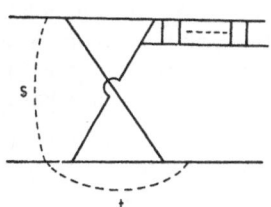

Fig. V.4. A Feynman diagram with a third double spectral function, behaving like t^{-1}.

(*Mandelstam*, 1963c). As we have mentioned previously, these cuts start at the same positions as those associated with the two particle t-channel

9*

unitarity contributions to Fig. (V.3). Note that the crucial difference between Fig. (V.5) and Fig. (V.3) is that Fig. (V.5) contributes to all three spectral functions

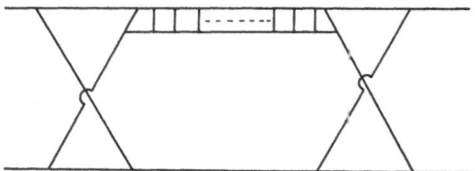

Fig. V.5. A Feynman diagram which leads to Mandelstam cuts.

We shall indicate in Section (V.6) our reasons for believing the cuts to exist.

V.4. Fixed J-Plane Poles

In this section we want to see under what conditions we can rule out fixed J-plane poles of the partial-wave amplitude. Firstly we consider a zero-spin problem satisfying elastic unitarity in some region of s, and we *assume* that there are no cuts. We use the fact that $B(l, s)$ is a real-analytic function to put

$$B^\pm(l^*, s + i\varepsilon)^* = B^\pm(l, s - i\varepsilon) \qquad (4.1)$$

for s real. Here, for notational convenience we are writing $B_l(s)$ as $B(l, s)$. We use this in (1.3) which becomes

$$B^\pm(l, s + i\varepsilon) - B^\pm(l, s - i\varepsilon) = 2i\,[q_{s\,13}\,q_{s\,24}]^l\varrho(s)\,B^\pm(l, s - i\varepsilon)\,B^\pm(l, s + i\varepsilon) \qquad (4.2)$$

for $s_0 < s < s_I$. The fact that $B(l, s)$ can have no fixed poles in the l-plane follows immediately from this equation, since such a pole would give rise to a double pole on the right-hand-side and only a single pole on the left. Now consider a problem where we do not have a one-channel elastic region, but where in $s_0 < s < s_1$ a finite number (n) of zero-spin two-particle channels are open. In this case (4.2) is replaced by the matrix equation (see Chapter II; the multi-channel formalism is discussed in more detail in Chapter VI)

$$\mathbf{B}^\pm(l, s + i\varepsilon) - \mathbf{B}^\pm(l, s - i\varepsilon) = 2i\,\mathbf{B}^\pm(l, s + i\varepsilon)\,\boldsymbol{\rho}_l\,\mathbf{B}^\pm(l, s - i\varepsilon) \qquad (4.3)$$

where the $\mathbf{B}$ are here $n \times n$ matrices, the element B_{ij} being the amplitude for scattering from channel i to channel j, and $\boldsymbol{\rho}_l$ is a diagonal matrix containing the appropriate phase-space factors.

We suppose there is a fixed pole in $\mathbf{B}^\pm$ at $l = \alpha$ (a constant) with residue $\boldsymbol{\beta}(s)$, then (4.3) tells us that

$$\boldsymbol{\beta}(s + i\varepsilon)\,\boldsymbol{\beta}(s - i\varepsilon) = 0\,, \qquad (4.4)$$

which for $n \geq 2$ does not allow us to conclude that $\boldsymbol{\beta} = 0$. Thus fixed poles cannot be ruled out here.

However, even in the many channel case, we can exclude fixed poles on the real l-axis. This can be seen by starting directly from the unitarity equation in the form

$$\mathbf{B}^{\pm}(l, s) - \mathbf{B}^{\pm}(l^*, s)^+ = 2i\, \mathbf{B}^{\pm}(l^*, s)^+\, \boldsymbol{\rho}_l\, \mathbf{B}^{\pm}(l, s)\,, \qquad (4.5)$$

where $+$ means hermitian conjugate. With a pole at $l = \alpha$, α real, this gives for the residue

$$\boldsymbol{\beta}(s)\, \boldsymbol{\beta}(s)^+ = 0\,, \qquad (4.6)$$

from which we *can* conclude that

$$\boldsymbol{\beta}(s) \equiv 0\,. \qquad (4.7)$$

All the previous derivations fail in regions of the J-plane where there are cuts. Such cuts full into two classes. First, there are the fixed cuts in the J-plane which occur when the particles have spin [see Section (IV.4)]. These cuts lie along the real J-axis from $J = -\sigma_T$ to $J = \sigma_T - 1$, where σ_T is the maximum total spin in any of the coupled channels. Instead of (4.6) we then have

$$\boldsymbol{\beta}_1(s)\, \boldsymbol{\beta}_2(s)^+ = 0 \qquad (4.8)$$

where the suffices $1, 2$ are introduced to distinguish whether we approach the real J-axis from above or below. It follows that we can no longer conclude that $\boldsymbol{\beta} \equiv 0$ and so cannot rule out fixed poles even on the real J-axis in the region of the cuts.

In all the cases we have considered so far, although the formal proof of the absense of poles may fail, there is no reason to believe that it does in fact fail, and it is unlikely that scattering amplitudes actually have fixed J-plane poles, either for $\operatorname{Im} J \neq 0$, or on the real axis associated with the fixed J-plane cuts.

We come now to the final reason for the breakdown of the "no-fixed poles" result, namely moving J-plane branch points. In this case it turns out that there are in fact fixed poles associated with the moving branch points. The possibility of such a pole at, say $J = J_0$ can be seen for example if, for all values of s for which the unitarity equation holds, the cut passes through $J = J_0$. Alternatively one could consider these branch point in the s-plane, since the proof using, for example, (1.2) would not work if, for a particular value of l, the moving cut completely covered the elastic unitarity cut. The reasons why we believe this actually happens in particular cases will be discussed below.

V.5. Singularities in the Froissart-Gribov Integral and Superconvergence Relations

The function $Q_l(z)$ has poles in the l plane at $l = -1, -2, -3$, etc., with residues which are polynomials, i.e.

$$Q_l(z) \simeq \pi \frac{\cos \pi l}{\sin \pi l}\, P_{-l-1}(z) \qquad (5.1)$$

for l near a negative integer [$B\,1,\ 3.3\,(3)$].

Thus if the Froissart-Gribov integral defining $A(l, s)$ for the scattering of two spinless particles (II.3.6) exists down to $\mathrm{Re}\, l = -1$, it follows that $B(l, s)$ has a fixed pole at $l = -1$. The only exception to this, which we return to below, is that the residue might vanish.

Now from the previous section we know that, except possibly through the existence of l-plane cuts, there cannot be a fixed pole at $l = -1$, so we deduce that the integral must diverge, for all s, somewhere to the right of $\mathrm{Re}\, l = -1$, i.e. there must be at least one Regge pole in $\mathrm{Re}\, l > -1$ for all s. This is quite familiar in potential scattering (e.g. see *Squires*, 1963a, or *Newton*, 1964) where, with a potential that behaves like $1/r$ at the origin, the leading Regge trajectory tends to -1 as the energy tends to infinity, and is always in $\mathrm{Re}\, l > -1$.

If the residue of the pole at $l = -1$ should vanish then the above discussion does not apply. In potential scattering this would occur if the coefficient of the $1/r$ term in the potential were zero and the potential behaved like r^0 at the origin; in which case the leading Regge trajectory would obey $\alpha(s) \to -2$ as $s \to \infty$. The general condition for the vanishing of the residue can be obtained from (5.1) and (II.3.6), and is

$$\int_{t_0}^{\infty} D_t^{\pm}(s, t)\, \mathrm{d}z_s = 0 . \tag{5.2}$$

A relation of this type is called a "superconvergence" relation. It must hold for all s for which the integral converges.

An alternative derivation of (5.2) is instructive. Suppose that, for some region of s, the region $\mathrm{Re}\, l \geq -1$, is free of singularities, then the amplitude $A(s, t)$ satisfies

$$t^{(1 + \varepsilon)}\, A(s, t) \xrightarrow[|t| \to \infty]{} 0 \tag{5.3}$$

for some positive ε. We now write a t dispersion relation for $A(s, t)$. Because of (5.3) no subtractions are required, so [c.f. (I.10.5)] we have

$$A(s, t) = \frac{1}{\pi} \int_{t_0}^{\infty} \frac{D_t(s, t')}{t' - t}\, \mathrm{d}t' + \frac{1}{\pi} \int_{u_0}^{\infty} \frac{D_u(s, u')}{u' - u}\, \mathrm{d}u' . \tag{5.4}$$

Now the limit in (5.3) must also hold for D_t and D_u, so we find for large t,

$$A(s, t) \to -\frac{1}{t}\left[\frac{1}{\pi} \int_{t_0}^{\infty} D_t(s, t')\, \mathrm{d}t' - \frac{1}{\pi} \int_{u_0}^{\infty} D_u(s, u')\, \mathrm{d}u' \right] \to$$

$$\to -\frac{1}{t}\left[\frac{1}{\pi} \int_{t_0}^{\infty} D_t^{-}(s, t')\, \mathrm{d}t' \right] \tag{5.5}$$

unless the term inside the square bracket is identically zero. Comparing (5.3) and (5.5) we see that this must indeed be the case, and we therefore obtain

$$\frac{1}{\pi} \int_{t_0}^{\infty} D_t^{-}(s, t')\, \mathrm{d}t' = 0 \tag{5.6}$$

i.e. equation (5.2) for the negative signature amplitude.

The reason why we obtained extra information in the first derivation, i.e. a superconvergence relation for both signatures rather than just negative signature, is that we used the unitarity equation to eliminate the possibility of fixed poles. If we had not used this constraint — and as we have seen it is not necessarily valid in the presence of moving cuts — then we would have again obtained (5.6) from (5.3). But we would not obtain (5.2) for the positive signature amplitude, since a fixed pole at $l = -1$ in the positive signature amplitude would not contribute to the full amplitude because of the signature factor which is zero at a "wrong" signature integral l (by "wrong" signature we mean $(\pm)$ signature for odd (even) l, and the appropriate generalization of this to the case of particles with spin).

It is of course possible to go further and remove the "barrier" at $\mathrm{Re}\, l = -2$ by imposing the condition that the pole at $l = -2$ of Q_l has zero residue when integrated in the Froissart-Gribov formula. In general it is clear that if we have the condition

$$t^{(N + \varepsilon)}\, A\,(s,\, t) \xrightarrow[|t| \to \infty]{} 0 \qquad (5.7)$$

for N an integer > 0, and ε any number > 0, then

$$\int_{t_0}^{\infty} t^n\, D^{\pm}\,(s,\, t)\,\mathrm{d}t = 0 \qquad (5.8)$$

for $n = 0, 1, \ldots, (N - 1)$, where the sign must be appropriate to the "right" signature [i.e. $(+)$ for n odd and $(-)$ for n even]. The "wrong" signature equations in (5.8) need not be true if there are cuts.

Although the satisfaction of the superconvergence relations given in (5.8) might appear to be accidental it has been conjectured by *Mandelstam*, 1966, that perhaps they are in fact true for all values of N. We shall discuss reasons (experimental and theoretical) why this might be so in Chapter VIII, and turn now to the case of particles with spin where it is possible to show rigorously that some superconvergence relations have to apply.

We start by considering the analogue of the second method used above. We use the fact that

$$\left(\frac{1 - z}{2}\right)^{\frac{-(\lambda - \lambda')}{2}} \left(\frac{1 + z}{2}\right)^{\frac{-(\lambda + \lambda')}{2}} \langle \lambda_3 \lambda_4 \,|A\,(s,\, t)|\, \lambda_1 \lambda_2 \rangle \qquad (5.9)$$

obeys a fixed-s dispersion relation, (IV.3.2). Thus, for large t, in the absence of any superconvergence relations,

$$\langle \lambda_3 \lambda_4 \,|A\,(s,\, t)|\, \lambda_1 \lambda_2 \rangle \sim t^{\lambda - 1}. \qquad (5.10)$$

It now follows from the Froissart bound (Section II.7) that if $\lambda > 2$ then there *must* be superconvergence relations. More generally if we postulate a Regge behaviour

$$\langle \lambda_3 \lambda_4 \,|A\,(s,\, t)|\, \lambda_1 \lambda_2 \rangle \sim t^{\alpha\,(s)} \qquad (5.11)$$

for large $|t|$, where $\alpha(s)$ is the leading s-channel Regge trajectory, then there will be superconvergence relations wherever

$$\lambda - 1 > \alpha(s) \tag{5.12}$$

for any s. In practical applications the $\alpha(s)$ is usually assumed from knowledge of known particles and their associated trajectories (see Chapter VIII).

The form taken by the superconvergence relations is very similar to that in the zero-spin case, namely

$$\int_{t_0}^{\infty} t^n D_{\overline{H}t}^{\pm}(s, t) \, \mathrm{d}t = 0 \tag{5.13}$$

for n equal to zero or any integer less than $\lambda - 1 - \alpha(s)$. The $\pm$ sign is chosen according to whether n is odd or even.

The alternative derivation of (5.13), using the Froissart-Gribov projection, requires a little care. Note first that, if the formula can be continued down to the appropriate value of J, then a sense-nonsense amplitude would behave like $(J - J_0)^{-1/2}$ in the absence of a superconvergence relation, and a nonsense-nonsense amplitude like $(J - J_0)^{-1}$. Now the square-root singularities at the sense-nonsense points in fact cancel exactly with corresponding factors in $d_{\lambda\lambda}^J$, and so no singularities occur in the integrand of (IV.5.1). However, as noted in Section (IV.5) the amplitudes have to be *zero* at the sense-nonsense points $(J = J_0)$ if a fixed power t^{J_0} is not to occur in the amplitude. For the wrong signature amplitude this zero is supplied by the signature factor; but for the right signature amplitudes we again need the superconvergence relation.

We close this section by noting that the superconvergence relations which have to hold for particular amplitudes involving particles with spin are unlikely to be true in certain "approximate" expressions for these amplitudes. The situation here is closely analogous to the discussion of the cancellation of the cuts of Section (V.3).Consider, for example, Fig. (V.2). As we stated in Section (V.3) this behaves like t^{-1} for large t. If we suppose there is a bound state in the ladder, and we continue the amplitude as a function of s_2 to the appropriate pole, then clearly the residue, which will be the scattering amplitude for this particle, will still go to zero as $|t| \to \infty$ like t^{-1} (or possibly faster). However, if we disperse in t and take just the single particle contribution to t-channel unitarity, then the usual Feynman rule would give a behaviour $t^{\sigma-1}$ (see Chapter VII), where σ is the spin of the bound state (here we assume that all the lines in the diagram correspond to particles of zero spin). Again this behaviour is cancelled by contributions coming from many particle intermediate states.

V.6. Gribov-Pomeranchuk Singularities

We turn now to a phenomenon which is closely related to that discussed in the previous section, the Gribov-Pomeranchuk (*Gribov* and *Pomeranchuk*, 1962b) singularities. Again we consider first the simple case of equal mass particles with zero spin. The left-hand discontinuity of the continued partial-wave amplitude is given by (II.11.9), and, as we have previously noted, the integrals in this equation converge for all values of l. It follows that the l-plane singularities of $\text{Im}\{B(l,s)\}_{\text{L.H}}$ are those of the integrands, i.e. those of the Q_l (since P_l is holomorphic in l for all finite l). From (5.1) this means that the only singularities are poles at the negative integers. The signature factor in (II.11.9) removes these poles for the right-signature amplitudes, so they occur only at wrong signature, negative-integral, values of l. Now contrary to the situation in Section (V.5), the residues of these poles can, for some range of s, be calculated exactly. This is because the residue is an integral over the third double spectral function (i.e. ϱ_{tu}) along a line at constant s. If we choose s suitably this integral can be calculated in terms of that part of the elastic double spectral function, which can be written down exactly [see Section (I.12)]. In particular ϱ_{tu} in this region does not change sign, and so the pole at $l = -1$ does not have zero residue. Similarly one can see that the other residues are not in general zero.

We are now faced with the following difficulty: $\text{Im}\{B_i^{\pm}(s)\}_{\text{L.H.}}$ has a fixed pole at $l = -1$ (among others), but, unless there are cuts in the l-plane, $B_i^{\pm}(s)$ cannot have a fixed pole at $l = -1$. There are thus two alternatives:

(a) there are cuts in the l-plane,

(b) $B_i^{\pm}(s)$ has a complicated type of singularity which does not violate the unitarity condition and which has a left-hand discontinuity having the required pole.

The nature of the singularity required in case (b) was examined by *Squires*, 1963b, and by *Atkinson* and *Contogouris*, 1965. The singularity is a branch point in the l-plane which changes its character with s, i.e. is of the form $(l+1)^{F(s)}$. For details we refer to the papers noted above. [Note that the singularity is usually referred to as an "essential" singularity although, whilst it has all the bad features of an essential singularity, it is in fact a branch point, and the term essential singularity is not strictly applicable (see e.g. *Titchmarch*, 1939, p. 93)]. We shall not pursue these points further here since possibility (a) now appears to be true.

In order to see why this is so we must consider the generalization of the above discussion to particles with spin. As in the previous section the fixed infinities of $\text{Im}\{B(J,s)\}_{\text{L.H.}}$ will now occur at values of $J = \sigma_T - n$, where σ_T is the largest of $(\sigma_1 + \sigma_2)$ and $(\sigma_3 + \sigma_4)$. The fact that in some cases these infinities may be of the square root form $(J - J_0)^{-1/2}$, rather than $(J - J_0)^{-1}$, does not alter the argument, so for simplicity we shall continue to refer to them as poles. These will again occur only in the wrong signature amplitudes, and, although the proof, based on unitarity, that they cannot have zero residue has not been given for the general spin case, it is likely to hold, and has been shown explicitly by *Charap*, 1963, for the spin 1/2- spin 1/2 case.

If we now assume that there are no cuts in the J plane, and adopt (b) above, the presence of these singularities would give rise to an asymptotic behaviour $t^{\sigma_T - 1}$, which is incompatible with the Froissart-bound if $\sigma_T > 1$, is probably unacceptable if $\sigma_T = 1$, and is certainly incompatible with the success of Regge phenomenology in other cases. Note that this type of singularity is *not* removed by the zero in the signature factor. This incompatibility with the Froissart bound was first noted by *Azimov*, 1963.

We are left with only one possible solution, namely that there are cuts in the angular momentum plane. These cuts must be such as to destroy the argument which prevents fixed poles at wrong signature amplitudes with $J = \sigma_T - 1, \sigma_T - 2$, etc. We then have no problem with the Froissart bound, since a fixed pole at a wrong signature value of J does not contribute to the high energy behaviour of the full amplitude because of the signature factor.

We recall from Section (IV.5) that the presence of these fixed poles means that sense-nonsense decoupling does not necessarily occur at wrong signature amplitudes, and hence that a given trajectory does not *necessarily* have to "choose", "sense" or "nonsense" when it passes through a wrong signature value of J.

Finally we note that our demonstration of the existence of cuts is conclusive if our analyticity postulates are accepted; or, to put this another way, we have shown that these postulates are incompatible unless there are moving cuts in the *J*-plane. *Azimov*, 1963, regarded this as evidence that some of the basic postulates are incorrect (in particular the assumption of a uniform power bound for the amplitude), but since, as we have seen in Section (V.3), it is perfectly possible that there are cuts which are at suitable places we need not take this pessimistic view. *Mandelstam*, 1963c, has shown that the arguments which remove cuts of the type discussed in Section (V.3) do not apply when more complicated diagrams are considered, and has shown that the Gribov-Pomeranchuk singularity present in a given set of diagrams can be removed by crossing to another sheet of the *J*-plane — thus proving that the cut must exist at least for this class of diagrams. We shall mention further supporting evidence for the existence of cuts in Chapter VII, where we study various classes of Feynman diagrams. All this makes it reasonable to suppose that the problem noted by *Azimov* is solved by the existence of cuts, and suggests that the presence of cuts should be allowed for in all phenomenological fitting. We return to this point in Chapter VIII.

For further details of the relation between the Gribov-Pomeranchuk singularities and the cuts in the *J*-plane, and of the nature of the singularities on other sheets of the *J*-plane we refer to papers by *Jones* and *Teplitz*, 1967b, and *Mandelstam* and *Wang*, 1967.

Chapter VI. Bootstraps

VI.1 The Bootstrap Hypothesis

According to maximal analyticity of the first kind, which we discussed in Chapter I, if we are given the bound-state and resonance poles of an amplitude, with their couplings, we can find all the other singularities by using the unitarity equations. In particular, for the four-line connected part, we can determine the double spectral functions. However, there is no obvious restriction on the number and types of the poles, though it seems likely, by analogy with potential scattering, that the presence of some particles, such as, say, the neutron and proton will necessarily imply the presence of other poles, such as the deuteron, as bound or resonant states of these particles. This ambiguity manifests itself in the undetermined subtractions of the Mandelstam representation discussed in section (I.11).

The hypothesis of maximal analyticity of the second kind, introduced in Chapter II, went some way towards resolving these difficulties by requiring that the amplitude at low angular momentum, where possible subtraction ambiguities can arise (we found they were limited to the S- and P-waves by the Froissart bound), should be obtainable by analytic continuation in l from the higher waves. Since the higher waves are completely determined by the double spectral functions, this implies that all the partial waves are determined by the form of the double spectral functions.

So, in summary, maximal analyticity of the first kind tells us that a knowledge of the poles determines the double spectral functions completely, and maximal analyticity of the second kind tells us that a knowledge of the double spectral functions enables us to determine the poles completely. Evidently there is a very strong self-consistency condition here. For if one were to add some arbitrary pole to an amplitude, one would generate, via maximal analyticity of the first kind, a whole new set of singularities, a whole new set of contributions to the double spectral functions. And these, because of maximal analyticity of the second kind, would probably imply the presence of further sets of poles, which would then imply further contributions to the double spectral functions, and so on ad infinitum, unless at some point self-consistency were reached, and the actual set of poles present in the amplitude implied the presence of just that set and no others.

The question then arises as to what set or sets of particles can satisfy this self-consistency condition. Clearly, if maximal analyticity of the first and second kinds are true in the real world of strongly interacting particles, then the actual particles which have been discovered must be part of such a set. There are probably an infinite number of particles in this set, though many are so unstable as to be undetectable. Are there, however, completely different sets of hypothetical particles which would satisfy our postulates? The conditions are so stringent that this seems unlikely, though we have no way of ruling it out rigorously at present.

The unlikelyhood has led to the proposal [see for instance *Chew* (1962b)] of the "bootstrap" hypothesis. We can express it as follows:

The Bootstrap hypothesis. The only set of particles (poles) consistent with the principles of maximal analyticity of the first and second kinds, is the actual set of strongly interacting particles found in nature.

The word "bootstrap", of course, derives from the proverbially impossible feat of "pulling oneself up by ones own bootstraps", which the self-consistency requirement (perhaps unfortunately) suggests. Originally the idea was that a single particle might be approximately self-consistent in some amplitude so bootstrapping itself [e.g. the ρ in $\pi - \pi$ scattering, *Chew* and *Mandelstam* (1961)], but strictly all the particles should be included in a global bootstrapping operation.

It should be stressed that this is not a new postulate, but an hypothesis as to the consequences of our six postulates. Unfortunately, solution of the unitarity equations involves solution of infinite sets of coupled, non-linear, singular, integral equations, so the hypothesis is impossible to test as a whole. To test it one would have to solve the entire strong interaction problem in one fell swoop. Instead many attempts have been made, with varying degress of success, to demonstrate that some small subset of the particles is approximately self-consistent, within some subset of the equations, which is believed to be approximately decoupled from the rest. These are called "bootstrap calculations". With so many approximations, no test can ever be crucial, but a sufficiently large number of partial successes have been achieved to make the more optimistic feel that the hypothesis may be true.

This chapter will be devoted to a discussion of bootstrap calculations, using the two principal techniques which have been applied to the problem; the N/D method for partial wave amplitudes, and the Mandelstam iteration. The literature on the subject is vast, and we refer the reader to some of the other surveys which have been made [e.g. *Zachariasen* (1965), *Udgaonkar* (1965)] for topics which we do not cover.

In rather few of the calculations has the Regge pole aspect of the problem been stressed, though of course any attempt to calculate bound or resonant states from the forces which are believed to generate them, involves the assumption that these particles are Regge poles. Here we shall concentrate on those types of calculation which bring out the importance of Regge poles in bootstrap dynamics, and in particular on the strip approximation.

Before doing this it may be worth saying something about the interdependence of the postulates on which we are basing our discussion, in particular about the interrelation of maximal analyticity of the first and second kinds. We have seen in section (II. 6) that, on the assumption that maximal analyticity of the first kind leads to the Mandelstam representation, we can prove maximal analyticity of the second kind wherever the Froissart-Gribov projection is defined, that is for $l > l_M(s)$. We also noted that the Froissart bound, required by cross-channel unitarity, demands that

$$l_M(s) \leqq 1 , \quad \text{for} \quad s \leqq 0 . \tag{1.1}$$

This limit has been somewhat extended into the positive elastic s-channel region by the work of *Bardakci* (1962) and *Prosperi* (1962), and in fact, by making plausible assumptions about the inelasticity, Prosperi has shown that the amplitude is meromorphic (i.e. has only poles) in $l > 1$ for all s. It seems possible that with sufficient accumen this might be proved. We are then left with possible arbitrary subtractions only in the S- and P-waves.

However, *Martin* (1962) has gone further. He shows that if we invoke crossing we can determine even these S- and P-waves. By a somewhat lengthy argument he demonstrates that if we have two amplitudes, $A(s, t)$ and $A'(s, t)$, which have the same double spectral functions in a region of s for which elastic unitarity holds in the s-channel, they can differ only in their subtractions in t. So

$$A(s, t) - A'(s, t) = \sum_{n=0}^{N} \beta_n(s) \, t^n , \tag{1.2}$$

for all s. If we go to $s < 0$ the Froissart bound requires $\beta_n(s) = 0$ for $n > 1$, and since $s < 0$ includes a finite region of the analyticity domain of $A(s, t) - A'(s, t)$ (since there is a region of $s < 0$ where neither $A(s, t)$ nor $A'(s, t)$ can have singularities), it must be true everywhere. So we have

$$A(s, t) - A'(s, t) = \beta_0(s) + \beta_1(s) \, t \tag{1.3}$$

only. Similarily, if we have a region in which t-channel elastic unitarity holds, there can only be arbitrary subtractions in s, and we have

$$A(s, t) - A'(s, t) = \gamma_0(t) + \gamma_1(t) \, s . \tag{1.4}$$

Comparing (1.3) and (1.4) we must have

$$\begin{aligned} \beta_0(s) &= a + b s \\ \beta_1(s) &= c + d s \end{aligned} \tag{1.5}$$

where a, b, c and d are constants. But unitarity of the s-channel partial waves requires that

$$|A_0'(s)| \quad \text{and} \quad |A_0(s)| < 1 , \quad \text{and} \quad |A_1'(s)| \quad \text{and} \quad |A_1(s)| < 1 , \tag{1.6}$$

even for large s, so we must have

$$\beta_0(s) < \text{const.}, \quad \text{and} \quad \beta_1(s) < \frac{\text{const.}}{s} . \tag{1.7}$$

Hence in (1.5)

$$b = c = d = 0 .$$

Since $A(s, t)$ and $A'(s, t)$ have a real region below all the thresholds, a must be real, so

$$\text{Im}\,\{A_0(s)\} = \text{Im}\,\{A_0'(s)\} \quad \text{for all } s , \tag{1.8}$$

and so, because in the s-channel elastic region $\text{Im}\,\{A\}$ determines A [see (II. 5.7)] (with a sign ambiguity discussed by *Martin*),

$$a = 0 .$$

In other words, if a scattering amplitude has a purely elastic region in two channels, and if the double spectral functions are known exactly in an arbitrarily small strip of one elastic region, then the subtractions are completely determined, and there is no arbitrariness in the amplitude. We do not of course know from this that the amplitude is analytically continuable for $l \leqq 1$, but we do know that the $l = 0$, 1 partial waves are completely determined by the higher ones.

Much work remains to be done on these matters, but it seems perfectly possible that eventually maximal analyticity of the second kind will be deduced from maximal analyticity of the first kind. If it is this will greatly strengthen the bootstrap hypothesis, which will then become just the statement that the only set of particles consistent with maximal analyticity of the first kind is the actual set of strongly interacting particles found in nature.

In the next section we begin the discussion of the techniques used for solving the bootstrap equations.

VI.2. The N/D equations

The N/D equations, introduced by *Chew* and *Mandelstam* (1960), are the most commonly employed method for bootstrap calculations. They have in fact been likened in importance to the Schroedinger equation of non-relativistic quantum mechanics.

Let us consider the amplitude A (s, t) for the scattering of two spinless particles, as discussed in Chapters I and II. We have seen in section (II. 11) that the s-channel reduced partial-wave amplitude [defined in (II. 5.17)]

$$B_l^{\pm}(s) = \frac{e^{i\,\delta_l(s)}\,\sin\delta_l(s)}{\varrho_l(s)} \tag{2.1}$$

is given by

$$B_l^{\pm}(s) = \frac{1}{16\pi^2} \int\limits_{t_0}^{\infty} Q_l(z')\, D_l^{\pm}(s, t')\, \frac{\mathrm{d}t'}{2q_s^{2l+2}}\ . \tag{2.2}$$

This amplitude has two cut regions, the right-hand s-channel unitarity cuts, and the left-hand cuts stemming from the crossed-channel singularities, and we can write a dispersion relation for $B_l^{\pm}(s)$ [c.f. (II 4.5)]

$$B_l^{\pm}(s) = \frac{1}{\pi} \int\limits_{s_0}^{\infty} \frac{\mathrm{Im}\{B_l^{\pm}(s')\}}{s' - s}\,\mathrm{d}s' + \frac{1}{\pi} \int\limits_{-\infty}^{s_L} \frac{\mathrm{Im}\{B_l^{\pm}(s')\}}{s' - s}\,\mathrm{d}s'\ , \tag{2.3}$$

where s_0 is the lowest threshold, and s_L is the beginning of the left-hand cut. Here, for simplicity, we have assumed that l is real (but not necessarily integral) so that the discontinuity across the cuts is just twice the imaginary part. For complex l we would simply have to replace $\mathrm{Im}\{B_l^{\pm}(s)\}$ in (2.3) by $\varDelta_s\{B_l^{\pm}(s)\}$, the discontinuity across the relevant s cuts (divided by $2i$), but for convenience we shall continue to write our equations as though they were only for real l.

Suppose that on the right-hand cut the amplitude satisfies elastic unitarity for $s_0 < s < s_I$, where s_I is the inelastic threshold, so (II. 5.18)

$$\text{Im}\{B_l^\pm(s)\} = \varrho_l(s)\,|B_l^\pm(s)|^2\,,\qquad s_0 < s < s_I\,. \tag{2.4}$$

If we make the rather drastic approximation that there is no coupling to inelastic channels, so $s_I \to \infty$, we can write (2.3) in the form

$$B_l^\pm(s) = B_l^L(s) + \frac{1}{\pi}\int\limits_{s_0}^{\infty}\frac{ds'}{s'-s}\,\varrho_l(s')\,|B_l^\pm(s')|^2\,, \tag{2.5}$$

where

$$B_l^L(s) = \frac{1}{\pi}\int\limits_{-\infty}^{s_L}\frac{\text{Im}\{B_l^\pm(s')\}}{s'-s}\,ds'\,. \tag{2.6}$$

Equation (2.5) tells us that

$$B_l^\pm(s)\xrightarrow[l\to\infty]{} B_l^L(s)\,, \tag{2.7}$$

since we know that both

$$B_l^\pm(s)\quad\text{and}\quad B_l^L(s)\,, \underset{l\to\infty}{\sim}\frac{e^{-l\xi(z_n)}}{\sqrt{l}} \tag{2.8}$$

by (II. 7.4)

If we suppose that the left-hand cut discontinuity is known, we can regard (2.5) as an integral equation for $B_l^\pm(s)$ given $B_l^L(s)$. The idea is to regard the left-hand cut as the "force" or "potential", and the imposition of unitarity as the solving of the scattering problem for the given potential. Indeed, were we to consider non-relativistic scattering with a Yukawa potential, the nearby part of the left-hand cut would be given directly by the potential (i.e. the first Born approximation). Solving the equations with just this cut would not give us an exact solution to the problem, because the elastic double spectral function, which contains the right-hand singularities, also contributes to the more distant parts of the left-hand cut. But it has been shown that the solution is sufficiently accurate for most purposes if the second Born approximation to the left-hand cut is used [*Luming* (1964)], and for weak coupling the first Born approximation may be enough. Good accounts of the analogy between scattering with a Yukawa potential and this approach to bootstrap calculations are given in the books by *Frautschi* (1963), and *Omnes* and *Froissart* (1963), and also in *Chew* (1965a).

To solve the equation it is simpler if we first linearize it by making the decomposition

$$B_l^\pm(s) = \frac{N_l(s)}{D_l(s)}\,, \tag{2.9}$$

where the numerator function $N_l(s)$ has the left-hand cuts, and the denominator function $D_l(s)$ has the right hand cuts of $B_l^\pm(s)$, that is

$$\text{Im}\{N_l(s)\} = \text{Im}\{B_l^\pm(s)\,D_l(s)\} = b_l(s)\,D_l(s)\,,\qquad s < s_L, \tag{2.10}$$

where
$$b_l(s) \equiv \text{Im} \{B_{\bar{l}}^{\pm}(s)\}, \quad s < s_L, \tag{2.11}$$
and
$$\text{Im} \{D_l(s)\} = N_l(s) \, \text{Im} \left\{ \frac{1}{B_{\bar{l}}^{\pm}(s)} \right\}, \quad s > s_0,$$

$$= - N_l(s) \frac{\text{Im}\{B_{\bar{l}}^{\pm}(s)\}}{|B_{\bar{l}}^{\pm}(s)|^2},$$

which from (2.4) gives
$$\text{Im} \{D_l(s)\} = - \varrho_l(s) \, N_l(s), \quad s > s_0. \tag{2.12}$$

Note also that since, from (2.1) and (2.9)
$$\frac{1}{B_l(s)} = \frac{D_l(s)}{N_l(s)} = \frac{e^{-i\delta_l(s)}}{\sin \delta_l(s)} \, \varrho_l(s), \tag{2.13}$$

and $N_l(s)$ is real for $s > s_0$, $D_l(s)$ must have the phase $e^{-i\delta_l(s)}$. We seek
to write dispersion relations for $N_l(s)$ and $D_l(s)$ with the conventional
normalization that
$$N_l(s) \to 0$$
and
$$D_l(s) \to 1, \quad \text{as} \quad s \to \infty. \tag{2.14}$$

We can use the method of *Wiener* and *Hopf* [*Titchmarsh* (1937) p.339]
to construct $D_l(s)$ in terms of its phase, and of its P_l poles at s_{il} and M_l
zeros at s_{jl} on the physical sheet, to give

$$D_l(s) = D_l(s_0) \prod_{i=1}^{P_l} \frac{s - s_{il}}{s_0 - s_{il}} \times$$
$$\times \prod_{j=1}^{M_l} \frac{s - s_{jl}}{s_0 - s_{jl}} \exp \left\{ - \frac{s - s_0}{\pi} \int_{s_0}^{\infty} \frac{\delta_l(s') - \delta_l(s_0)}{(s' - s)(s' - s_0)} \, ds' \right\}. \tag{2.15}$$

This evidently ensures that the phase of $D_l(s)$ is $e^{-i\delta_l(s)}$. We have
assumed that $\delta_l(s) \to$ constant, so that one subtraction is needed in the
integral. If we do not include any poles in $N_l(s)$ all the poles of the
amplitude will correspond to zeros of $D_l(s)$, either bound states on the
physical sheet at $s = s_{il}$, or resonances on unphysical sheets.
    As $s \to \infty$
$$D_l(s) \sim s^{\{M_l - P_l + \pi^{-1} [\delta_l(\infty) - \delta_l(s_0)]\}}, \tag{2.16}$$

but we have chosen the normalization (2.14), so the difference between
the phase shift at threshold and infinity must be
$$\delta_l(\infty) - \delta_l(s_0) = \pi \, [P_l - M_l]. \tag{2.17}$$

This result agrees with Levinson's Theorem [*Levinson* (1949)].
    We are free to make the conventional choice that
$$\delta_l(s_0) = M_l, \tag{2.18}$$

the number of bound state poles of the amplitude, since the magnitude of the phase shift is arbitrary to an integral multiple of π, so

$$D_l(s) \sim s^{(\delta_l(\infty)/\pi - P_l)} \tag{2.19}$$

and

$$\delta_l(\infty) = \pi\, P_l\,. \tag{2.20}$$

With the poles of (2.15) and the imaginary part given in (2.12), the dispersion relation for $D_l(s)$ may be written in the form

$$D_l(s) = 1 - \frac{1}{\pi} \int\limits_{s_0}^{\infty} \frac{\varrho_l(s')\, N_l(s')}{s' - s}\, ds' + \sum_{i=1}^{P_l} \frac{\gamma_{il}}{s - s_{il}}\,, \tag{2.21}$$

where we have taken the residues of the poles to be γ_{il}. Since we can specify γ_{il} and s_{il} arbitrarily, it is evident that $D_l(s)$ is not completely determined by our input, $B_l^L(s)$. This ambiguity is known as the CDD ambiguity [after the initials of its discoverers, *Castillejo, Dalitz* and *Dyson* (1956)], and the poles are called CDD poles. The CDD poles are zeros of the amplitude, and nearby, at a position depending on $N_l(s)$ as well as γ_{il} and s_{il}, there will be a zero of $D_l(s)$ in (2.21), corresponding to a (particle) pole of the amplitude. Hence the CDD poles introduce particles of arbitrary masses and couplings into the partial-wave amplitudes.

We have seen that for large enough l

$$B_l^{\mp}(s) \to B_l^L(s) \to 0 \tag{2.22}$$

[see (2.8)], and so

$$\delta_l(\infty) \to \delta_l(s_0)\,,$$

and there will be no bound states $(M_l = 0)$. Levinson's theorem (2.17) thus implies that for large enough l

$$P_l \to 0\,,$$

and there is no CDD ambiguity. And we are requiring that maximal analyticity of the second kind should hold, in which case the lower partial waves should be attainable from the higher by analytic continuation. We shall indicate below Mandelstam's argument [*Mandelstam* (1963 d)] that the solution of the N/D equations without CDD poles constitutes such a continuation, and hence we conclude that CDD poles must be absent from our amplitudes. We are working in the elastic unitarity approximation, however. In the real world, where there is coupling to other channels, the situation with regard to Levinson's theorem is much more complicated, and we shall see in the next section that, even with maximal analyticity of the second kind, CDD poles can occur in some channels. In the absence of such complications, however, if we were to find experimentally that

$$\delta_l(\infty) - \delta_l(s_0) = M_l$$

this would prove that maximal analyticity of the second kind was true, and that all the particles were composites. CDD poles correspond to elementary particles.

Assuming that there are no CDD poles, we can write the following dispersion relations for $N_l(s)$ and $D_l(s)$ using (2.10) and (2.12),

$$N_l(s) = \frac{1}{\pi} \int_{-\infty}^{s_L} \frac{ds'}{s'-s} \, b_l(s') \, D_l(s') \,, \tag{2.23}$$

and

$$D_l(s) = 1 - \frac{1}{\pi} \int_{s_0}^{\infty} \frac{ds'}{s'-s} \, \varrho_l(s') \, N_l(s') \,. \tag{2.24}$$

This pair of coupled integral equations can be solved by substituting one into the other. *Chew* and *Mandelstam* (1960) substituted (2.23) for $N_l(s)$ into (2.24), and obtained

$$D_l(s) = 1 - \frac{1}{\pi^2} \int_{s_0}^{\infty} \frac{\varrho_l(s')}{s'-s} \int_{-\infty}^{s_L} \frac{D_l(s'') \, b_l(s'')}{s''-s'} \, ds'' \, ds' \tag{2.25}$$

and, on performing the s' integral, found

$$D_l(s) = 1 + \frac{1}{\pi^2} \int_{-\infty}^{s_L} k_l \, (s, s'') \, b_l(s'') \, D_l(s'') \, ds'' \tag{2.26}$$

where

$$k_l \, (s, s'') = - \int_{s_0}^{\infty} \frac{\varrho_l(s')}{(s'-s) \, (s''-s')} \, ds' \,. \tag{2.27}$$

The function $k_l \, (s, s'')$ can easily be obtained for integer values of l, but this method is not very useful for continuation in angular momentum. Instead it is more convenient to follow *Uretsky* (1961), and *Mandelstam* (1963d), in defining a function

$$C_l(s) \equiv N_l(s) - B_l^L(s) \, D_l(s) \tag{2.28}$$

which has no left-hand cut, since

$$\mathrm{Im} \, \{B_l^L(s)\} = \mathrm{Im} \, \{N_l(s)\}/D_l(s) \,, \quad s < s_L \,. \tag{2.29}$$

Then, since $B_l^L(s)$ has no right-hand cut,

$$\mathrm{Im} \, \{C_l(s)\} = - \, B_l^L(s) \, \mathrm{Im} \, \{D_l(s)\} \,, \quad s > s_0 \,, \tag{2.30}$$

so the dispersion relation

$$C_l(s) = \int_{s_0}^{\infty} \frac{\mathrm{Im}\{C_l(s')\}}{s'-s} \, ds' \tag{2.31}$$

gives

$$N_l(s) = B_l^L(s) \, D_l(s) - \frac{1}{\pi} \int_{s_0}^{\infty} \frac{B_l^L(s') \, \mathrm{Im}\{D_l(s')\}}{s'-s} \, ds' \,, \tag{2.32}$$

and, substituting (2.24) for $D_l(s)$ and (2.12) for Im $\{D_l(s)\}$, we end up with

$$N_l(s) = B_l^L(s) + \frac{1}{\pi} \int\limits_{s_0}^{\infty} \frac{B_l^L(s') - B_l^L(s)}{s' - s} \, \varrho_l(s') \, N_l(s') \, \mathrm{d}s' \, . \qquad (2.33)$$

Equation (2.33) [or (2.26)] is a Fredholm integral equation [see e.g. *Smithies* (1962)] provided that the kernel is square integrable, that is provided that [see *Mandelstam* (1963d) for a more complete discussion]

$$\int\limits_{s_0}^{\infty}\!\!\int \mathrm{d}s \, \mathrm{d}s' \left| \frac{B_l^L(s') - B_l^L(s)}{s' - s} \right|^2 \varrho_l(s') \, \varrho_l(s) < \infty \, , \qquad (2.34)$$

and

$$\int\limits_{s_0}^{\infty} \mathrm{d}s' \, \varrho_l(s') \, |B_l^L(s')|^2 < \infty \, . \qquad (2.35)$$

We have already noted that, because only a finite range of integration is involved in (II. 11.9), $B_l^L(s)$ is analytically continuable in l apart from poles at the negative integers. The lower bound in l will depend on the behaviour of the kinematical factor $\varrho_l(s)$ at threshold where $B_l^L(s)$ is finite, and limits us to Re $\{l\} > -3/2$ unless, as is usually the case, there is a more restrictive bound due to the behaviour of $B_l^L(s)$. In particular if $B_l^L(s) \sim s^{\beta - l - 1}$, then (2.35) is satisfied only if Re $\{l\} > 2\beta - 1$. The upper bound in l depends on the asymptotic s behaviour of $\varrho_l(s)$, and requires $l < 1$, so we need

$$\max \{2\beta - 1, \, -3/2\} < \mathrm{Re} \, \{l\} < 1 \, . \qquad (2.36)$$

The solution of a non-singular integral equation is a meromorphic function of any given parameter, providing only that the kernel is analytic in that parameter, so (2.33) will be meromorphic in l in the range (2.36). This range is rather restrictive, but Mandelstam also shows that, if we solve the similar equations for

$$B_l^n(s) \equiv B_l^{\mp}(s) \, (q_s^2)^n$$

the range of validity is

$$n + \max \{2\beta - 1, \, -3/2\} < \mathrm{Re} \, \{l\} < n + 1 \, ,$$

and that the various $B_l^n(s)$ are analytic continuations of one another. So the N/D equations provide solutions which satisfy maximal analyticity of the second kind.

The only poles of $B_l(s)$ are the zeros of $D_l(s)$. Solving (2.24) for all values of l, we get a Regge trajectory, $\alpha(s)$, from the implicit equation

$$D_{\alpha(s_{\mathrm{R}})} \, (s_{\mathrm{R}}) = 0 \, , \qquad (2.37)$$

s_{R} being the position of the pole.

On expanding $D_l(s_{\mathrm{R}})$ about $l = \alpha(s_{\mathrm{R}})$ we get

$$B_l(s) \approx \frac{N_{\alpha(s_{\mathrm{R}})}(s_{\mathrm{R}})}{\left(\dfrac{\partial D_l(s)}{\partial l}\right)_{\substack{l = \alpha(s_{\mathrm{R}}) \\ s = s_{\mathrm{R}}}} [l - \alpha(s_{\mathrm{R}})]} \, , \quad s \approx s_{\mathrm{R}} \qquad (2.38)$$

which by comparison with (II. 9.15) gives, remembering (III. 1.7),

$$\gamma(s_R) = \frac{N_{\alpha(s_R)}(s_R)}{\left(\dfrac{\partial D_l(s)}{\partial l}\right)_{\substack{l=\alpha(s_R)\\s=s_R}}};$$ (2.39)

or, if we expand $D_l(s)$ about $s = s_R$ and use (2.12), we get

$$B_l(s) \approx \frac{N_l(s)}{\left(\dfrac{\partial \mathrm{Re}\{D\}}{\partial s}\right)_{s=s_R}(s - s_R) + i\varrho_l(s)N_l(s)}$$

$$\approx \frac{N_l(s)/2E_R D'_l(s)}{E - E_R + i\,\dfrac{\varrho_l(s)N_l(s)}{2E_R D'_l(s)}}$$ (2.40)

for $s \approx s_R$. This gives the usual Breit-Wigner form for a resonance pole at E_R of width

$$\Gamma = -\frac{\varrho_l(s)N_l(s)}{E_R D'_l(s)}$$ (2.41)

if $s > s_0$, $[D'_l(s)$ will be negative so Γ is positive: compare (II. 10.7)], and below s_0, where $\varrho_l(s) = 0$, there is the expected bound-state pole form. Comparing (2.41) with (II. 10.7) we find

$$\frac{\gamma(s_R)}{\alpha'(s_R)} = \frac{N_\alpha(s_R)}{D'_\alpha(s_R)},$$ (2.42)

where the prime denotes $\partial/\partial s$ in each case.

So solving (2.33) to find $N_l(s)$, and then (2.24) to find $D_l(s)$, allows us to obtain the Regge trajectory and residue functions, and they will have the analyticity properties which we discussed in section (III. 1) [see *Taylor* (1962)].

Since $D_l(s)$ is normalized to 1 at $s = \infty$, the only way of getting a zero of $D_l(s)$ at ∞, which will give the end point of a trajectory, $\alpha(\infty)$, is for $D_l(s)$ to have a fixed-l infinite singularity. Fixed singularities are present in the kernel of the integral equation (2.33) at the negative integers (due to the poles in l of Q_l at $l = -1, -2 \ldots$), and these will generate Fredholm poles in $N_l(s)$, at values of l which depend on the details of the solution [see *Chew* and *Jones* (1964)]. So, unlike potential scattering, where the end points of the trajectories must be integers [*Mandelstam* (1962)], our trajectories will take the form [*Jones* (1964)]

$$\alpha(s) = \alpha(\infty) + \frac{\text{const.}}{s} + \cdots \qquad s \to \infty,$$ (2.43)

with $\alpha(\infty)$ dynamically determined, and for the highest lying trajectories

$$\alpha(\infty) > -1.$$

Also, since [from (2.24) and (2.33) remembering that $D_\alpha(s_R) = 0$]

$$\frac{\gamma(s_R)}{\alpha'(s_R)} = \frac{N_\alpha(s_R)}{D'_\alpha(s_R)}$$

$$= \frac{\int\limits_{s_0}^{\infty} ds'\, \varrho_\alpha(s')\, N_\alpha(s')\, B_\alpha^L(s')/(s' - s_R)}{\int\limits_{s_0}^{\infty} ds'\, \varrho_\alpha(s')\, N_\alpha(s')/(s' - s_R)^2},$$ (2.44)

unless special moment conditions are satisfied by the integrals we can expect

$$\gamma(s) \sim \alpha'(s)\, s \sim 1/s, \qquad s \to \infty, \tag{2.45}$$

where the last step depends on (2.43). Such a behaviour has been found in the solutions of these equations [see section (VI. 6)].

Before solving the equations we need to know the left-hand cut function, and there are several ways of writing this.

One expression can be obtained from (II. 11.9), which gives the left-hand cut discontinuity in the form

$$b_{l}^{\pm}(s) = \frac{1}{16\pi^2} \int_{a(s)}^{b(s)} Q_l(z')\, \varrho_{tu}(t', u')\, [1 \mp e^{-i\pi l}]\, \frac{dt'}{2q_s^{2l+1}} +$$

$$+ \frac{1}{32\pi} \int_{t_0}^{-4q_s^2} P_l(-z')\, D_{\bar{l}}^{\pm}(s, t')\, \frac{dt'}{2(-q_s^2)^{l+1}}, \tag{2.46}$$

and we can obtain

$$B_l^L(s) = \frac{1}{\pi} \int_{-\infty}^{s_L} \frac{b_l^{\pm}(s')}{s' - s}\, ds'. \tag{2.47}$$

An alternative expression is found by substituting (II. 2.8) for $D_{\bar{l}}^{\pm}(s, t)$ into (II. 11.7). This gives

$$B_{\bar{l}}^{\pm}(s) = \frac{1}{16\pi^3} \int\int \left[\frac{\varrho_{st}(s', t'') \pm \varrho_{su}(s', t'')}{s' - s}\, ds' + \frac{\varrho_{tu}(t'', u') \pm \varrho_{tu}(u', t'')}{u' - u''}\, du' \right] \times$$

$$\times Q_l[z_s(s, t'')]\, \frac{dt''}{2q_s^{2l+2}}. \tag{2.48}$$

The discontinuity across the right-hand cut of $B_{\bar{l}}^{\pm}(s)$ is evidently

$$\frac{1}{16\pi^2} \int [\varrho_{st}(s, t'') \pm \varrho_{su}(s, t'')]\, Q_l[z_s(s, t'')]\, \frac{dt''}{2q_s^{2l+2}}, s > s_0, \tag{2.49}$$

and, if we subtract this contribution from (2.48), we are just left with

$$B_l^L(s) = \frac{1}{16\pi^3} \int\int \left[\frac{\varrho_{st}(s', t'') \pm \varrho_{su}(s', t'')}{s' - s} \right] \times$$

$$\times \left[\frac{Q_l[z_s(s, t'')]}{2q_s^{2l+2}} - \frac{Q_l[z_s(s', t'')]}{2q_{s'}^{2l+2}} \right] ds'\, dt'' + \tag{2.50}$$

$$+ \frac{1}{16\pi^3} \int\int \left[\frac{\varrho_{tu}(t'', u') \pm \varrho_{tu}(u', t'')}{u' - u''} \right] \frac{Q_l[z_s(s, t'')]}{2q_s^{2l+2}}\, du'\, dt''.$$

Another possible way of proceeding is to use (II. 11.3) and write

$$B_{\bar{l}}^L(s) = \frac{1}{16\pi^2} \int_{-\infty}^{0} \text{Im}\, \{Q_l[z_s(s, t'')]\}\, V(s, t'')\, \frac{dt''}{2q_s^{2l+2}} \tag{2.51}$$

where $V^{\pm}(s, t)$ is the amplitude $A^{\pm}(s, t)$, less that part which contributes to the right hand cut. We shall discuss this further in section (VI. 6).

All of these are of course exact expressions, and assume a complete knowledge of the double spectral functions. Our aim is to calculate, by means of the N/D equations, the contribution of one part of the double spectral functions (the elastic s-channel double spectral function) given a knowledge of the other parts, and evidently some approximations will be needed. We shall give some examples in the later sections.

So far we have been assuming that elastic unitarity holds for all $s > s_0$. Apart from being a rather poor approximation for many purposes this also often results in non-Fredholm equations because of the behaviour of $B_l^L(s)$ for large s. So it is necessary to find out how to put in-elasticity into our equations.

VI.3. Inelasticity in N/D equations

There are several methods of including the coupling of inelastic channels in the N/D equations. As long as all channels which are open at a given energy are two particle channels, the unitarity condition (II. 5.7) is simply replaced by (II. 5.12), and we have

$$\mathrm{Im}\,\{B_l^{ab}(s)\} = \sum_n \varrho_l^n(s)\, B_l^{bn}(s_+)\, B_l^{na}(s_-)\,, \qquad (3.1)$$

which we can write in matrix form [*Bjorken* (1960)]

$$(\mathrm{Im}\,\{\mathbf{B}_l\})_{ab} = (\mathbf{B}_l^* \cdot \boldsymbol{\varrho}_l \cdot \mathbf{B}_l)_{ab}\,, \qquad (3.2)$$

where $\boldsymbol{\varrho}_l(s)$ is a diagonal matrix,

$$\boldsymbol{\varrho}_l = \begin{pmatrix} \varrho_l^1 & \cdot & \cdot & \cdot & 0 \\ \cdot & \varrho_l^2 & \cdot & \cdot & \\ \cdot & \cdot & \varrho_l^3 & & \\ \cdot & \cdot & & \cdot & \\ 0 & \cdot & & & \cdot \end{pmatrix}, \qquad (3.3)$$

and

$$\mathbf{B}_l = \begin{pmatrix} B_l^{11} & B_l^{12} & \cdot & \cdot & \cdot \\ B_l^{21} & B_l^{22} & \cdot & \cdot & \cdot \\ \cdot & \cdot & & & \\ \cdot & \cdot & & & \end{pmatrix}. \qquad (3.4)$$

In these matrices the rows and columns correspond to the various channels.

Then

$$(\mathrm{Im}\,\{\mathbf{B}_l\})_{ab}^{-1} = -\,(\mathrm{Im}\,\{\mathbf{B}_l\} \cdot (\mathbf{B}_l^* \cdot \mathbf{B}_l)^{-1})_{ab} \qquad (3.5)$$

which from (3.2)

$$= -\,\delta_{ab}\, \varrho_l^a(s)\, \vartheta\,(s - s_a) \qquad (3.6)$$

where s_a is the threshold in channel a $[\varrho_l^a(s) = 0$ for $s < s_a]$. So if we write

$$\mathbf{B} = \mathbf{N} \cdot \mathbf{D}^{-1}\,, \qquad (3.7)$$

or

$$B_l^{ab}(s) = \sum_n N_l^{an}(\mathbf{D}_l(s)^{-1})_{nb}$$

$$= \sum_n \frac{N_l^{an}(s)\,\bar{D}_l^{nb}(s)}{\det\{\mathbf{D}_l(s)\}} \,, \tag{3.8}$$

where $\bar{\mathbf{D}}_l(s)$ is the co-factor matrix of $\mathbf{D}_l(s)$,

i.e. $\mathbf{D} \cdot \bar{\mathbf{D}} = \det\{\mathbf{D}\}$,

and if we follow the reasoning of section (VI. 2), we end up with

$$D_l^{ab}(s) = \delta_{ab} - \frac{1}{\pi}\int_{s_a}^{\infty}\frac{\varrho_l^a(s')\,N_l^{ab}(s')}{s'-s}\,ds' \tag{3.9}$$

and

$$N_l^{ab}(s) = B_l^{L\,ab}(s) + \sum_n \frac{1}{\pi}\int_{s_a}^{\infty}\frac{B_l^{L\,an}(s') - B_l^{L\,an}(s)}{s'-s}\,\varrho_l^a(s')\,N_l^{nb}(s')\,ds' \tag{3.10}$$

corresponding to (2.24) and (2.33).

From (3.8) we see that a trajectory function $\alpha(s)$, will be given by

$$\det\{\mathbf{D}_{\alpha(s_R)}(s_R)\} = 0 \,, \tag{3.11}$$

and if we separate $\det\{\mathbf{D}\}$ into its real and imaginary parts for real s, in the form

$$\det\{\mathbf{D}_l\} = \Delta_{lR}(s) + i\,\Delta_{lI}(s) \tag{3.12}$$

we have, corresponding to (2.40),

$$B_l^{ab}(s) \approx \frac{\sum_n N_l^{an}(s)\,\bar{D}_l^{nb}(s)/\Delta_{lR}'(s)}{(s-s_R) + i\Delta_{lI}(s)/\Delta_{lR}'(s)} \,, \tag{3.13}$$

(again the prime denotes $\partial/\partial s$) so the full width is

$$\Gamma = -\frac{\Delta_I}{\Delta_R'}\frac{1}{E_R} \,, \tag{3.14}$$

and the partial widths to the various channels can be obtained from the numerator.

A trajectory given by (3.11) occurs, as one would expect, in all the channels which communicate with each other (have the same quantum numbers), but its residue differs from one channel to another. Some examples of calculations using this formalism will be quoted in section (VI. 5). General questions concerning the validity of the many-channel formalism have been discussed by *Warnock* (1967).

Once the inelastic channels contain more than two particles, however, the unitarity equations become too complicated to use [see the discussion in section (I. 4)]. Instead we seek some method of parametrizing the inelasticity which will enable us to modify the single-channel equations in some simple way.

One such method, proposed by *Froissart* (1961 b), uses (II. 5.15) and (II. 5.16) to define an inelasticity parameter

$$R_l(s) \equiv \frac{\sigma_l^{tot}(s)}{\sigma_l^{el}(s)} = \frac{\mathrm{Im}\{B_l^{\pm}(s)\}}{|B_l^{\pm}(s)|^2}\frac{1}{\varrho_l(s)} \,. \tag{3.15}$$

Hence

$$\text{Im}\left\{\frac{1}{B_l(s)}\right\} = -\,\varrho_l(s)\,R_l(s)\,, \tag{3.16}$$

and using this in (2.12) gives the following modification of our previous equations:

$$D_l(s) = 1 - \frac{1}{\pi}\int_{s_0}^{\infty}\frac{\varrho_l(s')\,R_l(s')\,N_l(s')}{s'-s}\,\mathrm{d}s' \tag{3.17}$$

$$N_l(s) = B_l^L(s) + \frac{1}{\pi}\int_{s_0}^{\infty}\frac{B_l^L(s)-B_l^L(s)}{s'-s}\,\varrho_l(s')\,R_l(s')\,N_l(s')\,\mathrm{d}s'\,. \tag{3.18}$$

Here $D_l(s)$ has the phase of the amplitude, since $N_l(s)$ is still real, for all $s > s_0$, but the phase of the amplitude is not the same as the phase shift, since $\delta_l(s)$ becomes complex for $s > s_I$ [see (II. 5.12) et seq.] Clearly

$$R_l(s) = 1 \quad \text{for} \quad s_0 < s < s_I\,,$$
and
$$R_l(s) > 1 \quad \text{for} \quad s > s_I\,.$$

This method is simple to use, but as we expect $R_l(s) \to \infty$ as $s \to \infty$ the equations become undefined at high energies, and the divergence problems mentioned in the previous section are only aggravated.

A more sophisticated method has been proposed by *Frye* and *Warnock* (1963) [see also *Coulter* et al. (1964)]. They take

$$B_l^{\pm}(s) = \frac{\eta_l(s)\,\mathrm{e}^{2i\,\delta_l^R(s)}-1}{2\,i\,\varrho_l(s)}\,, \tag{3.19}$$

where the phase shift is divided into its real and imaginary parts (for real s),

$$\delta_l(s) = \delta_l^R(s) + i\,\delta_l^I(s)\,, \tag{3.20}$$
and
$$\eta_l(s) = \mathrm{e}^{-2\,\delta_l^I(s)}. \tag{3.21}$$

Thus
$$\eta_l(s) = 1 \quad \text{for} \quad s_0 < s < s_I\,,$$
and
$$\eta_l(s) \to 0$$

if the amplitude becomes wholly inelastic. We again put

$$B_l(s) = \frac{N_l(s)}{D_l(s)}\,,$$

but give $D_l(s)$ the phase $\mathrm{e}^{-i\,\delta_l^R(s)}$ rather than the phase of the amplitude, so that $N_l(s)$ has a cut for $s > s_I$. In fact, from (3.19),

$$N_l(s) = B_l(s)\,D_l(s) = \left[\frac{\eta_l(s)\,D_l^{*}(s)/D_l(s)-1}{2\,i\,\varrho_l(s)}\right]D_l(s)\,,$$

$$= \frac{\eta_l(s)\,D_l^{*}(s) - D_l(s)}{2\,i\,\varrho_l(s)}\,, \tag{3.22}$$

and, taking the real and imaginary parts of this expression, we get

$$\text{Im}\,\{N_l(s)\} = \frac{1 - \eta_l(s)}{2\varrho_l(s)}\,\text{Re}\,\{D_l(s)\}\,, \qquad s > s_\text{I} \tag{3.23}$$

and

$$\text{Im}\,\{D_l(s)\} = \frac{-2\varrho_l(s)}{1 + \eta_l(s)}\,\text{Re}\,\{N_l(s)\}, \qquad s > s_0\,. \tag{3.24}$$

From the latter we obtain, corresponding to (2.24),

$$D_l(s) = 1 - \frac{1}{\pi}\int\limits_{s_0}^{\infty} \frac{2\varrho_l(s')}{1 + \eta_l(s')}\,\frac{\text{Re}\{N_l(s')\}}{(s' - s)}\,\mathrm{d}s'\,. \tag{3.25}$$

Like (2.28), we define the function

$$C_l(s) \equiv N_l(s) - B_l^L(s)\,D_l(s) - \left[\frac{1}{\pi}\int\limits_{s_\text{I}}^{\infty} \frac{\text{Im}\{N_l(s')\}/\text{Re}\{D_l(s')\}\,\mathrm{d}s'}{s' - s}\right]D_l(s)\,. \tag{3.26}$$

This function has no left-hand cut, while on the right, using (3.23) its imaginary part is

$$\text{Im}\,\{(C_l(s)\} = -\,\bar{B}_l(s)\,\text{Im}\,\{D_l(s)\}\,, \qquad s > s_0\,, \tag{3.27}$$

where

$$\bar{B}_l(s) = B_l^L(s) + \frac{\text{P}}{\pi}\int\limits_{s_\text{I}}^{\infty}\left[\frac{1 - \eta_l(s')}{2\varrho_l(s')}\right]\frac{\mathrm{d}s'}{s' - s}\,, \tag{3.28}$$

and P implies taking the pricipal value. So, from (3.26),

$$C_l(s) = N_l(s) - \bar{B}_l(s)\,D_l(s) - i\,\frac{[1 - \eta_l(s)]}{2\varrho_l(s)}\,D_l(s)\,, \tag{3.29}$$

and, from (3.27) with (3.24).

$$C_l(s) = \frac{1}{\pi}\int\limits_{s_0}^{\infty}\frac{\bar{B}_l(s')\,\text{Re}\{N_l(s')\}}{s' - s}\left[\frac{2\varrho_l(s')}{1 + \eta_l(s')}\right]\mathrm{d}s'\,. \tag{3.30}$$

On substituting (3.25) into the right-hand side of (3.29), and equating it with (3.30), we end up with (after some manipulation)

$$\bar{N}_l(s) = \bar{B}_l(s) + \frac{1}{\pi}\int\limits_{s_0}^{\infty}\frac{\bar{B}_l(s') - \bar{B}_l(s)}{s' - s}\,\frac{\varrho_l(s')}{\eta_l(s')}\,\bar{N}_l(s')\,\mathrm{d}s' \tag{3.31}$$

where

$$\bar{N}_l(s) \equiv \frac{2\eta_l(s)}{1 + \eta_l(s)}\,\text{Re}\,\{N_l(s)\}\,. \tag{3.32}$$

As before, we can solve (3.31) and (3.25) to find the positions and residues of the Regge trajectories, given $B_l^L(s)$ and $\eta_l(s)$ as input. It is worth expressing the parameter $\eta_l(s)$ in terms of the discontinuity of $B_l^{\pm}(s)$ due to inelastic effects. If we write (II. 5.18) in the form

$$\text{Im}\,\{B_l^{\pm}(s)\} = \varrho_l(s)\,|B_l^{\pm}(s)|^2 + \text{Im}\,\{B_l^I(s)\} \tag{3.33}$$

we get, on substituting (3.19) for $\text{Im}\{B_l^\mp(s)\}$ and $B_l^\mp(s)$ and rearranging,

$$\eta_l(s) = [1 - 4\,\varrho_l(s)\,\text{Im}\{B_l^I(s)\}]^{1/2}\,. \tag{3.34}$$

Both the Froissart and the Frye-Warnock methods suffer from a difficulty if the forces in the inelastic channels are strong, however. For suppose we look at a two channel problem expressed in the multichannel formalism of (3.7), so that

$$B_{11} = \frac{N_{11}D_{22} - N_{12}D_{21}}{D_{11}D_{22} - D_{12}D_{21}}\,. \tag{3.35}$$

If the two channels are completely decoupled, we have the single channel solutions

$$B_{11} = \frac{N_1}{D_1} \quad \text{and} \quad B_{22} = \frac{N_2}{D_2} \quad \text{(say)}. \tag{3.36}$$

Suppose now that the two channels are weakly coupled by some parameter λ, so that

$$N_{11} = N_1 + O(\lambda^2)\,,$$
$$D_{11} = D_1 + O(\lambda^2)\,,$$

and the off-diagonal elements, N_{12}, D_{12} etc., are of order λ. Then

$$B_{11} = \frac{N_1}{D_1} + O(\lambda^2)$$

except where D_1 or D_2 have zeros. If D_2 has a zero at $s = s_p$ when $\lambda = 0$, and D_1 does not, then both the numerator and the denominator of (3.35) have zeros at s_p for $\lambda = 0$. But for $\lambda \neq 0$ these zeros will not coincide, and we shall find an adjacent zero and pole of B_{11}, near s_p, despite the fact that there is not sufficient force in channel 1 to produce such a pole. It will in fact correspond to a CDD pole of the B_{11} amplitude. This fact was observed by *Squires* (1964), *Bander* et al. (1965) and *Atkinson* et al. (1965). The absence of CDD poles is of course built into the treatment of inelasticity by the Froissart and Frye-Warnock methods, which are therefore unable to sope with this situation. In fact in the Frye-Warnock method such a pole corresponds to the vanishing of $\eta_l(s)$. This makes the integral equation (3.31) non-Fredholm, and it can be shown that the solution of the singular integral equation contains the CDD pole [*Bander* et al. (1965), *Atkinson* et al. (1966), *Jones* and *Hartle* (1965)]. Similarly in the Froissart method $R_l(s) \to \infty$ for such a pole.

This does not mean that we can not distinguish between elementary and dynamically produced particles, however. It remains true that there can be no CDD poles in any channel for large l, and the solution for low l is still a continuation of that for large l. *Jones* and *Hartle* (1965) have shown how a CDD zero of the amplitude can emerge from the inelastic cut as l is decreased. When this happens Levinson's theorem ceases to be a valid criterion for compositness, but *Atkinson* et al. (1966) have shown that if one diagonalizes the scattering amplitude by a suitable transformation matrix $\mathbf{O}_l$, such that

$$\mathbf{O}_l \cdot \varrho_l^{1/2} \cdot \mathbf{B}_l \cdot \varrho_l^{1/2} \cdot \mathbf{O}^{-1} = \mathbf{B}^D = \begin{pmatrix} B_{11}^D & \cdots & 0 \\ & B_{22}^D & \\ & & \ddots \\ 0 & & \end{pmatrix},$$

then the eigen phase shifts (the phase shifts of the eigen amplitude $\mathbf{B}^D$) will satisfy the no CDD pole criterion if maximal analyticity of the second kind is valid, and it is only the transformation matrix, $\mathbf{O}_l$, which produces the violation in the cases we have mentioned. Since, however, there are an infinite number of channels open at infinite energy, Levinson's theorem can not be regarded as a very practical test of maximal analyticity.

VI.4. Particle Exchange Forces

We have noted that the left hand cut, which takes its main contribution from crossed channel singularities, corresponds to the "force", or potential, in non-relativistic scattering, and that the solution of the N/D equations is analagous in many ways to the solving of Schroedinger's equation for the scattering problem. Thus a pole in the t-channel, corresponding to the exchange of a single particle, as in Fig. (VI. 1a), gives rise to a force which may generate an s-channel pole, as in Fig. (VI.1b). The well known analogy between the Yukawa potential and

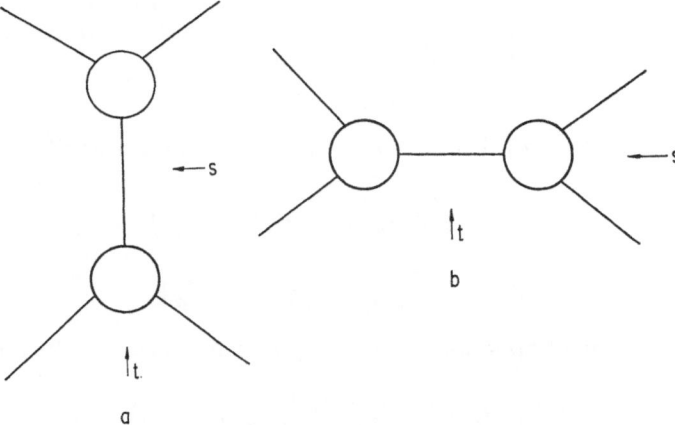

Fig. VI.1a and b. (a) A single particle exchange pole in the t-channel, which contributes part of the force required to generate (b), a pole in the s-channel.

scalar meson exchange is being invoked here. In fact potential scattering with a Yukawa potential satisfies the Mandelstam representation with just one double spectral function (ϱ_{st}, if s is the scattering channel, and the t-singularities represent the potential), as was shown by *Blanken-becler* et al. (1960), and the N/D method can be used for its solution [see *Luming* (1964)]. A discussion of this would take us too far from our subject, and we refer the reader to *Frautschi* (1963), *Omnes* and *Froissart*

(1963), and *Chew* (1965a) for further discussion. The analogy has been extended and turned into an approximation scheme for particle physics, by *Balazs* (1965a), (1965b), and followed up by *Finkelstein* (1966), (1967).

What concerns us in this section, however, is that single particle exchange in the t-channel corresponds to the first Born approximation and multiparticle exchanges to higher Born approximations. From our discussion in section (II. 7) one would expect the more distant multiparticle singularities to be less important because of their shorter range. We shall often refer loosely to single particle exchange as "the potential", but a more precise definition of "potential" will be given in section (VI. 7) when we discuss better approximations than single particle exchange. (Our use of the word "exchange" here should not be confused with "exchange potentials", which correspond to u-channel singularities. Since we are always working with amplitudes of definite signature we do not have to discuss them separately.)

The simplest form of a single particle exchange diagram is a bound-state, scalar (spin zero, even parity) particle in the t-channel, which contributes to the amplitude

$$V^{\pm}(s, t) = \frac{g}{m^2 - t} , \tag{4.1}$$

m being its mass, and g the residue of the pole. Evidently such a term has no s-singularities, and so can not contribute to the right-hand cut of a partial-wave amplitude. It does however contribute to the left-hand cut, since its contribution to the t-discontinuity is

$$D_t^{V\pm}(s, t) = \pi g \, \delta \, (m^2 - t) , \tag{4.2}$$

and, on substituting this in (2.46), we get, assuming for simplicity equal mass kinematics,

$$b_l(s) = \frac{1}{16\pi} \frac{g}{2q_s^{2l+2}} \frac{1}{2} P_l \left(1 + \frac{m^2}{2q_s^2} \right), \tag{4.3}$$

or in (II. 11.7)

$$B_l^L(s) = \frac{1}{16\pi} \frac{g}{2q_s^{2l+2}} Q_l \left(1 + \frac{m^2}{2q_s^2} \right). \tag{4.4}$$

Simiarly we might approximate a spin l_t resonance (such as the ϱ of spin one) by

$$V^{\pm}(s, t) = \frac{(2l_t + 1) g (q_t^2)^{l_t}}{m^2 - t - i\Gamma} P_{l_t} \left(1 + \frac{s}{2q_t^2} \right), \tag{4.5}$$

which, in the narrow width approximation $[\Gamma \to 0]$ gives

$$D_t^{V\pm}(s, t) = \pi (2l_t + 1) g (q_m^2)^{l_t} P_{l_t} \left(1 + \frac{s}{2q_m^2} \right) \delta (m^2 - t) \tag{4.6}$$

and

$$B_l^L(s) = \frac{1}{16\pi} \frac{g (2l_t + 1)}{2q_s^{2l+2}} Q_l \left(1 + \frac{m^2}{2q_s^2} \right) P_{l_t} \left(1 + \frac{s}{2q_m^2} \right) (q_m^2)^{l_t} \tag{4.7}$$

where q_m^2 is the value of q_t^2 corresponding to $t = m^2$.

We will discuss the solution of the N/D equations with such potentials in the next section, but here we want to note we have assumed that we are exchanging a fixed spin particle, or at least we have supposed that it will be a satisfactory approximation to hold the spin fixed in continuing $V^\pm (s, t)$ away from $t = m^2$. It would be more consistent with our analyticity postulates were we to take account of the Regge pole nature of the exchanged particles, and the fact that the spin alters as we vary t. At the same time it is worth noting that there can be an ambiguity in these potentials for particles with spin. For, if we use (2.51) to find $B_l^L(s)$, (4.1) still gives (4.4), but (4.5) does not give (4.7) except for $l_t = 0$. If we take $l_t = 1$ and $\Gamma = 0$ in (4.5)

$$V^\pm (s, t) = \frac{3g}{m^2 - t} \frac{2q_t^2 + s}{2} . \tag{4.8}$$

If the amplitude in question is, say, $\pi - \pi$ scattering, so that

$$q_t^2 = \frac{t - m_\pi^2}{4} \tag{4.9}$$

we can re-write this as

$$V^\pm (s, t) = \frac{3g}{4} \left[-1 + \frac{2s - 4m_\pi^2 + m^2}{m^2 - t} \right] . \tag{4.10}$$

The -1 term in (4.10), which has no t-dependence, will contribute only to the s-wave, $B_0^L (s)$, while the rest of (4.10) gives a result which agrees with (4.7). And the higher the exchanged spin the greater is the disagreement. These anomalies have been discussed by *Abers* and *Teplitz* (1965), and *Srivastava* and *Nath* (1966). The difference is simply that in (4.6) we have evaluated the angular-momentum factor $P_l (1 + s/q_t^2)$ with t fixed at m^2, whereas in (4.10) we have retained its functional dependence on t. *Chew* (1965 b) has shown that the use of the Regge representation of section (III. 4) helps to resolve this difficulty.

With (4.5), a single particle of fixed spin l_t contributes only to the l_t^{th} partial wave in the t-channel (again $\Gamma = 0$),

$$V_{l_t}^\pm (t) = \frac{g}{16\pi} (q_t^2)^{l_t} \frac{1}{m^2 - t} , \tag{4.11}$$

while the Khuri-Jones representation of (III. 4.8) gives (interchanging s and t)

$$V_{l_t}^\pm (t) = - \gamma (t) (-q_t^2)^{\alpha (t)} \frac{e^{- [l_t - \alpha (t)] \xi (t)}}{\alpha (t) - l_t} . \tag{4.12}$$

If we take a linear approximation to the trajectory function $\alpha (t)$, with α passing through l_t at $t = m^2$, i.e.

$$\alpha (t) = l_t + \alpha' (t - m^2) , \tag{4.13}$$

α' being the slope of the trajectory, we get

$$V_{l_t}^\pm (t) = \frac{\gamma (t) (-q_t^2)^{\alpha (t)} e^{- [l_t - \alpha (t)] \xi (t)}}{\alpha' (m^2 - t)} . \tag{4.14}$$

If we also assume that in $\xi (t)$ [see (III. 4.5), but with s and t interchanged]

$$s_1 \gg |t|$$

we can approximate

$$\xi(t) \approx \log\left(\frac{s_1}{q_t^2}\right)$$

giving   (4.15)

$$V_{l_t}^{\pm}(t) \approx \frac{\gamma(t)\,(q_t^2)^{\alpha(t)}}{\alpha'(m^2 - t)}\left(\frac{s_1}{q_t^2}\right)^{\alpha(t)-l_t}.$$

It is clear that in defining $\gamma(t)$ in (III. 2.6) by

$$\gamma(t)\,(q_t^2)^{\alpha(t)} \equiv \beta(t),$$

we have produced a function whose dimensions will vary as t varies, and which hence may vary very rapidly. It is better to define a function which, while having the required branch point at $q_t^2 = 0$, retains the dimensions of $\beta(t)$, viz.

$$\bar{\gamma}(t)\left(\frac{q_t^2}{\bar{q}_t^2}\right)^{\alpha(t)} \equiv \beta(t),$$ (4.16)

and to chose the constant $\bar{q}_t^2$ such that $\bar{\gamma}(t)$ is a slowly varying function. So we get

$$V_{l_t}^{\pm}(t) \approx \frac{\bar{\gamma}(t)}{\alpha'(m^2 - t)}\left(\frac{q_t^2}{\bar{q}_t^2}\right)^{l_t}\left(\frac{s_1}{\bar{q}_t^2}\right)^{\alpha(t)-l_t}.$$ (4.17)

Comparing this formula with (4.11) we see that the difference between the contribution of a Regge pole to the partial wave l_t, and that of an elementary particle, lies in the difference between the constant

$$\frac{g}{16\pi}$$

and the expression (called a "form factor" by *Chew*)

$$\frac{\bar{\gamma}(t)}{\alpha'}\left(\frac{s_1}{\bar{q}_t^2}\right)^{\alpha(t)-l_t},$$ (4.18)

and if $\bar{\gamma}(t)$ is a slowly varying function of t then the difference lies mainly in the factor

$$\left(\frac{s_1}{\bar{q}_t^2}\right)^{\alpha(t)-l_t}.$$

To estimate this we need to know $\bar{q}_t^2$. *Chew* and *Teplitz* (1964) have given dynamical arguments which favour

$$\bar{t} \approx s_1/2,$$

s_1, it being remembered, is the value of s at which Regge asymptotic behaviour has set in. Hence, since (I. 6.12)

$$\bar{q}_t^2 \approx \bar{t}/4$$

we have

$$s_1/\bar{q}_t^2 \approx 8.$$

The potential may be found by substituting (4.17) in the partial-wave series

$$V^{\pm}(s, t) = 16\pi \sum (2\,l_t + 1)\, V_{l_t}^{\pm}(t)\, P_{l_t}\left[z_t\,(t, s)\right],$$ (4.19)

(this will converge for $s < s_1$), and using (2.51). The integral (2.51) is over negative t only, and we know that $\alpha(t) < 1$ for $t < 0$ for all trajectories. So each successive physical angular momentum, l_t (and remember that these are alternate integers, even or odd depending on the signature of the trajectory), that is each successive particle up the trajectory, will have its contribution reduced over that of the next lowest by a factor $\approx (8)^{-2}$. Thus the effect of exchanging a trajectory will usually be more or less equivalent to exchanging just the lowest spin particle on that trajectory, and the higher spin particles will have a negligible effect. For the ϱ trajectory the lowest physical partial wave is $l_t = 1$, and the force turns out to be very similar to the fixed spin expression (4.7) [see *Collins* (1966)], but for higher spins (4.7) will certaintly be wrong. This explains why π-exchange is often a very important force, despite the fact that its trajectory is rather low lying. And the ambiguity which we noted previously, that the potential calculated from (4.5) contains parts which contribute to the lower angular momenta in the crossed channel which disagree with (4.7), can now be understood. When a trajectory is exchanged its high spin components will certainly give these extra contributions, but they are heavily damped down by the form factor, (4.18), and are usually unimportant.

It is also interesting to note that for Pomeranchon exchange we can expect $\alpha(t) > 0$, at least for the upper region of the integral in (2.51), and the lowest (and so dominant) partial wave in (4.19) will be the S-wave, $l_t = 0$. Hence $V_{l_t}^{\pm}(t)$ in (4.12) will be negative, and the force will be repulsive. This fact has been emphasized by *Chew* (1965c), who shows that the "range" of the repulsion for a given process, as given by the inverse of its logarithmic derivative with respect to t at $t = 0$, will be essentially the same as the width of the high-energy diffraction peak for that process. This is a considerably longer range than most of the attractive particle-exchange forces, and completely swamps the spin-two part of the P trajectory.

VI.5. Some Simple Bootstrap Calculations

The requirements for a complete bootstrap are very complicated, and our only real hope is to try and find approximations in which it is practicable to carry out calculations without too much of the physics being lost. The difficulty is that one can never be sure just how bad an approximation one is making, since a priori the corrections might well turn out to be larger than the effects included. But, as a start, let us try to see what can be done with some really drastic approximations.

As a very simple example of how one can try to test the bootstrap hypothesis, let us imagine the simplest possible world of strongly interacting particles which might satisfy the axioms, a world containing just one type of particle. This should be a meson of zero spin, since the choice of a higher spin would almost certainly result in lower spin particles being generated; scalar, rather than pseudoscalar like the pion; and of zero baryon number, and other additive quantum numbers, so that a three particle vertex is allowed. In the elastic unitarity approximation of section (VI. 2) we can regard the meson as a bound-state pole of two

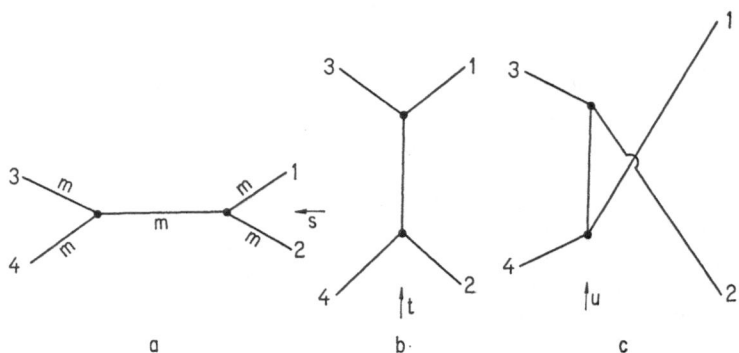

Fig. VI.2 a—c. (a) A scalar meson of mass m produced as a bound state, in the s-channel, of two other such mesons. The "potential" is given by the corresponding poles in the t- and u-channels, (b) and (c).

such mesons, as in Fig. (VI. 2a). The force to generate such a particle should, from our arguments in section (VI. 4), come, in the first approximation, from the corresponding pole terms in the t- and u-channels [Fig. (VI. 2b, c)].

So from (4.1) we have, when we remember to include both the t- and u-channel contributions with the $\pm$ signature,

$$V^\pm (s, t) = \frac{g}{m^2 - t} \pm \frac{g}{m^2 - t}$$
$$= \frac{2g}{m^2 - t} \quad \text{or} \quad 0 \tag{5.1}$$

m being the mass of our meson, and g the residue of the pole. For like particles Bose statistics forbids odd angular momentum states, and so the odd signature amplitude $A^-(s, t)$ vanishes identically. Equation (5.1) gives $b_l(s)$ and $B_l^\pm(s)$ as in (4.3) and (4.4), apart from the factor 2. We may then substitute these into the N/D equations, (2.26) and (2.23), or (2.33) and (2.24), and solve to see if there is a zero of the S-wave D function, $D_0(s)$, at $s = m^2$, corresponding to Fig. (VI. 2a). The residue, g, is a free parameter, so we can adjust it until such a zero occurs. This fixes g, but, if we have really succeeded in generating the particle, the residue of the output pole $(=g'$ say$)$, given by

$$\left(\frac{N_l(s)}{D_l'(s)} \right)_{\substack{s = m^2 \\ l = 0}} = \frac{g'}{16\pi} , \tag{5.2}$$

should be the same as g. In fact it was found by numerical computation [*Collins* (1964)] that $g/m^2 = 16.5$ is needed to get $D_0(m^2) = 0$, but that then $g'/m^2 = 105$. (We note that g/m^2 is dimensionless. The absolute mass scale can not of course be determined; only mass ratios are significant in S-matrix theory.) Evidently the bootstrap does not work.

The input coupling is rather large so one might expect that the g^2 term, corresponding to two particle exchange, would make a significant difference, but it does not [*Collins* (1965)]. The magnitude of the discrepancy is such that we can safely conclude, without undue surprise, that a universe which consisted of nothing but scalar mesons (as its only type of strongly interacting particle; there might of course be leptons etc. as well) could not satisfy our analyticity postulates. The corresponding field theory, with an interaction Lagrangian [see for example *Bogoliubov* and *Shirkov* (1959) p. 352]

$$\mathscr{L}_I = \lambda \, \phi^3 \tag{5.3}$$

is renormalizable, and the solutions must presumably have a CDD pole at $s = m^2$ corresponding to Fig. (VI. 2a). It represents a minor advantage of bootstraps over field theory that such a hypothetical world can be eliminated.

Having already looked at a situation in which we do not expect the bootstrap to work, let us now try an almost equally simple one which might, the ρ in $\pi - \pi$ scattering. Since $\pi - \pi$ scattering is crossing symmetric, one might hope, as a first approximation, to represent the scattering amplitude as just a sum of ρ poles in the s-, t- and u-channels. The f being of spin 2, will probably give a negligible force in view of the arguments of the previous section, and the possible S-wave resonances, σ and ε, whose existence is still uncertain, will also probably give weak effects because of their low spin.

One complication over our previous calculation is the need to work with the complete isotopic multiplet (π^+, π^0, π^-) rather than just the individual pions. We assume degeneracy of the multiplet, and isotopic spin invariance, so it is more convenient to work with states of definite isospin, rather than with physical states. But a process such as

$$\pi^+ \, \pi^+ \to \pi^+ \, \pi^+$$

which is pure $I = 2$, crosses to

$$\pi^+ \, \pi^- \to \pi^+ \, \pi^-$$

which is a combination of $I = 0, 1$ and 2 amplitudes. This is readily taken into account by using the isospin crossing matrix, which for $\pi - \pi$ scattering is [*Chew* and *Mandelstam* (1960)]

$$\beta\,(I_s, I_t) = \begin{matrix} I_t \quad 0 \qquad 1 \qquad 2 \quad I_s \\ \begin{pmatrix} \dfrac{1}{3} & 1 & \dfrac{5}{3} \\[2mm] \dfrac{1}{3} & \dfrac{1}{2} & -\dfrac{5}{6} \\[2mm] \dfrac{1}{3} & -\dfrac{1}{2} & \dfrac{1}{6} \end{pmatrix} \begin{matrix} 0 \\[2mm] 1 \\[2mm] 2 \end{matrix} \end{matrix}, \tag{5.4}$$

I_s being the s-channel isospin, etc. So, remembering that the ρ has spin 1, we find, from (4.7), that in the narrow width approximation the force from ϱ-exchange in the t- and u-channels is

$$B_l^{L\,I_s}(s) = \frac{1}{16\pi}\,\frac{3\,\Gamma m_\rho}{q_s^{2l+2}}\,\beta(I_s,\,1)\,Q_l\left(1+\frac{m_\rho^2}{2q_s^2}\right)P_1\left(1+\frac{s}{2q_\rho^2}\right),\qquad(5.5)$$

where we have replaced $(g\,q_\rho^2)$ by (Γm_ρ), Γ being the width of the ρ, using the Breit-Wigner relation between the width and the residue of an elastic resonance pole [see (2.40)].

This can be used to solve the N/D equations. There is the difficulty, however, that

$$B_l^{L\,I_s}(s) \underset{s\to\infty}{\sim} \frac{s\,\log s}{s^{l+1}}\qquad(5.6)$$

and the kernel of (2.33) is not square integrable [see (2.36)]. The problem stems from our assumption that elastic unitarity holds for all s, and one way of avoiding it is simply to cut off the integrals at some value of s ($=s_1$ say) so that

$$N_l(s) = B_l^L(s) + \frac{1}{\pi}\int_{s_0}^{s_1}\frac{B_l^L(s') - B_l^L(s)}{s'-s}\,\varrho_l(s')\,N_l(s')\,ds'\qquad(5.7)$$

and similarly for $D_l(s)$.

The assumption we are making is that elastic unitarity holds approximately up to s_1, but the precise meaning of the cut-off as an inelastic effect is obscure. There is also the disadvantage that s_1 becomes a new parameter of the equations, and if the solution is strongly dependent on the choice made for s_1, the bootstrap is rather unconvincing.

In fact the original calculation of *Zachariasen* (1961) introduced the cut-off in a rather different way by using the first determinantal approximation of *Baker* (1958), which involves simply setting

$$N_l(s) = B_l^L(s)\,,\qquad(5.8)$$

and making a subtraction in the equation for $D_l(s)$, (2.24), at some point s_a, setting

$$D_l(s_a) = 1\,,$$

so that

$$D_l(s) = 1 - \frac{s-s_a}{\pi}\int_{s_0}^{\infty}\frac{\varrho_l(s')\,N_l(s')}{(s'-s)\,(s'-s_a)}\,ds'\,.\qquad(5.9)$$

The extra power of s in the denominator makes the integral converge, and the subtraction point s_a is the new free parameter. Zachariasen fixed s_a to lie at the end of the left hand cut of $B_l^{\pm}(s)$ (i.e. at s_L), on the grounds that this makes $B_l^{\pm}(s) = B_l^L(s)$ near s_L, as it should if the crossed channel ϱ is dominant. The result he obtained was that the ϱ was self-consistent if its mass was $m_\varrho = 350$ Mev (see the erratum to Zachariasen's paper), and its width $\Gamma = 300$ Mev, compared to the experimental $m_\varrho = 750$ Mev, $\Gamma = 110$ Mev. Many other apperoximate calculations have been performed, for example by *Balazs* (1962), (1963), but an exact solution of the N/D equations with a cut-off does not give a self-consisent solution. If we fix the exchanged ρ mass at the experimental number, and take $s_1 = 200 m_\pi^2$, an input width $\Gamma = 2.25 \, m_\pi$ (≈ 320 Mev) is needed to get the output ϱ in the correct place, but then its output width is very large [see *Teplitz* and *Teplitz* (1965), and *Collins* (1966)]. A glance at the crossing matrix (5.4) shows that the force due to ϱ echange in $I_s = 0$ is stronger than it is in $I_s = 1$ by a factor 2, so in this channel there is a higher trajectory than the ϱ which we might like to identify with the P. However the P has not been included in the input so true self-consistency is not possible. Trajectories from this type of calculations are shown in Fig. (VI. 3).

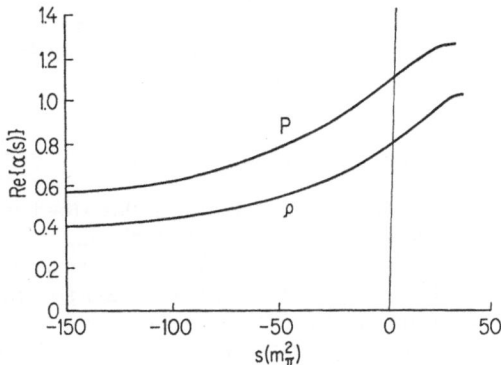

Fig. VI.3. The ϱ and P trajectories generated by the exchange of a ϱ of physical mass, and width $\Gamma = 2.25\,m_\pi$, with $s_1 = 200$. These values are chosen so that the output ϱ has the correct mass, but its width turns out to be very large.

More ambitious calculations, still with fixed-spin pole exchange but including several two-particle channels, have been undertaken, in the determinantal approximation by *Zachariasen* and *Zemach* (1962), and using (5.7) by *Fulco* et al. (1965). The latter authors included the $\pi - \pi$, $\pi - \omega$, and $K \bar{K}$ channels, exchanging ρ, K^* and ϕ, as appropriate. They fixed the input widths at the physical values, and increased s_1 until the output ρ appeared in the right place. However, this procedure gives very flat trajectories, and large output widths, and their results were very similar to the single channel calculations. The final conclusion must be that although the ρ produces a strong force, it is not able to bootstrap itself.

An apparently more successful example of a bootstrap calculation is the N, N* reciprocal bootstrap, proposed by *Chew* (1962c), in which it is found that, in $\pi - N$ scattering, the exchange of a nucleon in the $\pi \bar{N}$ crossed channel results in a force which generates the N*, while N* exchange generates the N. More detailed calculations have been carried out by *Abers* and *Zemach* (1963), and *Fulco* et al. (1965). The output widths in these calculations are much better than for the ρ, but the results are very cut-off dependent, so that the cut-off has to be fixed to get the output pole in the required place. Evidently the high energy

11*

behaviour is crucial to the success or failure of such calculations. In any case these calculations are in serious disagreement with more recent work on the determination of the $\pi - $ N phase shifts. It is found that the P_{11} phase shift is negative just above threshold, but passes through zero at about 180 Mev, probably reaches 90° at about 600 Mev (the Roper resonance), and then flattens off. But Levinson's theorem (2.17) applied to these calculations tells us that

$$\delta_l(\infty) - \delta_l(s_0) = -\pi,$$

and the phase shift will certainly not change sign. It thus seems more likely that the nucleon is really a CDD pole in the $\pi - $ N channel, and is bound in some other channel or channels. A discussion of this problem, with references, is given by *Atkinson* and *Halpern* (1966).

VI.6. The New Form of the Strip Approximation

It should be clear from the calculations described in the previous section that fixed-spin pole exchange, with a single, or small number of two-particle channels, is rather unsatisfactory even as an approximate bootstrap scheme. The simplifications made were so drastic, however, that we need not be too disconcerted by this.

There are two important ways in which the calculations can be improved by invoking our knowledge of Regge poles. Firstly, they enable us to give a better description of the particle poles themselves, and the way in which these should be analytically continued from one region of the variables to another. In the calculations of the previous section the output poles were Regge poles, but not the input ones, and we would like to make things more obviously crossing symmetric. The other feature of Regge poles, the fact that they represent the subtractions required by the divergences of the amplitude, also gives us a means of discussing the inelasticity. For we expect the high energy region in, say, the s-channel to be controlled by the highest lying t- and u-channel Regge poles, so these poles must in some way represent the opening up of more and more inelastic channels.

The new form of the strip approximation was devised by *Chew* and *Jones* (1964), following work by *Chew* (1963a), to build both these features into a calculational framework. It was called the "new form of the strip approximation" to distinguish it from an earlier proposal based on the elastic double spectral functions. This older form has been revived more recently, however, and will be discussed in the next section. Calculations with the new form of the strip approximation have been carried out for $\pi - \pi$ scattering, and we shall continue to use the simplifications of equal mass kinematics and zero spin external particles, in our discussion.

The method involves dividing the amplitude into three contributions,

$$A(s, t) = A^s(s, t) + A^t(s, t) + A^u(s, t), \tag{6.1}$$

where the s-channel poles and two-particle thresholds are contained in $A^s(s, t)$, the t-channel ones in $A^t(s, t)$, etc. So in the language of section (VI. 4)

$$V^s(s, t) \equiv A^t(s, t) + A^u(s, t) \tag{6.2}$$

will constitute the "potential" for generating the s-channel poles. The approximation consists of writing each term in the form

$$A^t(s, t) = \sum_j A^t_j(s, t) \tag{6.3}$$

where $A^t_j(s, t)$ is the contribution of the jth t-channel Regge pole, and we include in the summation all those poles which reach the right-half l_t plane. We are thus neglecting the possibility of singularities other than poles in the right-half l_t plane. For each such pole term we require that it include just the one Regge pole, and no other poles, though it will contain the various t-channel threshold branch points. It must also be constructed such that

$$A^t_j(s, t) \to 0 \quad \text{as} \quad t \to \infty, \tag{6.4}$$

since we expect the s- and u-channel poles to dominate at high t, not the t-channel ones. And we also require that the amplitude should satisfy the Mandelstam representation, which with this decomposition, means that each pole term individually must satisfy the Mandelstam representation. We have already described such terms in Chapter III, and the Chew-Jones representation of a t-channel pole is, from (III. 3.7),

$$R^{s_1}_j(s, t)$$
$$= \frac{1}{2} G_j(t)(-q_t^2)^{\alpha_j(t)} \left[-\frac{P_{\alpha_j}(t)(1 + s/2q_t^2)}{\sin \pi \alpha_j(t)} - \frac{1}{\pi} \int_{-4q_t^2}^{s_1} \frac{P_{\alpha_j}(t)(-1 - s'/2q_t^2)}{s' - s} \, \mathrm{d}s' \right]. \tag{6.5}$$

This contains the Regge pole in the first term on the right-hand side, and the various t-channel thresholds are present in the functions $\alpha(t)$ and $G(t)$. It vanishes as $t \to \infty$ if $G(t)$ vanishes suitably, and satisfies the Mandelstam representation with double spectral function boundaries along $s = s_1$ and $t = t_0$. Remembering the signature of the poles, the complete contribution will be

$$A^t_j(s, t) = R^{s_1}_j(s, t) + (-1)^{I_t} R^{u_1}_j(u, t) \tag{6.6}$$

where the second term has its double spectral function bounded by $t = t_0$, and $u = u_1$. Including a set of such poles for each channel gives us a Mandelstam representation as shown in Fig. (VI. 4). It is convenient, but not essential, to take $s_1 = t_1 = u_1$, so that there is no overlap of the various double spectral functions.

The obvious defect of such an approximation is that we have neglected the "corners" of the double spectral functions, such as $s < s_1, t < t_1$, and also the interior, $s > s_1, t > t_1$. Despite this the dominant features of scattering amplitudes which are found in experiments would seem to have been included fairly well. These are the resonance poles in each channel together with the lower thresholds, and, at higher energies

$(s > s_1)$, in the regions adjacent to the forward and backward directions ($t = 0$ and $u = 0$ for the s-channel), the peaks controlled by the crossed-channel poles. Any representation which gets these two features right should be a good approximation to the experimental amplitude, and

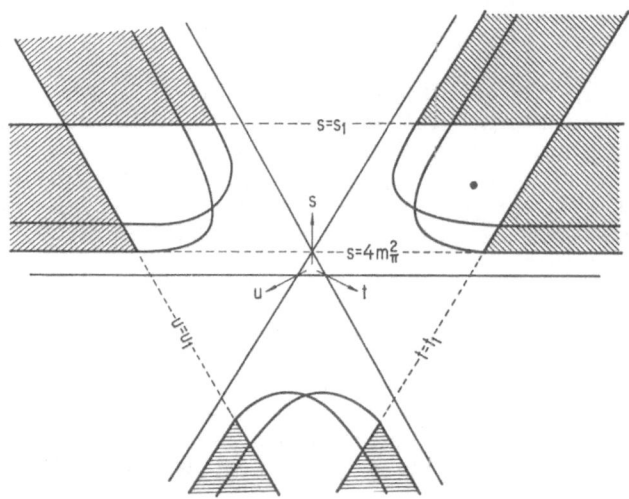

Fig. VI.4. The Mandelstam representation for the new form of the strip approximation, showing the strips (shaded), and the correct double spectral function boundaries. The region enclosing ● is the s-channel elastic double spectral function.

this is the justification of the strip approximation. But of course we can not be sure that the approximation will be valid in the nearby unphysical regions, where we have assumed special forms for the double spectral functions, and it is essential to the calculations that this should be the case.

Having obtained an amplitude which by construction satisfies maximal analyticity of the first and second kinds, performing the bootstrap calculation consists simply of imposing unitarity, and again we try to achieve this with the N/D equations. The first step is to construct amplitudes of definite signature in the s-channel from $A (s, t)$. We know that the double spectral function corresponding to (6.5) is [from (III.3.4)]

$$\varrho_{st} (s, t) = \varDelta_t \left\{ \frac{1}{2} G (t) (- q_t^2)^{\alpha (t)} P_{\alpha (t)} \left(-1 - \frac{s}{2 q_t^2} \right) \right\} \vartheta (s - s_1) \, \vartheta (t - t_0) \quad (6.7)$$

and putting such expressions into (II. 2.6), we find after some calculation that

$$A^\pm (s, t) = A^{s \pm} (s, t) + V^{s \pm} (s, t) \qquad (6.8)$$

where

$$A^{s \pm} (s, t) = \sum_i R_i^{t_1} (s, t) \pm (- 1)^{I_i} R_i^{u_1} (s, t) \qquad (6.9)$$

and

$$V^{s\pm}(s,t) = \sum_j \beta(I_s, I_j) \, G_j(t) \, (-q_t^2)^{\alpha_j(t)} \times$$

$$\times \left\{ \frac{1}{\pi} \int\limits_{-4q_t^2}^{s_1} du' \, P_{\alpha_j(t)}\left(-1 - \frac{u'}{2q_t^2}\right) \left[\frac{1}{u'-s} + \frac{(-1)^{I_j}}{u'-u}\right] + \right.$$

$$+ \left. \frac{(-1)^{I_j} P_{\alpha_j(t)}(-1-s/2q_t^2) + P_{\alpha_j(t)}(1+s/2q_t^2)}{\sin \pi \alpha_j(t)} \right\}$$

$$+ (-1)^{I_j}\beta(I_s, I_j)\frac{1}{\pi} \int\limits_{s_1}^{\infty} du' \, G_j(t')\,(-q_{t'}^2)^{\alpha_j(t')} \times$$

$$\times P_{\alpha_j(t')}\left(-1 - \frac{u'}{2q_{t'}^2}\right) \left[\frac{1}{u'-u} + \frac{(-1)^{I_j}}{u'-t}\right] \tag{6.10}$$

with $s + t + u = s + t' + u' = 4\,m_\pi^2$.

In writing this we have remembered to include the isospin crossing matrix for $\pi - \pi$ scattering, and have been able simply to add the t- and u-channel contributions of each pole, which by crossing symmetry must occur in both channels.

The partial wave projection of $V^{s\pm}(s,t)$, which we will denote by $B_l^V(s)$, can be obtained from (2.51). We can see that it will contain not only the expected left-hand cut, but also the right-hand cuts of the amplitude for $s > s_1$. The right-hand cut for $s_0 < s < s_1$ is contained in $A_l^{s\pm}(s)$, the partial wave projection of $A^{s\pm}(s,t)$. However, $A^{s\pm}(s,t)$ will also contribute to the left hand cut for $s < -t_1 + 4\,m_\pi^2$ since it contributes to $D_t^\pm(s,t)$ for $t > t_1$. In fact, using (III. 3.5),

$$D_t^{s\pm}(s,t) = \sum_i G_i(s)\,(-q_s^2)^{\alpha_i(s)} P_{\alpha_i(s)}\left(-1 - \frac{t}{2q_s^2}\right), \tag{6.11}$$

and from (2.46) we get

$$B_l^{SL}(s) = \frac{1}{16\pi^2} \int\limits_{-\infty}^{s_L^1} ds' \int\limits_{t_1}^{-4q_s'^2} dt'' \frac{D_t^{s\pm}(s',t'')\,P_l(-1-t''/2q_s'^2)}{(s'-s)\,4\,(-q_s'^2)^{l+1}}, \tag{6.12}$$

where

$$s_L^1 = -t_1 + 4\,m_\pi^2.$$

So the function

$$B_l^L(s) \equiv B_l^{SL}(s) + B_l^V(s) \tag{6.13}$$

contains the complete left-hand cut of $B_l^\pm(s)$, at least in the strip approximation, and also the right-hand cut for $s > s_1$.

So we want to impose unitarity in the region $s_0 < s < s_1$. We can, in principle, include as many two body channels as we like, but for the moment let us stick to just one, the $\pi - \pi$ channel, so that we have elastic unitarity. As before we put

$$B_l^\pm(s) = \frac{N_l(s)}{D_l(s)} \tag{6.14}$$

but now $N_l(s)$ has all the cuts of $B_l^L(s)$, including the right-hand cut for $s > s_1$, and $D_l(s)$ has just the unitarity cut for $s_0 < s < s_1$. So corresponding to (2.12) we have

$$\text{Im} \{D_l(s)\} = - \varrho_l(s) \, N_l(s) \,, \qquad s_0 < s < s_1 \,, \tag{6.15}$$

and if maximal analyticity of the second kind is valid, and only the one channel is important, it can be shown that there will still be no CDD poles [*Jones* (1965)], and we can write the dispersion relation [c. f. (2.24)]

$$D_l(s) = 1 - \frac{1}{\pi} \int\limits_{s_0}^{s_1} \frac{\varrho_l(s') \, N_l(s')}{s' - s} \, \mathrm{d}s' \,. \tag{6.16}$$

Corresponding to (2.28), the function

$$C_l(s) \equiv N_l(s) - B_l^L(s) \, D_l(s) \tag{6.17}$$

has only a right cut for $s_0 < s < s_1$, where

$$\text{Im} \{C_l(s)\} = - B_l^L(s) \, \text{Im} \{D_l(s)\} \,, \tag{6.18}$$

and we obtain

$$N_l(s) = B_l^L(s) + \frac{1}{\pi} \int\limits_{s_0}^{s_1} \frac{B_l^L(s') - B_l^L(s)}{s' - s} \, \varrho_l(s') \, N_l(s') \, \mathrm{d}s' \,, \tag{6.19}$$

which is just the same as (2.33) apart from the finite range of integration, and indeed looks the same as (5.7). It has the advantage over (2.33) that there is no problem about the square integrability of the kernel, but, unlike (5.7), the equations (6.16) and (6.19) are exact (given our assumptions) and s_1 is not a cut-off, but represents the point at which Regge behaviour takes over from elastic unitarity. The approximation is to represent the region in which this will occur by a sharp transition. There is, however, a price to be paid for this advantage, which is that $B_l^L(s)$ has, from the first term of (6.10), a logarithmic singularity at s_1. This singularity at the end point of integration means that (6.19) is not a Fredholm equation. It arises because we have to match the phase shifts obtained from elastic unitarity, below s_1, with those given by the asymptotic behaviour for $s > s_1$. Below s_1 we have

$$\sin^2 \delta_l(s) = \varrho_l(s) \, \text{Im} \{B_l^\pm(s)\} \,, \tag{6.20}$$

and at s_1 we are forcing the phase shift to

$$\sin^2 \delta_l(s_1) = \varrho_l(s_1) \, \text{Im} \{B_l^Y(s_1)\} \,, \tag{6.21}$$

with $\text{Im} \{B_l^Y(s_1)\}$ given by

$$\text{Im} \{B_l^Y(s_1)\} = \sum_j \frac{\beta(I_s, I_j)}{16\pi} \int\limits_{-\infty}^{0} \frac{\mathrm{d}t}{q_{s_1}^{2l+2}} \text{Im} \left\{ Q_l \left(1 + \frac{t}{2 q_{s_1}^2}\right)\right\} \times$$
$$\times G_j(t) \, (-q_t^2)^{\alpha_j(t)} \, P_{\alpha_j(t)} \left(-1 - \frac{s_1}{2 q_t^2}\right) \tag{6.22}$$

from (6.10). Clearly there can be no solution unless

$$\varrho_l(s_1) \, \mathrm{Im}\, \{B_l^V(s_1)\} \lesssim 1 \,, \tag{6.23}$$

a restriction imposed by s-channel unitarity on the t-channel pole contributions. If this condition is satisfied it can be shown [*Chew* (1963b), *Teplitz* (1965)] that despite the singularity the integral equation can be solved by making a suitable transformation to remove the singular part, and that then both $N_l(s)$ and $D_l(s)$ are singular at s_1, in that

$$D_l(s) \quad \text{and} \quad N_l(s) \sim \frac{1}{(s_1 - s)^{\delta_l(s_1)/\pi}} \tag{6.24}$$

but in the ratio, N/D, this singularity cancels, and is not present in $B_l^\pm(s)$. It turns out that (6.19) can be solved numerically by the usual matrix inversion technique, despite the singularity [*Jones* and *Tiktopoulos* (1966)].

The equations have been solved with just a ϱ trajectory in each channel [*Collins* and *Teplitz* (1965)]. The input Regge functions, $\alpha(t)$, $\gamma(t)$ are parametrized in a suitable way, and $B_l^\pm(s)$ calculated from (6.10), (6.12) and (6.13). Note that in these equations we are only integrating over negative t, so we only require $\alpha(t)$ and $\gamma(t)$ for $t < 0$, where they are real (assuming no crossing of trajectories). We then solve (6.19) and (6.16), and find the output $\alpha(s)$ and $\gamma(s)$ from (2.37) and (2.42). The parameters are adjusted until input and output agree. The results are not especially sensitive to the choice of s_1, because reducing s_1 increases the potential in (6.10), and decreases the range of integration in (6.16) and (6.19). To some extent these compensate. The effect of making s_1 very large is to flatten the trajectories, thus giving large output widths, but also to make the trajectories more tightly bound, because of the increased range of integration in (6.16). [$N_l(s)$ is positive so increasing s_1 lowers the value of s at which the zero of $D_l(s)$ occurs].

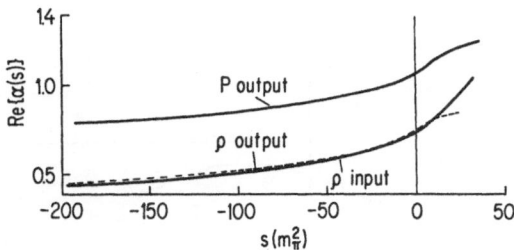

Fig. VI.5. The self-consistent ϱ trajectory with $s_1 = 200$. A P trajectory is also generated. (From *Collins* and *Teplitz*, 1965).

With $s_1 = 200 \, m_\pi^2$, the ϱ trajectory

$$\alpha_\varrho(t) = 0.33 + 0.42/(1 - t/75) \tag{6.25}$$

was found to be self-consistent [see Fig. (VI. 5)] but the output trajectory did not pass through $l = 1$ to generate the ϱ particle. [Remember only

$\alpha(t)$ and $\gamma(t)$ for $t < 0$ can be made self-consistent, since $B_l^{\Gamma}(s)$ is independent of $t > 0$.] The fact that $\alpha(t)$ turns over just above threshold means, given the dispersion relation (III. 1.30), that Im $\{\alpha(t)\}$ must be large above threshold.

The ρ exchange force also gives rise to a P trajectory, as we explained in the previous section, so it is inconsistent not to include both ρ and P in our calculations. If the P is included we have to cope with the long range repulsion which it produces. At first sight it seems that this repulsion must be incorrect, or at least that it must be cancelled by some other contribution to $I = 0$ exchange. For, since we have taken the double spectral functions to be zero outside the strips of Fig. (VI. 4) we can expand $D_i^{\pm}(s, t)$ in a convergent partial wave series for $s_0 < s < s_1$ and, if we use the Froissart-Gribov projection rather than the Wong form, we find

$$B_l^{V}(s) = \frac{\beta(I_s, 0)}{16\pi} \frac{1}{\pi} \int_{t_0}^{t_1} \frac{dt}{2q_s^{2l+2}} Q_l \left(1 + \frac{t}{2q_s^2}\right) \times$$

$$\times \sum_{l_t} (2l_t + 1) \, \text{Im} \, \{A_{l_t}(t)\} \, P_{l_t} \left(1 + \frac{s}{2q_t^2}\right). \tag{6.26}$$

Since, with elastic unitarity in the t channel strip, Im $\{A_{l_t}(t)\}$ must be positive, it is evident that $B_l^{V}(s)$ be positive at least for $s \approx 0$. As this is not the case for P exchange, if the strip approximation is to be correct there must be some contribution, apart from P (or P′ which is similar), which is positive. This might be provided by trajectories which do not

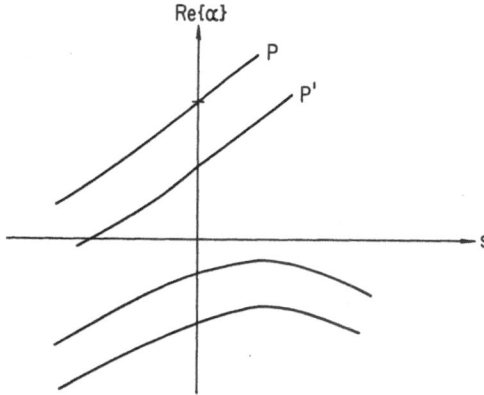

Fig. VI.6. The $I = 0$ trajectories, including two hypothetical trajectories which do not reach the right-half l plane, but produce an attractive force.

reach the right-half l plane, like those shown in Fig. (VI. 6). Since such trajectories would not be manifest, either physically, or in the output of this type of calculation, there would be no point in including them individually, but *Chew* and *Teplitz* (1965) showed how to represent their effect by "normalizing" the potential. This consists of simply subtracting from the P potential, $V^P(s, t)$, the part $V^P(0, t)$, and then adding back $V^P(0, t)$ partial wave by partial wave. So

$$B_l^V(s) = \frac{1}{16\pi} \frac{1}{\pi} \int\limits_{-\infty}^{0} \frac{dt}{2q_s^{2l+2}} \operatorname{Im}\left\{Q_l\left(1 + \frac{t}{2q_s^2}\right)\right\} [V^P(s, t) - V^P(0, t)] +$$

$$+ \frac{\beta(I_s, 0)}{16\pi} \frac{1}{\pi} \int\limits_{t_0}^{t_1} \frac{dt}{2q_s^{2l+2}} Q_l\left(1 + \frac{t}{2q_s^2}\right) \sum_{l_t} (2l_t + 1) \operatorname{Im}\{A_{l_t}(t)\}.$$

$$(6.27)$$

$\operatorname{Im}\{A_{l_t}(t)\}$ can be made consistent in the s, t, and u-channels, and this ensures the positivity of $B_l^V(s)$. This procedure was used [*Collins* (1966)] to obtain self-consistent ϱ and P trajectories, but the results were not remarkably different from those with just ρ exchange, because the normalized P contribution is very small. There was no sign of a secondary P' trajectory in the output. A further problem arises in that because the P makes a large contribution to the amplitude at large s, there is often difficulty in satisfying the unitarity bound (6.23) for the lower values of l. This is because the self-consistent ρ and P both have large residues.

Further work has been done on coupling in the $K\bar{K}$ channel, and including K* and ϕ exchange, but the results were very similar to the single channel case [*Bali* and *Chiu* (1967)]. The general conclusion is that trajectories corresponding to the physical ρ and P do not bootstrap themselves in the new form of the strip approximation, and that though self-consistent trajectories can be obtained they are defective in having small slopes, not rising to high enough l to produce physical particles, having large residues which correspond to particles of excessively large width, and, in the case of the P, requiring the rather arbitrary "normalization".

In fact normalizing may be quite the wrong thing to do, because the argument given above, that the $I = 0$ exchange force should be positive, assumes that elastic unitarity holds for $t_0 < t < t_1$, whereas in fact it will hold only for $t_0 < t < t_I$, where $t_I = 16\, m_\pi^2$ for $\pi - \pi$ scattering. The double spectral functions are really non zero within the boundaries shown in Fig. (VI. 4), and these give the true limit to the domain of convergence of the partial-wave series for $D_t^\pm(s, t)$ used in (6.26). The region indicated with a dot in Fig. (VI. 4) contains the elastic s-channel double spectral function which, by definition, should not form part of the potential, but which does contribute to $D_t^\pm(s, t)$ where it represents inelastic t-channel processes. There is thus no reason why the signs of $D_t^\pm(s, t)$ and $B_l^V(s)$ should be the same [and, with strong coupling, they are not the same in potential scattering, *Collins* (1965)].

If, on the other hand, we accept the repulsion at its face value, it is found that the combination of attractive ϱ and repulsive P trajectories results in nonsensical solutions of the N/D eqations, in that we get resonance poles on the physical sheet [see *Collins* (1966)]. This is also true in potential scattering where it is found that a combination of attractive and repulsive potentials, taken in the first Born approximation, and solved by the N/D method, gives similarly meaningless results. Since the full solution of the potential problem guarantees that all resonances are on unphysical sheets, this can only stem from the inadequacy of the first Born approximation to the left-hand cut. The Born series is essentially a series in the coupling strength, and it is evident that for a repulsion the terms of the series will alternate in sign, and that too much repulsion will be obtained from the first Born approximation.

   Chew (1965c) has argued that the Pomeranchon, representing as it does the many coupled channels open to the $\pi - \pi$ system at high energies, should produce a narrowing of the output widths, since a resonance which is strongly coupled to such channels spends less time in its decay channel. In terms of N/D equations the long range P repulsion should reduce the N function near threshold, and hence the width of low energy resonances, without altering much the position of the zero of the D function, which depends on the shorter range ρ force. Thus it is essential that we should be able to include the P repulsion properly, and we need some equivalent in relativistic scattering to the Born series in non-relativistic scattering. This is provided by the Mandelstam iteration.

VI.7. The Mandelstam Iteration

   The "old" form of the strip approximation proposed by *Chew* and *Frautschi* (1961a) made use of the equation for calculating the double spectral function which we derived in section (I. 12), that is (I. 12.11)

$$\varrho^{\text{els}}\,(s,\,t) = \frac{1}{16\pi^2 q_s \sqrt{s}}\,\int\limits_{t_0}^{K=0}\!\!\int\,dt_1\,dt_2\,\frac{D_t^{\pm}\,(s_+,\,t_1)\,D_t^{\pm}\,(s_-,\,t_2)}{K^{1/2}\,(t,\,t_1,\,t_2,\,s)} \qquad (7.1)$$

[Note that with amplitudes of definite signature there is no D_u, so (I. 12.12) does not apply.]

   If we make a division of the amplitude similar to (6.8), so that

$$A^{\pm}\,(s,\,t) = A^{s\pm}\,(s,\,t) + V^{s\pm}\,(s,\,t) \qquad (7.2)$$

where

$$A^{s\pm}\,(s,\,t) = \frac{1}{\pi^2}\,\int\!\!\int^{\infty}\,\frac{\varrho^{\text{els}}\,(s',\,t'')}{(s'-s)\,(t''-t)}\,ds'\,dt'' \qquad (7.3)$$

and $V^{\pm s}\,(s,\,t)$ contains the rest of the amplitude, then we can write [from (II. 2.8)]

$$D_t^{\pm}\,(s,\,t) = \frac{1}{\pi}\,\int^{\infty}\,\frac{\varrho^{\text{els}}\,(s',\,t)}{s'-s}\,ds' + D_t^{V}\,(s,\,t)\,, \qquad (7.4)$$

where $D_t^{V}\,(s,\,t)$ is the t-discontinuity of $V^{s\pm}\,(s,\,t)$. Equations (7.1) and (7.4) give us an iterative procedure for calculating $\varrho^{\text{els}}\,(s,\,t)$. Assuming that $D_t^V(s,t)$ is known we can calculate $\varrho^{els}(s,t)$ from (7.1), then substitute this in (7.4) to give a new value of $D_{\bar{t}}^{\pm}(s,t)$, and so on. In fact because of the form of K, and the finite range of integration in (7.1), to find $\varrho^{els}(s,t)$ for a given value of t (say t_x) we only need $D_{\bar{t}}^{\pm}(s,t)$ for values of t less than t_x, and (7.4) only uses $t < t_x$. Thus for $\pi - \pi$ scattering, a knowledge of $D_{\bar{t}}^{\pm}(s,t)$ for $t_0 < t < 16\,m_\pi^2$, where $\varrho^{els}(s,t)$ does not exist [see Fig. (VI. 4)], enables us to find $\varrho^{els}(s,t)$ for $t < 64\,m_\pi^2$. Using this in (7.4) we are able to calculate $D_{\bar{t}}^{\pm}(s,t)$ up to $64\,m_\pi^2$, which with (7.1) allows us to find $\varrho^{els}(s,t)$ out to $t = 256\,m_\pi^2$, and so on. This is the Mandelstam iteration [*Mandelstam* (1958), (1959a), (1959b)].

The difficulty with such calculations is that in general $A^{s\,\pm}(s,t)$ given by (7.3) will not become small with increasing s. In fact if we put a t-channel Regge pole $\alpha(t)$ into $D_t^V(s,t)$ we have

$$D_t^V(s,t) \sim s^{\alpha(t_1)}$$

for some $t = t_1$, and in (7.1) this gives

$$\varrho^{els}(s,t) \sim s^{2\alpha(t_1)-1}$$

and, after n iterations, ϱ will behave (for large enough t) like

$$\varrho^{els}(s,t) \sim s^{2^n\alpha(t_1)-n}\,.$$

We have already noted this phenomenon in connection with the Amati-Fubini-Stanghellini cuts in section (V. 3), and we know that the problem arises because of the assumption that elastic unitarity holds for large s. There must be really some delicate cancellations, caused by multi-particle unitarity, which will remove this behaviour, and make $\varrho^{els}(s,t)$ vanish as $s \to \infty$, but these are hard to build into the strip approximation.

One way round this difficulty which has been used recently is simply to modify the unitarity relation, (7.1), to read

$$\varrho^{els}(s,t) = \frac{g(s)}{16\pi^2\,q_s\sqrt{s}} \int_{t_0} \int^{K=0} dt_1\,dt_2\,\frac{D_{\bar{t}}^{\pm}(s_+,t_1)\,D_{\bar{t}}^{\pm}(s_-,t_2)}{K^{1/2}(t,t_1,t_2,s)} \qquad (7.5)$$

where

$$g(s) = 1 \quad \text{for} \quad s < s_I\,,$$

and

$$g(s) \to 0 \quad \text{as} \quad s \to \infty\,.$$

It has been shown [*Mandelstam* (1963d), *Bali* et al. (1966)] that an iteration with such a cut-off will lead to a $\varrho^{els}(s,t)$ which is power bounded in t, so we can expect to get Regge asymptotic behaviour at high t in the s-channel strip region. In fact once we have calculated $D_{\bar{t}}^{\pm}(s,t)$, from (7.5) and (7.4), out to sufficiently large values of t we shall find

$$D_{\bar{t}}^{\pm}(s,t) \xrightarrow[t\to\infty]{} \overline{G}(s)\,t^{\alpha(s)} \qquad (7.6)$$

where, from (6.11),

$$\overline{G}(s) = \pi^{-1/2}\,G(s)\,\frac{\Gamma[\alpha(s)+1/2]}{\Gamma[\alpha(s)+1]}\,, \qquad (7.7)$$

so that
$$\log |D_i^{\pm}(s, t)| = \log |\bar{G}(s)| + \mathrm{Re}\,\{\alpha(s)\}\log t\,, \qquad (7.8)$$
and
$$\arg(D_i^{\pm}(s, t)) = \arg(\bar{G}(s)) + \mathrm{Im}\,\{\alpha(s)\}\log t\,. \qquad (7.9)$$

So a fit of $\log|D_i|$ and $\arg(D_i)$ to $\log t$ will enable us to find $\alpha(s)$ and $\gamma(s)$.

Such methods have been used by *Bransden* et al. (1963) (using a rather different cut-off scheme), and more recently by *Bali* (1966), to calculate the ϱ and P trajectories in $\pi - \pi$ scattering, using a potential produced by fixed spin ρ exchange (and an f in the case of the former group). The trajectories obtained were a good deal better than those obtained with the N/D method, and the first Born approximation to the left hand cut, in that large input widths were not required to get the output trajectories in the right place, and the output trajectories rose somewhat higher in l. An example from Bali's paper is shown in Fig. (VI. 7). The trajectories are still too flat however, and the output widths too large ($\sim 3\,m_\pi$ in Bali's calculation) though the latter might be improved if the P contribution were added. Considerable numerical accuracy is needed to obtain trajectories in this way, and it may turn out to be better to use the N/D equations, but employ the iterative method to calculate the corners of the double spectral functions [i.e.

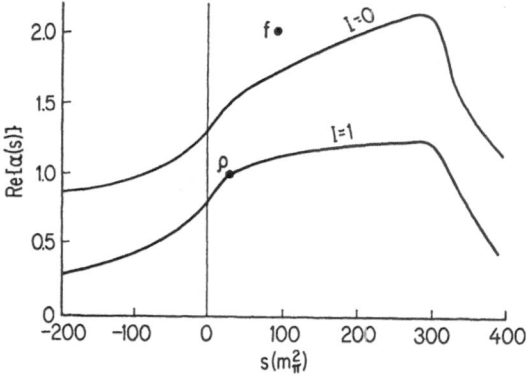

Fig. VI.7. The ϱ and P trajectories obtained from the exchange of a fixed spin ϱ of width $1.1\,m_\pi$, and with $s_1 = 400\,m_\pi^2$ (From *Bali*, 1966).

$s < s_1$ and $t < t_1$ in Fig. (VI. 4)], so that (say) the third or fourth Born approximation to Regge pole exchange is included in the left-hand cut. Because of crossing symmetry a knowledge of this part of the double spectral function would also give us the inelasticity in the s-channel, and the corresponding contribution to $B_i^{\pm}(s)$ could be fed into the equations by the Frye-Warnock method, and (3.34). But the work involved in carrying out such a programme would be very great, especially if several channels were included as well.

VI.8. Some Problems

There remain some quite formidable difficulties with the strip approximation quite apart from the failure in numerical detail of the calculations described above. Perhaps the most serious is the fact that, whereas the rate at which one would expect an amplitude to fall in t away from the forward peak, if it satisfied

$$A^{\pm}(s, t) = \frac{1}{\pi} \int_{t_0}^{\infty} \frac{D_t^{\pm}(s, t')}{t' - t} \, \mathrm{d}t' , \tag{8.1}$$

(at high s above the s-channel resonances, where we can expect the integral to converge) would be

$$A^{\pm}(s, t) \sim 1/t , \tag{8.2}$$

in fact, as we shall discuss in section (VIII. 6), the experiments show quite clearly that it generally falls much faster than this, and more like

$$A^{\pm}(s, t) \sim e^{bt} \quad \text{for} \quad 0 > t > -1 \ (\text{Gev}/c)^2 , \tag{8.3}$$

turning to

$$A^{\pm}(s, t) \sim e^{-a\sqrt{-t}} \quad \text{for} \quad t < -1 \ (\text{Gev}/c)^2 . \tag{8.4}$$

The latter is the fastest fall off permitted by analyticity and unitarity requirements [*Martin* (1965)]. Such a behaviour requires that an infinite number of moment conditions should be satisfied by $D_t^{\pm}(s, t)$ in (8.1). Some attempt has been made to incorporate such a behaviour into the strip approximation [*Collins* (1967)], but detailed calculations will require a much more sophisticated approach to the problem.

A further, and perhaps related, difficulty is that there is fairly strong experimental evidence [see section (VIII. 1)] that trajectories rise to quite large values of Re $\{\alpha\}$, perhaps even $\approx 19/2$ for baryons, with a particle mass of 3230 Mev. If $\alpha(t)$ obeys a dispersion relation such as (III. 1.30), it can only keep on rising as long as Im $\{\alpha(t)\}$ has not passed its maximum in t, and, since the double spectral function is proportional to Im $\{\alpha(t)\}$ [see (6.7)], this means that the double spectral function must be large out to $t > (3230)^2$ Mev^2, which is well outside the sort of strip width expected in our equations.

One way of keeping trajectories rising so high would be for the dynamics to be controlled by really high mass particles. One might speculate that the low mass particles which we observe are really bound states of two and three quark systems, and so, by the arguments of section (VI. 3), CDD poles in the amplitudes we are looking at [see for instance *Squires* and *Watson* (1967) and section (VIII. 9)]. Since, however this presumably puts the main weight of the double spectral function out at $t \approx 4M_Q^2$ (M_Q is the quark mass) it makes the sharp fall of of the forward and backward peaks very hard to understand. One would expect

$$A(s, t) \sim \frac{1}{t - 4M_Q^2} . \tag{8.5}$$

An alternative mechanism would be for the trajectories to swap from one channel to another with a higher threshold as the energy increases. If always $\alpha(t) \to -\infty$ as $s \to -\infty$, as *Mandelstam* (1966) has suggested, then the coupling to other channels must be important in determining the dynamics of trajectories [this is discussed further in section (VIII. 9)]. The asymptotic forms (2.43) and (2.45) are built into the single channel calculations, and a dynamical scheme involving only a very few channels could, at best, be valid only for a limited region of a trajectory.

We have described, in the last two sections, proposals for boot-strapping trajectories using the strip approximation. There have of course been many other proposals which we lack space to discuss in detail. An interesting method of using unitarity to obtain integral equations for the Regge parameters directly has been given by *Frautschi* et al. (1964), based on methods proposed by *Cheng* and *Sharp* (1963a), (1963b). Some results for potential scattering have been given by *Hankins* et al. (1965). More recently methods of determining Regge parameters which involve the construction of a unitarity S-matrix as a sum of Regge poles, followed by the imposition crossing symmetry, have been developed by *Abbe* et al. (1965), (1966), (1967).

Many further topics remain to be discussed. In particular there are the problems of the existence of the internal isospin symmetry, $SU(2)$, the partial symmetry including strangeness, $SU(3)$, which is broken in such a regular manner, and the conjectured $SU(6)$ dynamical symmetries. Much work has been done on all these aspects, not usually working out the dynamics in detail, but simply trying to see if such symmetries seem likely to be self-sustaining in light of the crossing matrices of the symmetry groups, and the observed multiplets [*Chan Hong-Mo* et al. (1964), *Cutkosky* (1963)]. But Regge poles feature hardly at all in these accounts, and we shall content ourselves with referring the reader again to the reviews by *Zachariasen* (1965) and *Udgaonkar* (1965), and to more recent work on $SU(6)$ by *Capps* (1966) and *Carruthers* (1967), and references therein.

Despite the problems which we have found the bootstrap concept retains its appeal, partly because it is clear that, if our analyticity postulates are correct, something of the sort has got to work, and partly because so many qualitative results support it. It is evident that the simple dynamical models which have been tried so far will need considerable refinement and sophistication before their suggestive features can become established facts. But the main point, which has been particularly emphasized by *Chew* (1965a), is that because S-matrix theory is so closely related to the experimental numbers which it is intended to explain, it will continue to develop, providing that we go on incorporating into it as much of our experimental information as possible. We shall see in Chapter VIII that Regge poles summarize a tremendous amount of such information, and it will certainly be necessary to include them in any future dynamical scheme.

Chapter VII. Perturbation Theory and Elementary Particles

VII.1. The Relevance of the Perturbation Expansion

In the previous sections we have given reasons for believing that the high energy properties of scattering amplitudes are dominated by crossed channel Regge poles or Regge cuts. The arguments have been indirect in that we have not discussed any specific calculations which would actually produce such poles and cuts, but it is not, of course, possible to do exact calculations, and so one either has to make approximations (as in chapter VI), or to introduce "models" which have at least some of the features of the real situation. We have already referred to one such model, namely non-relativistic potential scattering, within which many of the ideas discussed in this book were first introduced. Here, as a somewhat more realistic model for particle physics, we shall discuss the perturbation expansion of a Lagrangian field theory. In a formal sense, i.e. ignoring questions of convergence, this satisfies the analyticity (of the first kind) and crossing postulates, and it is also relativistically invariant. We defer the question of whether it is compatible with maximal analyticity of the second kind until section (VI. 5).

To give a general introduction to Lagrangian field theory and its formal perturbation solution is outside the scope of this book, and (for this chapter only) we shall assume that the reader is familiar with the basic ideas, and in particular with the way in which the terms of the perturbation series can be represented as Feynman diagrams. We recall that in a Lagrangian field theory one starts with a set of field operators which correspond to a set of "elementary" particles. Only these field operators appear in the Lagrangrian, and all of the lines in the Feynman diagrams correspond to such elementary particles.

In general, as we shall see, individual Feynman diagrams do not have Regge behaviour for high t, in fact they behave like $t^N (\log t)^M$ where N and M are some *constants* (i.e. not a function of s), so it is interesting to see how Regge behaviour (i.e. $t^{\alpha(s)}$) can arise when one sums infinite classes of such terms. We shall see how the various J-plane singularities which have been mentioned previously, namely Regge poles, Regge cuts, and fixed (Gribov-Pomeranchuk) singularities, can arise. Of course at best the results can only be regarded as suggestive, even if we accept the validity of the model, since we shall be restricting ourselves to particular classes of diagrams, and making the assumption that the processes of summing an infinite sequence of terms, and of taking the large t limit, can be commuted. Nevertheless the results are useful as a supplement to the unitarity arguments used earlier, since an individual Feynman diagram generally includes part of the contribution of a large number of "unitarity diagrams", and implicitly includes some of the effects of many-particle unitarity. As we shall see, the fact that we are interested only in the high t limit simplifies considerably the contributions of the individual Feynman diagrams, and enables many formal summations to be carried out.

Discussions of the high energy behaviour of perturbation theory have been given by *Polkinghorne* (1963a, b, c), *Halliday* (1963), *Halliday* and *Polkinghorne* (1963), *Federbush* and *Grisaru* (1963a), and *Tikto-poulos* (1963a, b). An early review is given by *Oehme* (1964), and a later and very comprehensive one by *Eden* et al. (1966).

In our discussion we shall confine our comments to Φ^3 type inter-actions, i.e. those in which only three particles meet at a vertex. Some consideration of the Φ^4 interaction is given in the above references. In general rather anomalous behaviour occurs, as might be expected, since a Φ^4 interaction in effect corresponds to a potential which is singular at the origin.

VII.2. High Energy Behaviour of Feynman Diagrams

The contribution to the amplitude $A\,(s,\,t)$ of any Feynman diagram involving only zero spin particles can be written in the form

$$I = \lim_{\varepsilon \to 0^+} \int \frac{d^4 k_1 \ldots d^4 k_l}{\prod\limits_{r=1}^{n} (q_r^2 - m^2 + i\varepsilon)} \tag{2.1}$$

where the q_r are the 4-momenta of the internal lines, the k_i are a set of independent loop momenta (so that all the q_r are linear combinations of the k_i and the external momenta), and we have, for simplicity, chosen all masses to be equal. It is convenient to use the Feynman identity

$$\frac{1}{f_1 f_2 \ldots f_N} = (N-1)! \int_0^1 \frac{\prod d\alpha_i\, \delta(\Sigma \alpha_i - 1)}{[\Sigma \alpha_i f_i]^N} \tag{2.2}$$

to write (2.1) as

$$I = (n-1)! \int_0^1 \prod d\alpha_i\, \delta\,(\Sigma\,\alpha_i - 1) \int \frac{\prod d^4 k_j}{[\Sigma\alpha(q_r^2 - m^2)]} \tag{2.3}$$

where here, and in future, we do not bother to write the $(i\,\varepsilon)$ terms. The k integrals can now be done explicitly [*Chisholm* (1952)], and I written as an integral over the α's, the integrand being a function of the α's and of the scalar invariants which can be constructed from the external four-momenta. In particular, for the four-point function (i.e. the two-particle to two-particle scattering amplitude), we can write the result in the form

$$I = \int_0^1 \frac{\prod d\alpha_i\, \delta(\Sigma \alpha_i - 1)\, N\,(\alpha)}{[g\,(\alpha)\,t + d\,(s,\,\alpha)]^m}\,, \tag{2.4}$$

where

$$m = n - 2l > 0\,. \tag{2.5}$$

Now the integrand of (2.4) has the behaviour t^{-m} for large t, unless $g(\alpha) = 0$. Hence, if in the region of integration of the α's $g(\alpha) \neq 0$, the integral itself will have this form. However, it is possible to obtain contributions which tend to zero less fast with increasing t, from regions of the α integration near where $g(\alpha) = 0$. It is clear that in order to obtain such a dominant contribution at high t it must not be possible to distort the α-integration contour away from the point where $g(\alpha) = 0$. This may be the case for one of two reasons:

1. If $g(\alpha) = 0$ at the boundary of the integration region of the α's.

2. If $g(\alpha) = 0$ at an interior point of the α-integration, and this point, for $t \to \infty$, is "pinched" by singularities of the integrand, so that the contour cannot be distorted away from it.

Dominant asymptotic contributions arising from (1) are known as "end-point" contributions, and from (2) as "pinch" contributions. End-point contributions will occur on all sheets of the amplitude, whereas it is a characteristic of pinches that they occur on some sheets and not on others. Thus on one sheet of the amplitude two singularities may approach the point $g(\alpha) = 0$ from opposite sides of the contour; on another sheet they may approach from the same side, in which case they do not prevent distortion of the contour away from the point.

As might be expected, the end-point contributions are the easiest to treat, and it is therefore useful to define a particular class of Feynman diagrams which have the property that they do not have any pinch contributions on the physical sheet. To this end we define "planar" diagrams. A planar diagram is a Feynman diagram which can be drawn in a plane without any of the internal or external lines crossing each other, and with the external lines occurring in the sequence p_1, p_2, p_4, p_3, in

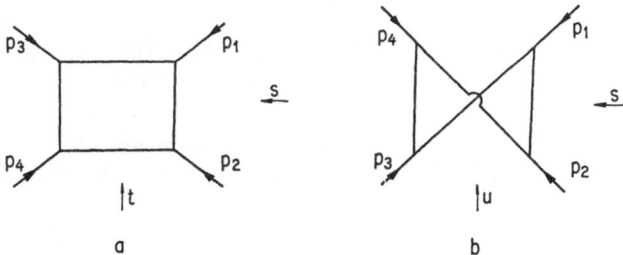

Fig. VII.1. A Feynman diagram which is planar if we consider fixed s and high t, but "non-planar" for fixed s, and high u.

clockwise order around the diagram [see Fig. (VII. 1a)]. Because of the latter restriction, a diagram which is planar when we are considering high $t = (p_1 + p_3)^2$, at fixed $s = (p_1 + p_2)^2$ (as we are here) might become non-planar if we required to consider high u at fixed s, say, and vice-versa [Fig. (VII. 1) gives a simple example of this].

For a planar diagram it can be shown [see *Eden* et al. (1966)] that $g(\alpha)$ in (2.4) is a sum of products of the α_i, each term having a positive

12*

coefficient. If follows that there are no zeros of $g(\alpha)$ for positive α's, and, since on the physical sheet no distortion of the α-contour is necessary, we can conclude that pinch contributions do not occur for planar diagrams.

VII.3. End-Point Contributions

We first consider the s-channel ladder diagrams shown in Fig. (VII.2). These diagrams are analogous to non-relativistic potential scattering by a Yukawa potential [*Blankenbecler* et al. (1960), see also e. g. *Omnes* and

Fig. VII.2. The s-channel ladder diagrams.

Froissart (1963)], and might therefore be expected to yield similar results, i.e. to have a Regge-pole behaviour. That this is indeed the case has been shown by direct computation of the partial-wave amplitude in the s-channel using the Bethe-Salpeter equation [*Lee* and *Sawyer* (1962)]. Here we shall see how this behaviour arises from a consideration of the high energy behaviour of Feynman diagrams.

If we denote the Feynman parameters (all previously called α) by α_i for the rungs of the ladder, and by β_i for the sides, then (2.4) takes the form, for the $(2n)^{\text{th}}$ order diagram,

$$I_n = g^2 \left(\frac{-g^2}{16\pi}\right)^{n-1} (n-1)! \int \frac{\Pi \, d\alpha_i \, \Pi \, d\beta_j \, \delta(\Sigma \alpha_i + \Sigma \beta_j - 1) \, [C(\alpha, \beta)]^{n-2}}{[(\Pi \alpha_i)t + d(\alpha, \beta, s)]^n}.$$

$$(3.1)$$

Clearly the coefficient of t is zero if any of the α_i are at the end-point, $\alpha_i = 0$; and the dominant contribution arises from the neighbourhood of the point where all α_i are near zero. Thus

$$I_n \sim g^2 \left(\frac{-g^2}{16\pi}\right)^{n-1} (n-1)! \int_0^\varepsilon \Pi \, d\alpha_i \int_0^1 \frac{\Pi \, d\beta_i \, \delta(\Sigma \beta_i - 1) \, [C(0, \beta)]^{n-2}}{[(\Pi \alpha_i) \cdot t + d(0, \beta, s)]^n}. \quad (3.2)$$

The α-integrations can now be carried out explicitly [*Polkinghorne* (1963a)] provided we keep only the leading term for large t, and we obtain

$$I_n \sim \frac{g^2}{t} \frac{(\log t)^{n-1}}{(n-1)!} \left[\left(\frac{-g^2}{16\pi^2}\right)^{n-1} (n-2)! \int_0^1 \frac{\Pi \, d\beta_i \, \delta(\Sigma \beta_i - 1) \, [C(0, \beta)]^{n-2}}{[d(0, \beta, s)]^{n-1}}\right].$$

$$(3.3)$$

The term in square brackets is exactly what would be obtained for the diagram in Fig. (VII. 3), with $(n-1)$ loops, except that the powers of d and C are reduced by one; in fact it corresponds to the contracted diagram of Fig. (VII. 3) evaluated with 2-dimensional vectors q (rather than the usual 4-dimensional vectors). From the form of Fig. (VII. 3),

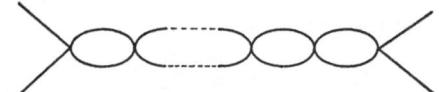

Fig. VII.3. The reduced diagram which gives the leading term of Fig. VII.2.

or by direct computation [*Polkinghorne* (1963a)], it follows that the square bracket has the form $(K(s))^{n-1}$, where $K(s)$ is the Feynman integral associated with a single loop (evaluated in 2-dimensions), so we have finally

$$I_n \sim \frac{g^2}{t} \frac{[K(s) \log t]^{n-1}}{(n-1)!}, \tag{3.4}$$

where

$$K(s) = -\frac{1}{16\pi^2} \int_0^1 \frac{d\beta_1 \, d\beta_2 \, \delta(\beta_1 + \beta_2 - 1)}{[\beta_1 \beta_2 s - (\beta_1 + \beta_2)^2]}. \tag{3.5}$$

We now sum the leading contributions of all ladders and assume that this gives the leading contribution of the sum of all ladders. Then

$$I_{\text{Ladders}} \sim \frac{g^2}{t} \sum_{n=1}^{\infty} \frac{[K(s) \log t]^{n-1}}{(n-1)!} \tag{3.6}$$

$$\sim g^2 \, t^{\alpha(s)} \tag{3.7}$$

where

$$\alpha(s) = -1 + K(s). \tag{3.8}$$

Thus we have obtained Regge behaviour, with a trajectory that tends to -1 as $|s| \to \infty$ or $g^2 \to 0$. We see that $\alpha(s)$ is a real analytic function of s with a right-hand cut starting at the physical threshold.

To illustrate the direct method of obtaining trajectories by summing perturbation theory diagrams we describe a method due to *Oehme* (1964), which gives the form of the trajectory in the weak coupling limit. The first diagram in Fig. (VII. 2) (i.e. the single particle exchange diagram) gives a contribution to the s-channel partial-wave amplitude [see (II. 4.4)]

$$A_l^{(1)}(s) = \frac{1}{16\pi} \frac{1}{2} \int_{-1}^{+1} P_l \left(1 - \frac{t}{2q_s^2}\right) \frac{g^2}{t - m^2} \frac{dt}{2q_s^2} \tag{3.9}$$

$$= \frac{1}{16\pi} \frac{g^2}{2q_s^2} Q_l \left(1 + \frac{m^2}{2q_s^2}\right). \tag{3.10}$$

In the neighbourhood of $l = -1$ this gives

$$A_l^{(1)} \simeq \frac{1}{16\pi} \frac{g^2}{2q_s^2} \frac{1}{(l+1)}. \tag{3.11}$$

We now seek a function $A_l(s)$ which tends to $A_l^{(1)}$ as $g^2 \sim 0$, and which satisfies the unitarity equation

$$\operatorname{Im} A_l(s) = \varrho(s)\, A_l(s)^* \, A_l(s)\ . \qquad (3.12)$$

Such a function is

$$A_l(s) = \frac{1}{16\pi} \frac{g^2}{2q_s^2} \left[\frac{-1}{\alpha(s)-l} + \frac{1}{l+1}\left\{1 + \frac{g^2\alpha_1(s)}{16\pi(l+1)} + \cdots\right\}\right], \quad (3.13)$$

where we have made the expansion

$$\alpha(s) = -1 + \frac{g^2}{16\pi}\,\alpha_1(s) + \cdots, \qquad (3.14)$$

and where

$$\operatorname{Im}\alpha(s) = \frac{1}{16\pi}\frac{g^2}{2q^2}\,\varrho(s). \qquad (3.15)$$

From the last equation we have

$$\alpha(s) = -1 + \frac{g^2}{16\pi^2} \int\limits_{4m^2}^{\infty} \frac{2\,ds'}{[s'(s'-4m^2)]^{1/2}(s'-s)} + \cdots, \qquad (3.16)$$

which is readily seen to be the same as (3.8) and (3.5).

We note from (3.16), or (3.8) and (3.5), that the expansion of $\alpha(s)$ in powers of the coupling constant is not valid near threshold. This should not surprise us since, as we saw in section (III. 1), $\alpha(s)$ is not analytic in s at threshold.

In the literature a variety of results are available for more complicated graphs, and special techniques have been used to calculate the leading t behaviour [e.g. *Federbush* and *Grisaru* (1963a, b), *Bjorken* and *Wu* (1963), *Polkinghorne* (1964)]. *Tiktopoulos* (1963a, b) and *Halliday* (1963) have shown that all planar graphs behave as $t^{-n} (\log t)^m$ where n and m are integers ($n \geq 1$, $m \geq 0$), and have given general rules for determining n and m. These rules are complicated, and a variety of "exceptional cases" have to be treated with great care, so we do not

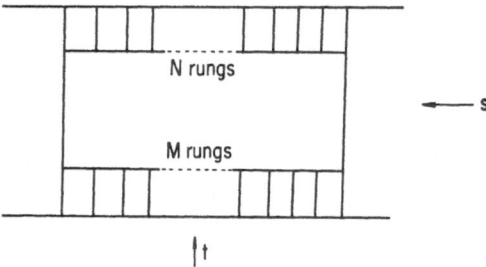

Fig. VII.4. A Feynman diagram which behaves as $t^{-2} \log t$ for large t, for all M, N.

quote them here. Their usefulness is limited by the fact one is usually interested in summing infinite sets of diagrams and for this purpose (see above) the coefficient of the leading term must be known. One important example where this is not necessary however is the diagram of Fig.(VII.4) for which the rules give an asymptotic behaviour proportional to $t^{-3} \log t$. Since this is independent of M and N, this behaviour will hold also when the contribution is summed over M and N (even if the ladders formally sum to give resonances). In particular, as we noted in chapter V, we do not obtain Regge cuts on the physical sheet from diagrams of this type.

It has been shown [see *Eden* et al. (1966) for details] that the Regge pole behaviour found above for ladder graphs also holds when a much wider class of diagram is included in the sum, the other diagrams giving higher order contributions to the residue and trajectory functions. In addition *Halliday* and *Polkinghorne* (1963) and *Polkinghorne* (1965) have considered production processes using these methods. They again find Regge-like behaviour for particular classes of diagrams. These authors also discuss the singularities of the functions $\alpha(s)$ and $\beta(s)$.

VII.4. Pinch Contributions

As might be expected, the analysis of pinch contributions is considerably more difficult than that of end-point contributions, and in fact no general rules for the form of the high t behaviour of non-planar diagrams have been given. However, it has been shown that particular classes of such diagrams sum to give cuts in the angular-momentum plane.

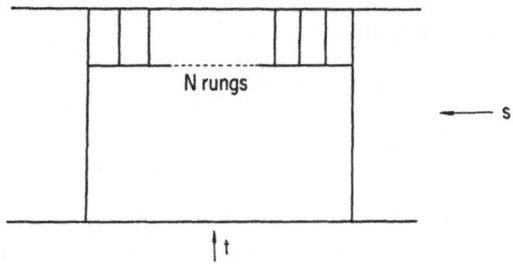

Fig. VII.5. A Feynman diagram with a pinch contribution on an unphysical sheet.

To illustrate this we consider first the diagram of Fig. (VII. 5). Since this is a planar diagram it has only end-point contributions and behaves like $t^{-2} \log t$ for large t. However we saw in section (V. 3) that if the sum over N yields a Regge trajectory, then the two-particle t-channel discontinuity of Fig. (VII. 5) summed over N contains a Regge cut contribution. Since, as we have seen, this cut is not present on the physical sheet, it must be present on the sheet reached by passing through the 2-particle t-channel unitarity cut. Here we want to see how this Regge-cut occurs in perturbation theory.

First we consider the lowest order example of Fig. (VII. 5), i.e. Fig. (VII. 6). For this diagram we have

$$g(\alpha) = \beta_1\beta_2\,(\alpha_1 + \alpha_2 + \delta + \gamma) + \alpha_1\alpha_2\,(\beta_1 + \beta_2 + \gamma + \delta_1) +$$
$$+ \beta_1\gamma\,\alpha_2 + \alpha_1\gamma\,\beta_2\,, \tag{4.1}$$

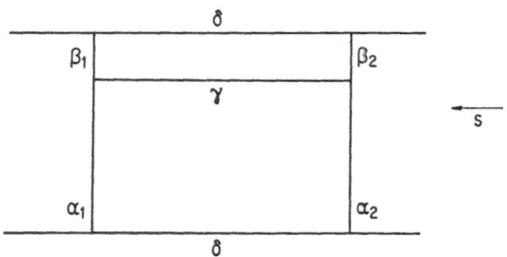

Fig. VII.6. The simplest example of Fig. VII.5.

which we write as

$$g(\alpha) = (\beta_1 + \beta_2 + \gamma + \delta_1)\left[\alpha_1 + \frac{\beta_1\beta_2 + \beta_1\gamma}{\beta_1 + \beta_2 + \gamma + \delta_1}\right]\left[\alpha_2 + \frac{\beta_1\beta_2 + \beta_2\gamma}{\beta_1 + \beta_2 + \gamma + \delta_1}\right] +$$
$$+ \beta_1\beta_2\,(\delta + \gamma) - \frac{\beta_1\beta_2\,(\beta_2 + \gamma)\,(\beta_1 + \gamma)}{(\beta_1 + \beta_2 + \gamma + \delta)} \tag{4.2}$$

We see that $g(\alpha)$ is equal to zero whenever β_1 or β_2 are zero (these are end points), and also either

$$\alpha_1 = -\frac{\beta_1\beta_2 + \beta_1\gamma}{\beta_1 + \beta_2 + \gamma + \delta_1} \tag{4.3}$$

or

$$\alpha_2 = -\frac{\beta_1\beta_2 + \beta_2\gamma}{\beta_1 + \beta_2 + \gamma + \delta_1}\,, \tag{4.4}$$

which are not end points. However, it is clear from the general form (2.4) that, whenever $g(\alpha)$ is equal to zero, there is a singularity in the limit $t \to \infty$. Clearly this cannot pinch the contour if t is on the physical sheet, since the Feynman parameters are then positive (this is an example of the general rule for planar diagrams); but the normal threshold corresponds to end point singularities in the α_1 and α_2 integrations, and so, on passing through the normal threshold to the second sheet, we distort the α contours. The result is that the singularities do pinch the contour on this sheet. To calculate the leading t behaviour we write the Feynman integral in the form

$$I = \text{const.} \int dx \int dy \int dz_1\,dz_2 \int_0^1 d\delta_1\,d\delta\,d\gamma \times$$
$$\times [(xy + z_1\,z_2)\,t + d\,(s_1\,\delta_1,\,\delta_1\,\gamma)]^{-3}\,, \tag{4.5}$$

where we have written

$$x = \alpha_1 + \frac{\beta_1 \beta_2 + \beta_1 \gamma}{\beta_1 + \beta_2 + \gamma + \delta_1} \tag{4.6}$$

$$y = \alpha_2 + \frac{\beta_1 \beta_2 + \beta_2 \gamma}{\beta_1 + \beta_2 + \gamma + \delta_1} \tag{4.7}$$

$$z_{1,2} = \beta_{1,2} \left[\delta + \gamma - \frac{\gamma^2}{\gamma + \delta_1} \right] \tag{4.8}$$

and have made use of the fact that the leading contribution will come from the neighbourhood of β_1 and β_2 equal to zero.

The x and y integrations in (4.5) can be performed explicitly, and yield a logarithmic function. If we choose the principal value of the logarithm, then we obtain the physical sheet value for which the integral behaves like t^{-3}. Other branches yield an extra term given by

$$I \sim \text{const.} \int_0^1 d\delta \, d\delta_1 \, d\gamma \int_0^{+ve} \frac{dz_1 \, dz_2}{t(z_1 z_2 t + d)^2} \tag{4.9}$$

$$\sim \text{const.} \, t^{-2} (\log t) \int_0^1 \frac{d\delta \, d\delta_1 \, d\gamma}{d} \,. \tag{4.10}$$

Clearly the final integral in (4.10) corresponds to the diagram Fig. (VII.6), with the lines corresponding to the Feynman parameters α_1, α_2, β_1, β_2 removed, i.e. to the diagram of Fig. (VII. 7). As before this diagram must be evaluated in *two dimensions* [because of the power of d in the denominator of (4.9)]. It is convenient to write the contribution of Fig. (VII. 7)

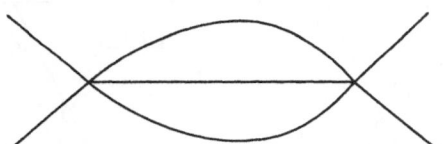

Fig. VII.7. The reduced diagram corresponding to Fig. VII.6.

as an integral over invariants [*Drummond* (1963)]. In this case this can easily be done, and we obtain

$$\int \frac{ds_1 \, ds_2 \, K(s_1)}{K^{1/2}(s, s_1, s_2, t) \, (s_2 - m^2)} \tag{4.11}$$

where $K(s)$ is the function defined above (3,5) and $K(s, s_1, s_2, t)$ is the function defined in (I. 12.7).

The above discussion can now be extended to the case where there is an arbitrary number of rungs in the ladder. The important differences are that in general we obtain a factor $z_1 z_2 \ldots z_N$ in the denominator of

(4.9), rather than $(z_1 z_2)$, which means that $\log t$ in (4.10) becomes $(\log t)^{N-1}$, and also that the coefficient corresponds to the Feynman diagram of Fig. (VII. 8) evaluated in two dimensions. It is clear that this coefficient, written in the form of (4.11), becomes

$$\int \frac{\mathrm{d}s_1 \, \mathrm{d}s_2 \, [K(s_1)]^{N-1}}{K^{1/2}(s, s_1, s_2, t) \, (s_2 - m^2)} \, . \tag{4.12}$$

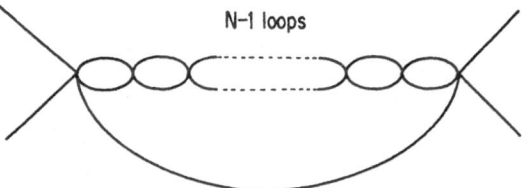

Fig. VII.8. A higher order example of Fig. VII.7.

The numerical factors work out such that, when we sum over N, we obtain for the asymptotic behaviour the form

$$\text{const.} \int \frac{\mathrm{d}s_1 \, \mathrm{d}s_2 \, t^{\alpha(s_1)-1}}{K^{1/2}(s, s_1, s_2, t) \, (s_2 - m^2)} \, , \tag{4.13}$$

where

$$\alpha(s) = 1 - K(s) \, . \tag{4.14}$$

This agrees exactly with the Amati-Fubini-Stanghellini cut contribution of section (V. 3) (remember of course that it is *not* on the physical sheet).

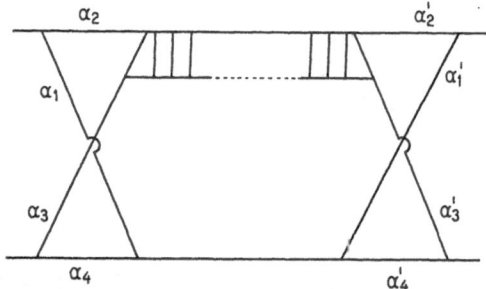

Fig. VII.9. A Feynman diagram which gives a physical sheet Regge cut when the ladder sum is carried out.

In order to obtain physical sheet cuts we must go to non-planar diagrams such as Fig. (VII. 9). The discussion of these diagrams is very similar to the above except that, for example, α_1 of Fig. (VII. 7) is now replaced by $(\alpha_1 \alpha_3 - \alpha_2 \alpha_4)$, which *can* be zero on the physical sheet i.e. for positive α's). The contracted diagram turns out to be exactly the same as in the previous case [namely Fig. (VII. 8)], so we now obtain the contribution of (4.13) even on the physical sheet.

Note that this discussion does not prove that the cut contributions are not cancelled by similar contributions coming from other sets of diagrams. However this is extremely unlikely, and, as we have seen, some cuts must be present if the Gribov-Pomeranchuk singularities are not to violate the Froissart-Hara bound [Section (V. 6)].

VII.5. Particles with Spin and the Reggeization of Elementary Particles

So far in our discussion of perturbation theory, we have taken all the particles to have zero spin. It is possible to generalize many of the results previously obtained, but the arguments are more complicated because the numerator function in the Feynman integral now contains powers of t. It is these powers of t which would cause trajectories to tend to limits above -1 for scattering of particles with spin, in the absence of superconvergence relations which cause cancellations.

The most interesting feature which arises from a study of perturbation theory for particles with spin is the suggestion that a genuine elementary particle might become Reggeized by the addition of extra Feynman diagrams. To unterstand this we must be clear what is meant by an "elementary particle" in this context. We have so far assumed that all the particles of strong interaction physics lie on Regge trajectories (this is part of the postulate of maximal analyticity of the second kind), which means that in principle they should be calculated from a knowledge of the double-spectral-functions. It is possible, however, that some particles might *not* lie on Regge trajectories but might correspond to CDD poles inserted in specific partial-wave amplitudes. These would not be contained in the double-spectral-functions. Clearly the presence of such a particle would not affect the analytically continued partial-wave amplitude (since this is uniquely defined by knowledge of the $A_J(s)$ for physical J greater than any arbitrary value), and so the continued amplitude would not be equal to the physical amplitude at the value of J corresponding to the spin of the elementary particle. Thus the amplitude is not analytic at that point, but contains a Kronecker delta contribution.

In Lagrangian field theory, as we noted in Section (VII. 1), the "elementary particles" are those which appear in the Lagrangian as field operators, and consequently as lines in Feynman diagrams. Thus if a particle A (say) were elementary, and if it coupled to two other elementary particles, B and C, then the perturbation expansion would include the diagram shown in Fig. (VII. 10), which would contribute only to the partial-wave with $J = \sigma_A$, σ_A being the spin of the particle. If, on the other hand, A was not elementary, but a composite of B and C etc., then this Feynman diagram would not exist, and the pole corresponding to particle A would arise from the sum of BC $\to$ BC scattering diagrams continued to the position of the A particle pole.

Returning to the case of the elementary particle, the diagram Fig. (VII. 10) clearly gives a non-analytic contribution to the partial-wave amplitude of the form $\delta_{J\sigma_A}$. In general, therefore, the existence of such a

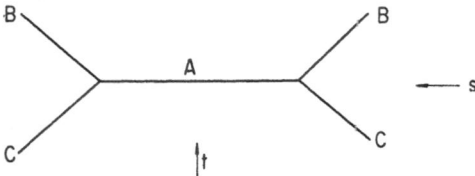

Fig. VII.10. A Feynman diagram which gives a Kronecker-delta singularity in the s-channel partial-wave amplitude.

particle violates the condition of maximal analyticity of the second kind. Since the existence of at least *some* elementary field is essential to a Lagrangian field theory, it would appear that such a theory must inevitably be in conflict with maximal analyticity, and could be shown to be inconsistent with experiment if amplitudes do in fact have Regge asymptotic behaviour with all the particles lying on Regge trajectories [see Chapter VIII].

The suggestion of *Gell-Mann* and *Goldberger* (1963) was that perhaps a sum of other diagrams might give a Kronecker term which would exactly cancel the above $\delta_{J\sigma_A}$, and replace it by a Regge trajectory. If this happened, the distinction between "elementary" and "non-elementary" particles, based on the asymptotic behaviour of amplitudes, would not be possible.

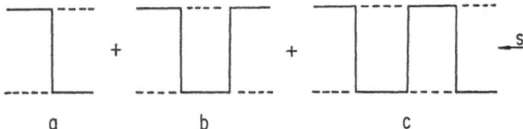

Fig. VII.11. Ladder diagram in vector-spinor scattering.

To see how this might work in practice we consider [*Gell-Mann* et al. (1963), (1964b)] the case of vector-spinor scattering. Then diagram of Fig. (VII. 10) behaves like t^0 for large t. Now we include diagrams analogous to the diagrams of section (VII. 3), i.e. those of Fig. (VII. 11). The diagram Fig. (VII. 11a) behaves as t^{-1} for large t, whereas Fig. (VII. 11b) behaves like $t^0 \log t$. We see therefore that Fig. (VII. 11a) and Fig. (VII. 11b) do not form the starting terms of a series which might sum to a Regge-pole behaviour. However Fig. (VII. 10) and Fig. (VII. 11b) do have the right behaviour to start such a series. Provided the higher order diagrams behave in the right way to complete the series the apparently non-Regge behaviour of Fig. (VII. 10) is cancelled exactly and replaced by a Regge-type behaviour. The treatment of the higher order terms is not easy, but at least the lower orders have been shown to have the required behaviour. This method is, however, extremely cumbersome so we refer to the original papers for further details, and turn now to a more elegant treatment given by *Mandelstam* (1965).

VII.6. Elementary Particles

As we saw in the previous section an elementary particle of spin σ contributes a Kronecker delta singularity $\delta_{J\sigma}$ to any amplitude in which it can occur. Kronecker delta singularities can also occur in the following way. Consider the partial-wave amplitude at a value of J for which some of the helicity states are "nonsense" states (see Chapter IV). Clearly in the evaluation of the physical amplitudes at such a value of J the "nonsense" helicity states should be ignored. However, for all other values of J these helicity states are to be included, so, unless their effect automatically becomes zero at the physical J considered, the continued amplitude will not agree with physical amplitude at that point. It is therefore important to know whether the sense-nonsense coupling affects the sense amplitude.

To study this question we recall that the function $e_{\lambda\lambda'}^J$ in the Froissart-Gribov projection behaves like $(J - J_0)^{-1/2}$ for J near a sense-nonsense value J_0. This behaviour cannot occur in the amplitude, which is bounded by unitarity, and is replaced by the form $(J - J_0)^{1/2}$ [see Section (V. 5)]. Note that we are here concerned only with the *right-signature* amplitude (a Kronecker delta in a wrong-signature amplitude is irrelevant) so sense-nonsense decoupling occurs in the way explained in Chapter IV and V. However, suppose we calculate the continued partial-wave amplitude by a perturbation expansion involving an effective "potential" (see Chapter VI for an explanation of this concept). The potential consists of the partial-wave projection of the integral of the left-hand cut of A (s, t), and so it is not bound by the unitarity restriction. Thus we may expect (from the Froissart-Gribov projection) that the *potential* will behave like

$$\langle s \, |V_J| \, n \rangle \sim \frac{b}{(J - J_0)^{1/2}} \tag{6.1a}$$

$$\langle n \, |V_J| \, n \rangle \sim \frac{c}{(J - J_0)} \tag{6.1b}$$

$$\langle s \, |V_J| \, s \rangle \sim \text{constant} \tag{6.1c}$$

for the sense-nonsense, nonsense-nonsense, and sense-sense amplitudes respectively. As an example one might approximate the potential by the simplest one particle exchange diagram. Specializing to the case of spin $1/2 -$ spin 1 scattering this would be as given in Fig. (VII.11a). This behaves as $t^{1/2}$ for large t, corresponding to the presence of the fixed singularities in (6.1) at $J = 1/2$ (the highest sense-nonsense value in this case).

At this stage we must emphasise that we are regarding the external particles of spins 1/2 and 1 as *elementary particles*; otherwise, as we saw in section (V. 3), the $t^{1/2}$ behaviour, and hence the infinities at $J = 1/2$ in (6.1), would be removed by other contributions to the potential. In the case of elementary particles, however, it appears unlikely that any such cancellation could occur, and from here on we assume that it does not.

Now we write a formal perturbation expansion (in V_J) for the sense-sense amplitude. This can be re-arranged to read (*Calogero* et al. (1963b)]

$$\langle s\, |A^J|\, s\rangle = \langle s\, |A^{J\,nn}|\, s\rangle \left[1 + \sum_n \langle s\, |V_J|\, n\rangle \langle n\, |A^J|\, s\rangle\right], \qquad (6.2)$$

where the suffix nn means "no nonsense", i.e. nonsense states are not included as intermediate states, and correspondingly the sum in (6.2) is over nonsense states only. In the term of (6.2) the V part behaves like $(J - J_0)^{-1/2}$, and the A part like $(J - J_0)^{1/2}$ so the product is finite but *non-zero*. Thus, for elementary particles, the nonsense states do affect the continued sense-sense amplitude, which is not equal to the physical amplitude (the physical amplitude is in fact just $\langle s\, |A^{J\,nn}|\, s\rangle$).

We must now see under what conditions the difference between the physical amplitude at $J = J_0$ and the continued amplitudes can exactly correspond to an elementary particle δ-function, so that we have the equality

$$\langle s\, |A^{J\,nn}|\, s\rangle + "\delta_{J J_0}" = \langle s\, |A^J|\, s\rangle. \qquad (6.3)$$

Such a behaviour would appear unlikely; and indeed it does not happen in general. However for some cases it has to occur.

To see this we first consider the origin of the difference between $A^{J\,nn}$ and A^J in a conventional N/D type calculation. It is clear that the left-hand cuts of these two amplitudes are identical, i.e. they have the same "potentials". For A^J the unitarity equation contains all helicity states but, as we have seen, at $J = J_0$ the nonsense states decouple exactly. Thus the only difference between the two solutions arises from the fact that, with a given left-hand cut, and given unitarity equation, the solution is not unique, but possesses the CDD ambiguity, i.e. it is possible to insert additional poles in the D (or N) function. In fact what happens is that the amplitude A^J, for general J, which is obtained by solving the N/D equations with *no CDD poles*, when continued to $J = J_0$, appears as a solution with CDD poles if only sense states are considered. Here we have an example of the property discussed in section (VI. 3) that the existence or otherwise of a CDD pole in a given amplitude depends upon the coupling to other channels. The CDD poles arise because in the coupled channel problem there are Regge poles in the nonsense channels which are in the neighbourhood of $J = J_0$, for weak potentials, and which move from $J = J_0$ as the coupling is increased. These Regge poles are absent when the nonsense channels are not included explicitly and must then be inserted in the sense channels as CDD poles. The number of CDD poles required is equal to the number of nonsense channels for the value of J being considered.

Next, we note that the partial-wave amplitudes have to satisfy various threshold conditions at the physical thresholds. These conditions impose restrictions on the parameters of the elementary particles, or CDD poles, which can be introduced, and *Mandelstam* (1965) has pointed out that in certain circumstances these parameters are determined uniquely i.e. for a given left-hand cut there is a unique amplitude containing no more than one CDD pole (or elementary particle) which satisfies the threshold conditions. In this case (6.3) must be satisfied, both sides being equal to this unique solution.

This situation occurs, for example, in the spin $1/2 -$ spin 1 scattering problem referred to earlier, when we consider scattering in the s-channel as in Fig. (VII. 11). We are then interested in the positive-parity state with $J = 1/2$, and for this there are two sense and one nonsense states. In the (l, S) representation, where l is the orbital angular momentum and S is the total spin, these are

Sense states:
$$l = 1, \quad S = 3/2;$$
$$l = 1, \quad S = 1/2.$$

Nonsense states:
$$l = -1, \quad S = 3/2.$$

In order to remove the kinematic singularity at $s = 0$, we use the variable $w = \sqrt{s}$. The generalisation of the MacDowell symmetry (Section IV. 10) relates amplitudes for negative w to amplitudes of opposite parity with positive w. There are again three of the latter, i.e.

Sense states:
$$l = 2, \quad S = 3/2;$$
$$l = 0, \quad S = 1/2.$$

Nonsense states:
$$l = 0, \quad S = 3/2.$$

The threshold conditions have to be satisfied at $w = \pm 2\,m$ (we take both particles to have mass m for the moment). Since the two positive parity sense states have $l = 1$, the corresponding three sense-sense amplitudes have a simple zero at $w = 2\,m$ i.e. behave like $(w - 2\,m)$. This gives three threshold conditions. For the negative parity sense states we have a behaviour $(w + 2\,m)^2$ near $w = -2\,m$, for the $l = 2$ to $l = 2$ amplitude, a behaviour $(w + 2\,m)$ for the cross term, and no condition for the $l = 0$ to $l = 0$ amplitude; thus are again three conditions.

If the amplitudes are normalized to have no kinematic singularities at $w = 0$, they will have a constant asymptotic behaviour at infinity; thus there will be three unknown subtraction constants for the three sense-sense amplitudes. It is clear that with a given left-hand cut it will not in general be possible to satisfy the six threshold conditions with these three parameters. However if we add a CDD pole we get three extra parameters (the position of the pole and a factorizable residue) corresponding to the mass and couplings of an elementary particle. We then have 6 parameters to fit the six conditions, and we obtain a *unique* solution. Note that this statement is *independent* of the left-hand cut; the values of the 6 parameters depend on the left-hand cut of course, but for a given left-hand cut they will be unique. It follows that (6.3) holds exactly for this problem; the Kronecker deltas cancel, and the solution exhibits pure Regge behaviour. This confirms the results suggested by analysis of the low orders of perturbation theory.

Fortunately this behaviour (which would greatly obscure the question of how to determine experimentally whether or not there are elementary particles) is not found in general. It does not happen in a spin 0-spin 1 amplitude, nor does it happen in the cross-channel (i.e. spin $1 +$ spin $1 \to$ $\to$ spin $^1/_2 +$ spin $^1/_2$) of the above process. Of course the fact that the amplitudes are not unique does not by itself prove that (6.3) is false, but it is possible to prove that it is false in these cases simply by looking at the perturbation series; though in saying this we are assuming that if (6.3) is not true in some order of perturbation theory, then it is not true; ignoring the possibility that (6.3) might be taken as a "self-consistency" condition for the coupling constants.

Abers and *Teplitz* (1967) have given an analysis on the lines of the above for the general spin, general mass, case and have discussed the circumstances under which the solution is unique, so that Reggeization can occur.

In conclusion, then, we have found that if elementary particles do occur, they might, in some particular amplitude, fail to yield the expected constant power behaviour, but they would certainly show up in this way in other amplitudes. Whether or not a given "elementary" particle contributes a constant power behaviour to a particular amplitude can be discovered by the method of *Mandelstam*, as generalised by *Abers* and *Teplitz*. Our attitude in this book has been to adopt the "maximal analyticity" principle, and so assume that there are no elementary particles. So far there is no evidence to contradict this, and much to support it, as we shall see in the next chapter.

Chapter VIII. A Survey of the Experimental Situation

VIII.1. The Regge Trajectories

a) Introduction

In this chapter we shall review the experimental situation in light of the ideas about Regge poles, etc. which have been discussed in previous chapters. Mainly we shall be concerned with how well the assumption that Regge poles dominate the high energy behaviour of scattering amplitudes fits the available data, but first we must look at the current evidence for Regge trajectories in the region where they cross physical J values, and show up as bound states or resonant states. We shall use the words "particle" or "state" indiscriminately to refer to bound states or resonances.

As we saw in section (II. 10) a trajectory corresponds to a physical particle when it crosses alternate integral (or, in the case of fermions, half-odd-integral) values of J for positive s. Thus Regge trajectories connect particles which differ in spin by two units, but have otherwise the same quantum numbers. The energy separation between two such particles depends on the slope of the trajectory, but if we invoke the potential scattering argument of section (III. 1), that $d\alpha(s)/ds$ is related to the "radius" of the state and hence to the range of the strong interaction forces, we might anticipate a slope of the order of $1\,(\text{Gev})^{-2}$, corresponding to a range of about 10^{-13} cm. This appears to agree with the experimental evidence, both for $s > 0$, and, as we shall see later, $s < 0$.

Since most of the resonances which have been positively identified (i.e. have had all their quantum numbers uniquely determined) have masses of at most a few Gev., it is not surprising that many trajectories have been seen only at one physical value of J. There are, however, some which have been observed to cross two or more physical values, and several are conjectured to rise to quite large values of J. Of course there is no reason why any given trajectory should not turn over before it has passed more than one physical value, and indeed we saw in chapter VI that in current bootstrap models, as in potential scattering, trajectories do appear to turn over very rapidly. In fact it is the apparent linearity of trajectories over a large energy range which is hard to understand.

b) Meson Trajectories

Let us discuss first the well established meson trajectories. It is convenient to use the $SU(3)$ classification scheme, according to which all the firmly identified mesons can be grouped into "nonets", consisting of an $SU(3)$ octet, and an $SU(3)$ singlet. The three best established nonets which are physical at J^P values 0^-, 1^-, 2^+, respectively, are given in Table (VIII. 1), together with the two less well established 0^+ and 1^+ nonets.

The evidence that these mesons do fit into the $SU(3)$ octets, as indicated, is reasonably good, at least for 0^-, 1^-, and 2^+, but depends heavily on there being a mixing between the isotopic singlet member of the octet, and the $SU(3)$ singlet state. This mixing is particularly important for the ϕ and ω particles, and the f and f'. For a discussion of the experimental status of $SU(3)$ we refer to the articles by *Leitner* (1966), and *Goldhaber* (1967). If these particles lie on Regge trajectories, we can make a so called "Chew-Frautschi plot" [*Chew* and *Frautschi* (1962)] in which $\mathrm{Re}\,\{\alpha(s)\}$ is plotted against s. Such plots are given in

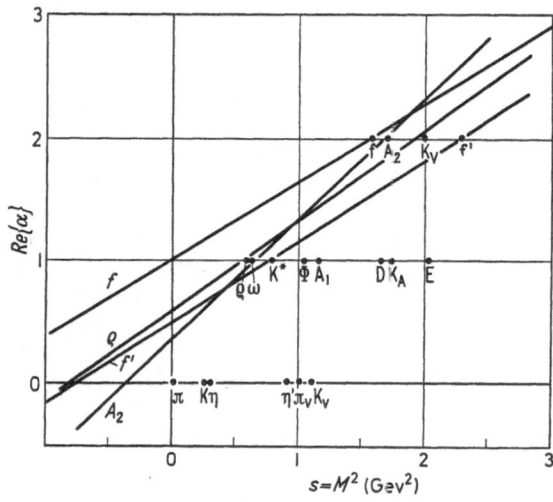

Fig. VIII.1. A Chew-Frautschi plot for the mesons. The particle masses are taken from *Rosenfeld* et al., 1967, and trajectories have been drawn for those cases in which $\alpha(0)$ is known from table (VIII.1).

Fig. (VIII. 1). In drawing these curves we have born in mind the values for $\alpha(0)$ which are given in Table (VIII. 1). The evidence for these numbers will be given in later sections.

If we assume that the slopes of the trajectories are positive below threshold, which they must be unless trajectories cross [see section (III. 1)], then the 0^- nonet will have $\alpha(s) < 0$, for $s < 0$. This means that for many processes the contributions of these trajectories to high energy scattering in the crossed channel will be small compared to that of the other higher-lying meson trajectories, and so they play rather a small part in most phenomenological fits. This is despite the fact that as forces these trajectories may be quite important, as our discussion in section (VI. 4) showed. In addition it turns out that there are sometimes constraints which may prevent, say, the pion (which will have $\alpha(0)$ only a very little below zero) from contributing in the forward direction to a given process. The validity of these constraints is an important, and as yet unresolved question, which we shall take up in detail in sections (VIII. 6) and (VIII. 7).

Table VIII.1. *The Meson Nonets*

J^P	Name	G	C	I	(Mass)^2 Gev^2	$\alpha(0)$	$\alpha'(0)$
0^-	π	—	+	1	0.018		
	K($\overline{\text{K}}$)			1/2	0.248		
	η	+	+	0	0.301		
	η'	+	+	0	0.918		
1^-	ρ	+	—	1	0.593	0.58[a]	1.00[a]
	K*($\overline{\text{K}}$*)			1/2	0.796		
	ω	—	—	0	0.614	0.45[c]	0.31[c]
	φ	—	—	0	1.039		
2^+	A_2	—	+	1	1.70	0.34[b]	0.35[b]
	$K_V(\overline{K}_V)$			1/2	1.99		
	f	+	+	0	1.57	1.0(= P)	0.12[c]
	f'	+	+	0	2.29	0.73[c](= P')	1.50[c]
0^+	π_V	—	+	1	1.01		
	η_V	+	+	0	1.10		
1^+	A_1	—	+	1	1.16		
	$K_A(\overline{K}_A)$			1/2	1.74		
	D	+	+	0	1.65		
	E	+	+	0	2.03		

Data from *Rosenfeld* et al., 1967. The masses are quoted for the neutral members of the isospin multiplets, and the values of C given also refer to the neutral members of the I = 1 multiplets. References are (a) *Höhler* et al., 1966, (b) *Phillips* and *Rarita*, 1965 b case 2, and (c) *Rarita* et al., 1967, solution 1.

The trajectories of the 1^- nonet, particularly the ϱ, have been extensively studied, and play an important part in many phenomenological analyses, particularly for inelastic processes. But the highest lying trajectories belong to the 2^+ nonet, and these generally give the most important contributions to those processes in which they can play a part. In particular the f and f' trajectories have the quantum numbers of the vacuum, and will therefore be involved in the high energy behaviour of all elastic processes. We shall see in section (VIII. 3) that, if high energy scattering is really dominated by crossed-channel Regge poles, then the leading trajectories at $s = 0$ must have the quantum numbers of the vacuum, and it was for this reason that a trajectory of this type was originally postulated by *Chew* and *Frautschi* (1961b). It is generally called the "Pomeranchon" trajectory (sometimes designated by P), after *Pomeranchuk* (1958), whose theorems on high energy behaviour it serves to explain in Regge pole terms.

We see from Fig. (VIII. 1) that the f trajectory is very likely to be the leading one (i.e. have the largest $\alpha(0)$), and so the identification of the f with the Pomeranchon is natural. We shall discuss this further in section (VIII. 3). The 2^+ trajectories presumably cross $J = 0$ at negative values of s, where they cannot correspond to particles, so the residues must vanish at these points.

The lowest "recurrences" of the three well established nonets would appear at J^P values 2^-, 3^-, and 4^+, respectively, but there are no completely identified particles which might be such recurrences. However, *Focacci* et al. (1966) have plotted the distribution of "missing masses" in the reaction

$$\pi^- + p \to p + \text{(missing mass)}^-$$

by measuring the recoil energy of the proton. The distribution shows a great deal of structure, and, in addition to the ρ and A_2, which are the only single particles from the above which can be produced, there are a large number of other peaks. *Foccacci* et al. number these in order of increasing mass, and plot the numbers against $(\text{mass})^2$. The points lie in a straight line [see Fig. (VIII. 2)], and it is tempting to regard the

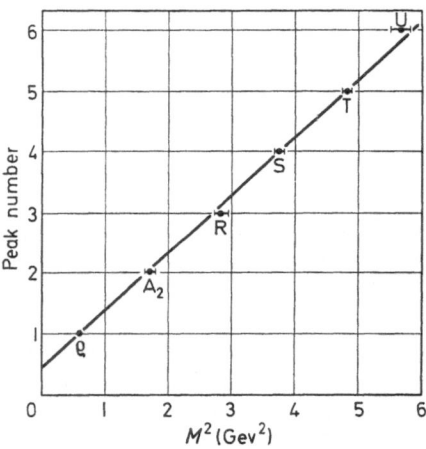

Fig. VIII.2. The data of *Focacci* et al., 1966, in a plot of peak number against $(\text{mass})^2$. The line may be interpreted as a trajectory $\alpha(s) = 0.45 + 1.05\ s$.

figure as a Chew-Frautschi plot, and the line as a pair of superimposed Regge trajectories of opposite signatures, by assigning spins to the particles as indicated. We must emphasize, however, that apart from the spins of the ρ and A_2 there is no direct evidence in favour of these (or indeed any other) assignments, although the hypothesis of high spin would help to explain the narrow widths of the higher states observed by *Focacci* et al. The R peak is certainly not just a single peak, but probably contains three resonances. Other evidence for non-strange particles in the region of energy covered by *Focacci* et al. is discussed by *Goldhaber* (1967).

If the line in Fig. (VIII. 2) really does represent a Chew-Frautschi plot, then, as we noted in chapter VI, it poses problems for dynamical theories. Its main characteristic, its lack of curvature, is hard to understand from the point of view of the bootstrap hypothesis, where the "scale" of energies is usually taken to be about 1 Gev. Note that the straight-line plot extrapolates to $\alpha(0) = 0.45$, so some slight curvature is needed if the curve is to pass through the value of $\alpha_p(0)$ given in Table (VIII. 1). This curvature is in the expected direction, since we know from section (III. 1) that

$$\frac{d^2 \alpha(s)}{ds^2} > 0, \quad \text{for } s < s_0,$$

unless two trajectories (of the same quantum numbers) cross, and there is no sign of this. Later we shall see further evidence for trajectories continuing as approximately straight lines through several physical values of J, and some possible reasons for this behaviour will be mentioned in section (VIII. 9).

A second remarkable feature of Fig. (VIII. 2) is the lack of any signature effect; that is trajectories of opposite signature appear to coincide almost exactly. Recalling the discussions of signature in sections (II. 2) and (II. 10), we see that the force generating even signature trajectories depends on the sum of the t- and u-channel singularities, whereas for negative signature it is the difference which matters. This must mean that there is no significant "exchange force", i.e. only one discontinuity, $D_t(s, t)$, contributes significantly to the fixed s dispersion relation. It is interesting that the quark model [see section (VIII. 9)], can offer a ready explanation of both these features of Fig. (VIII. 2).

The five hypothetical nonets, and the non-strange recurrences discovered by *Focacci* et al., include almost all the mesons which are sufficiently well established to have found their way into the wallet cards (*Rosenfeld* et al., 1967). Exceptions are the $\delta(965)$, B(1210), $\pi(1640)$, and $K_A(1800)$, but none of these is very certain, and we refer to *Goldhaber* (1967), for a discussion of their status. There is, however, special interest in the δ despite the fact that none of its quantum numbers, except $I = 1$, are known at present.

We have noted in section (III. 8) that there is no real evidence for "daughter" meson trajectories. The first daughter would have opposite signature to the parent trajectory, but otherwise the same quantum numbers, and would be one unit of angular momentum lower than the parent at $s = 0$. The δ-meson, which has been seen as a small bump in the missing-mass spectrum referred to above [*Focacci* et al. (1966)], at about 965 Mev., is the only candidate for a boson daughter available at present. It is in roughly the right place to be the first daughter of the ρ meson, for which it would need $J^P = 0^+$ and positive G-parity [see *Freedman* and *Wang* (1967a)]. It will be very interesting to see whether daughter trajectories do show up as particles, and they will doubtless be eagerly sought, but, as we have indicated in section (III. 8), evidence may be hard to come by. Of course their absence would not necessarily imply that the daughter idea is wrong, because there is no need to assume that the daughter trajectories are similar to their parents [see section (III. 8)]. They might for example be extremely flat, and so never reach physical J values, or alternatively they might have much the same sort of slopes, but their residues might vanish at physical J. In either of these eventualities, the only hope of detecting them would be as crossed-channel poles controlling high energy behaviour. We mentioned some of the difficulties of finding them in this way in Chapter III.

c) Baryon Trajectories

We turn now to the baryon trajectories, and consider first those which have zero strangeness. These occur as resonances in $\pi - N$ scattering, and have therefore been studied extensively. There are two

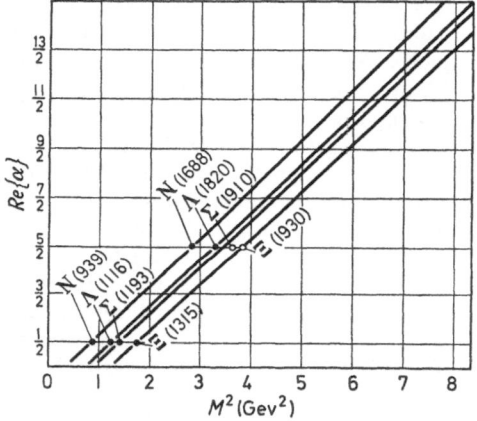

Fig. VIII.3. Trajectories of the α (even parity, even signature) baryon octet. Solid dots represent firmly identified particles; open dots are bumps for which only I is known. The nucleon trajectory is approximately $\alpha(s) = -0.39 + 1.01\,s$.

well established $I = 1/2$ trajectories, one having as its lowest spin particle $1/2^+$ (the nucleon), and the other $3/2^-$, and one with $I = 3/2$, and lowest spin particle $3/2^+$. These are plotted in Figs. (VIII. 3), (VIII. 4), and (VIII. 5) along with their $SU(3)$ partners. The lower states have been uniquely identified as having all the appropriate quantum numbers, and the higher states have been observed as peaks with the correct isotopic spin, but their spins and parities have not been determined, and as with the mesons they have been inserted in the diagrams at the positions appropriate to their masses if straight line trajectories are assumed. Again it is remarkable that particles do seem to occur just where straight Regge trajectories would cross physical spin values, and it seems quite probable that the assignments given in Figs. (VIII. 3–5) are correct. In section (VIII. 8) we shall see confirmatory evidence for the assignments in some cases, when we discuss the work of *Barger* and *Cline*.

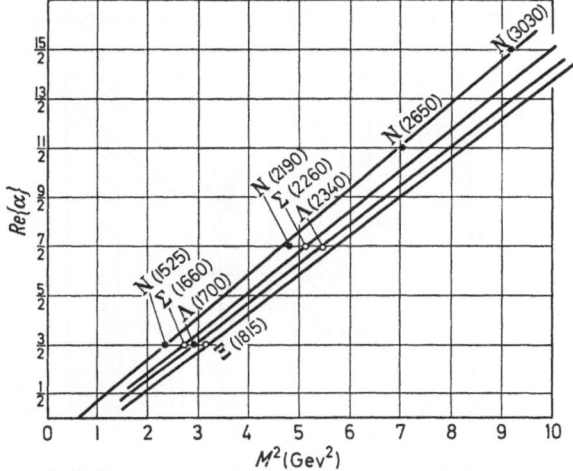

Fig. VIII.4. Trajectories of the γ (odd parity, odd signature) baryon octet. The N_γ trajectory is approximately $\alpha(s) = -0.46 + 0.88\,s$.

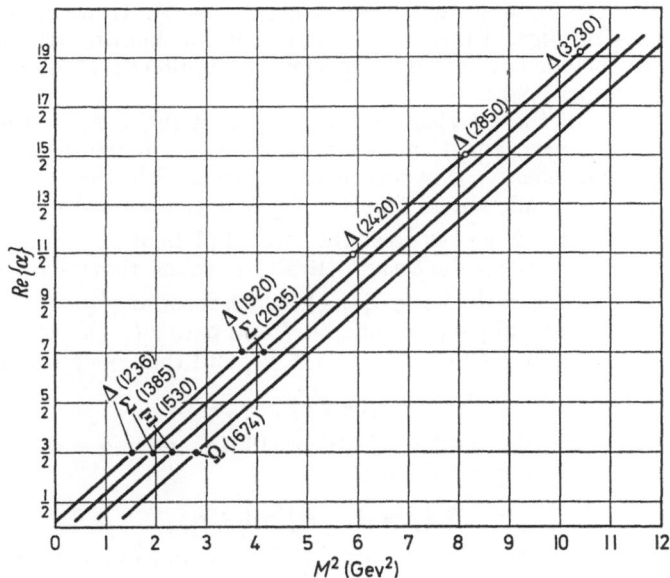

Fig. VIII.5. Trajectories of the δ (even parity, odd signature) baryon decuplet. The Δ trajectory is approximately $\alpha(s) = 0.15 + 0.90\,s$.

The nucleon is well known to be a member of an $SU(3)$ octet, and the Δ (3—3 resonance) a member of a decuplet. Less certain is the assignment of the other $I = 1/2$ trajectory to an octet. Some of the recurrences of the strange particle trajectories have been observed but the general situation is not so clear for strange baryons. The $SU(3)$ singlets are plotted in Fig. (VIII. 6). For further details as to the status

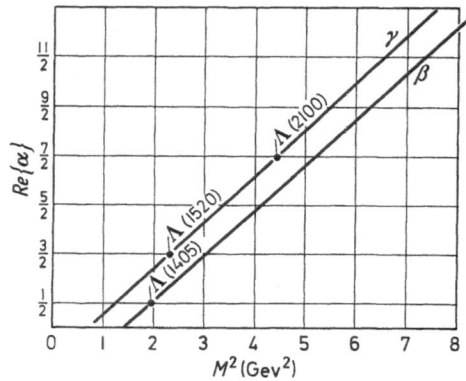

Fig. VIII.6. Trajectories of the γ (odd parity, odd signature) and β (odd parity, even signature) baryon singlets. The λ trajectory is approximately $\alpha(s) = -0.70 + 0.95\,s$.

of the various states we refer to the reviews by *Murphy* (1967) and *Ferro-Luzzi* (1967), and to *Rosenfeld* et al. (1967). If we accept the assignments of Figs. (VIII. 3—6), almost all the known baryon resonances are included. The only important exceptions are the P_{11}, and some S-wave, $\pi - N$ states.

The slopes of these trajectories are all about the same as those for mesons, $\approx 1\,(\text{Gev})^{-2}$, and there is no evidence of curvature over quite a wide range of energy. It is all the more remarkable that this should be so when we recall [see section (IV. 7)] that the natural variable for fermion trajectories is $\sqrt{s}$, whereas thes are all plots of $\alpha(s)$ versus s. The MacDowell symmetry [*MacDowell* (1960)] requires that there should be two trajectories, $\alpha_1(\sqrt{s})$ and $\alpha_2(\sqrt{s})$, which meet at $\sqrt{s} = 0$. The trajectory α_1 has physical particles of spin J and parity $(-1)^{J-1/2}$, while α_2 has particles of spin J and parity $(-1)^{J+1/2}$, and they obey the relation

$$\alpha_1(\sqrt{s}) = \alpha_2(-\sqrt{s})\,, \quad s > 0\,.$$

This has been discussed by *Sakmar* (1964), and more recently by *Desai* (1966), who suggests the form

$$\alpha(\sqrt{s}) = A + B\sqrt{s} + C\,s \tag{1.1}$$

which would connect $1/2^+$ states with $1/2^-$ ones, and so on, as shown in Fig. (VIII. 7) [see also *Chiu* and *Stack* (1967) and section (VIII. 6)]. However, most of the required states are not known, and none have been seen for the corresponding decuplets. It is of course possible that the residues vanish at all the physical integers for $\sqrt{s} < 0$, so that no particles result. But unless B in (1.1) is very small, we shall not get straight line trajectories, and if it is small, then the minimum of $\alpha(\sqrt{s})$ will be very sharp. This problem is so far unresolved.

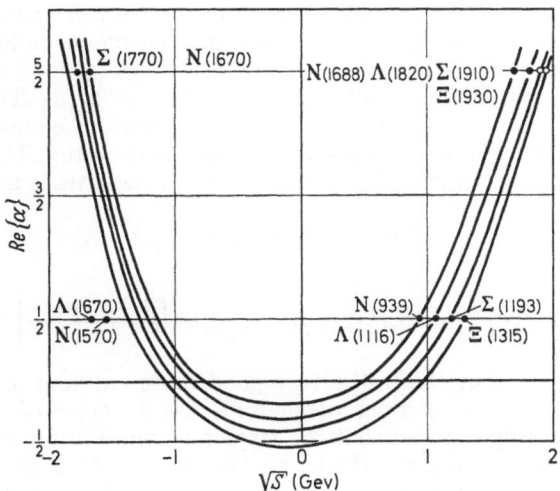

Fig. VIII.7. A possible form for the α baryon octet trajectories, connecting them to $^1/_2{}^-$ and $^5/_2{}^-$ states. The known $^1/_2{}^-$ states are of too high a mass to fit onto any simple curve joining the two regions of $\sqrt{s}$.

The main conclusion from this section is that essentially all the known particles can be classified by combining $SU(3)$ with straight Regge trajectories. In view of the very large number of states involved this indicates a remarkable simplicity. But the evidence is not yet so strong that one must accept this classification, and the overall picture may look quite different in a few years' time.

VIII.2. Regge Poles and High Energies

In section (II. 9) we found that there was a direct connection between the singularities in the J_s-plane and the asymptotic t behaviour as exemplified by (II. 9.8). This presents us with another way of finding the J_s-plane singularities, in that we can look at the amplitude at high t, in the t-channel physical region, and determine $\alpha(s)$ for $s \lesssim 0$. The usual convention in high energy physics, which we shall adopt in the remainder of this chapter, is to take s as the square of the energy, and $(-t)$ as the square of the momentum transfer. So we can invoke crossing, and say instead that the high s behaviour will tell us about the J_t-plane singularities for $t \lesssim 0$. And of course if we look in the backward direction, $u \lesssim 0$, we can also find the J_u singularities.

We have seen in the previous section that there are several high lying Regge trajectories, and since our discussion in chapter V indicated that cuts require the exchange of at least two poles (though of course this does not necessarily mean that they will lie lower in J_t), it is natural to begin by trying to fit the high energy data with just poles. If cuts are also important we shall presumably find this out by our failure to get an adequate fit.

The first thing is to decide what particles can be exchanged in the crossed channel for a given process, given the various quantum number restrictions, and this determines the available trajectories. The amplitude is then written as a sum of the contributions of these trajectories, e.g. for spinless particles

$$A\,(s,\,t) = \sum_i \Gamma_i(t)\, P_{\alpha_i(t)}(z_t)\,, \tag{2.1}$$

where

$$\Gamma_i(t) = -\,16\pi^2 \left[2\,\alpha_i(t) + 1\right]\beta_i(t) \left[\frac{1 \pm e^{-i\pi\alpha_i(t)}}{\sin \pi \alpha_i(t)}\right]\,, \tag{2.2}$$

and it is hoped that this will give a good representation of the experimental data. In later sections we shall discuss the types of test that are available, bearing in mind that the trajectory and residue functions are of course free parameters, which can be varied until a fit is achieved. But, once a Regge trajectory has been determined by a fit to one amplitude, this same trajectory must be used in fits to other amplitudes, since the trajectory functions are common to all processes. Similarly, the residues are constrained by the factorization theorem, and also by the kinematical constraints discussed in section (IV. 6).

At first we shall make the most simple assumption, that all the Regge poles have real $\alpha(t)$ and $\gamma(t)$ below threshold. This assumption is compatible with a considerable amount of the data, but there are some discrepancies which may lead us to abandon this, or some other, assumption later.

It is important, before comparing the high energy theory with experiment to try and decide what is meant by high energy. There are two aspects to this, one trivial, and the other not.

The trivial point is that it is usual to replace $P_\alpha(z_t)$ in (2.1) by its leading term for high z_t. Obviously for this to be valid it is necessary that z_t should be large. From [B 1, 3.2 (23)], we can expand

$$P_\alpha(z_t) = z_t^\alpha [a + b z_t^{-2} + \cdots] \tag{2.3}$$

and, since a and b are of the same order, the replacement can be made if

$$z_t \gg 1\,. \tag{2.4}$$

For equal masses this requires

$$s \gg |q_t^2|\,,$$

so in the forward direction, where $t = 0$ and $q_t^2 = -m^2$ (m being the mass of the particles involved), we need

$$s \gg m^2 . \tag{2.5}$$

Similarly if the masses are unequal we need

$$s \gg m_i^2 , \tag{2.6}$$

where m_i is the heaviest mass involved. There is of course the complication, discussed at some length in chapter III, that the given Regge pole term may not itself be dominant near $t = 0$, because of the divergence of q_t at $t = 0$. If daughter trajectories are used to cancel this singularity, then the strength of the second term in the series corresponding to (2.3) depends on the parameter of the non-singular part of the first daughter, and there will also be logarithmic terms, as in (III. 7.4), unless the slopes of the parent and daughter are identical.

On the other hand, if the cancellation is achieved by the background as in (III. 6.9), then the condition (2.6), which holds for the Regge pole term for $t = 0$, will presumably still hold for the full amplitude at $t = 0$. In either event the normal Regge asymptotic behaviour is maintained.

The non-trivial problem is to decide at what energy a given Regge pole will dominate over others, whose trajectories are lower, and over the background. The relative importance of two trajectories (1 and 2, say) in the sum (2.1) is approximately given by

$$\frac{\Gamma_1(t) \, (z_t)^{\alpha_1(t)}}{\Gamma_2(t) \, (z_t)^{\alpha_2(t)}} , \tag{2.7}$$

so clearly, if the Γ's are of a similar order of magnitude, the appropriate requirement is again that z_t should be large, leading to (2.6). But this assumes that the Γ's are an appropriate measure of the coupling. In fact we know that $\Gamma(t)$ has a branch point at threshold, and in section (VI. 4) we re-wrote the residues in the form

$$\beta(t) \equiv \gamma(t) \, (q_t^2)^{\alpha(t)} \equiv \bar{\gamma}(t) \left(\frac{q_t^2}{q_i^2}\right)^{\alpha(t)} \tag{2.8}$$

on the grounds that then $\bar{\gamma}(t)$ has only dynamical singularities, and is of constant dimension. The constant factor q_i^2 is chosen such that $\bar{\gamma}(t)$ is not a rapidly varying function of t. Putting this in (2.7), we find the ratio of the contributions to be

$$\frac{\bar{\Gamma}_1(t) \left(\dfrac{s}{2q_{t_1}^2}\right)^{\alpha_1(t)}}{\bar{\Gamma}_2(t) \left(\dfrac{s}{2q_{t_2}^2}\right)^{\alpha_2(t)}} \tag{2.9}$$

where

$$\bar{\Gamma}_i(t) = \Gamma_i(t) \left(\frac{q_t^2}{q_{t_i}^2}\right)^{\alpha_i(t)} , \quad i = 1, 2 . \tag{2.10}$$

If we assume that $\bar{\Gamma}$ is a reasonable measure of the coupling, then the dominance of the given trajectory depends on having

$$s \gg q_{t_i}^2 , \quad i = 1, 2 , \tag{2.11}$$

which thus becomes the criterion of high energy. The energy scale is therefore determined by $q_{t_i}^2$, rather than by the masses which appear in (2.6). It has become usual, following *Hadjioannou* et al. (1962), to take

$$q_{t_i}^2 \approx m_N^2 \approx 1 \text{ Gev}^2, \qquad (2.12)$$

where m_N is the nucleon mass, for all trajectories. We have already noted that this is the scale of strong interactions as measured by the slopes of trajectories, and now we have further confirmation, based on the rate of variation of the residue functions. Since the slopes of the trajectories determine the masses of the low spin particles, (2.5) and (2.11) are presumably connected in some deeper way.

It is of course possible to check roughly that (2.12) is right for a given trajectory, since the trajectory will not dominate over the background until (2.11) is satisfied. We shall see that the experimental evidence is that Regge behaviour has set in by the time the energy has reached a few Gev., so our estimate seems to be about right. In section (VIII. 8) we shall discuss attempts to link the low and high energy regions by representing the amplitude as a sum of direct channel resonances and crossed channel Regge pole terms.

VIII.3. High Energy Total Cross-Sections

a) Regge Poles and Total Cross-Sections

The spin-averaged total cross-section, $\sigma^T(s)$, for the scattering of two particles, 1 and 2, is related to the forward elastic cross-section by the "optical theorem"

$$\sigma^T(s) = (2q_{s12}s^{1/2})^{-1}(2\sigma_1+1)^{-1}(2\sigma_2+1)^{-1} \sum_{\lambda_1\lambda_2} \text{Im}\{\langle\lambda_1\lambda_2|A(s,0)|\lambda_1\lambda_2\rangle\}$$

$$(3.1)$$

[c.f. (I. 7.17)]. Thus there is a linear relation between the total cross-sections and Regge-pole residues in the t-channel, and no cross-terms are involved. Because of this, and the simplicity of the kinematics and lack of spin complications, total cross-sections are very readily analysed in terms of Regge-poles [*Udgaonkar* (1962)].

From (2.1) and (II. 9.17) we have

$$\text{Im}\{A(s,t)\} \equiv D_s(s,t) = \sum_i 16\pi^2 [2\alpha_i(t)+1]\beta_i(t) P_{\alpha_i(t)}(z_t), \qquad (3.2)$$

which with (2.8) gives

$$D_s(s,t) \xrightarrow[s\to\infty]{} \sum_i \frac{16\pi^2 \Gamma\left[\alpha_i(t)+\dfrac{1}{2}\right]}{\sqrt{\pi}\,\Gamma[\alpha_i(t)+1]} [2\alpha_i(t)+1]\,\bar{\gamma}_i(t) \left(\frac{s}{s_0}\right)^{\alpha_i(t)}, \qquad (3.3)$$

where we have put $s_0 \equiv 2q_t^2$. So (3.1) tells us that

$$\sigma^T(s) \xrightarrow[s\to\infty]{} \sum_i G_i(0) \left(\frac{s}{s_0}\right)^{\alpha_i(0)-1}, \qquad (3.4)$$

where $G_i(t)$ includes the various factors from (3.1) and (3.3). It is worth noting that, since the laboratory momentum of particle 1 (2 is the target) is

$$|p_{lab}|^2 = \frac{[s - (m_1 + m_2)^2]\,[s - (m_1 - m_2)^2]}{4\,m_2^2}$$

it is at good approximation at high energy ($s \gg m_1^2, m_2^2$) to use

$$|p_{lab}| \approx \frac{s}{2\,m_2}$$

in place of (s/s_0) in (3.4). Or, since

$$E_{1\,lab} = \frac{1}{2\,m_2}\,(s - m_1^2 - m_2^2)\,,$$

to use E_{lab} instead of (s/s_0). In all the experiments with which we shall be concerned the target particle will be the nucleon, so $m_2 \approx 1$ Gev, and if we measure p_{lab} in (Gev/c), or E_{lab} in Gev, this replacement corresponds to taking $s_0 \approx 2\,(\text{Gev})^2$, as in (2.12).

There are five processes for which data is available to be used in these analyses, namely NN, $\overline{\text{N}}$N, KN, $\overline{\text{K}}$N and πN scattering, and each of these occurs in two isospin states, so there are ten experimental total cross-sections in all.

b) The Pomeranchuk Limit

An important experimental property of total cross-sections is that they apparently tend to constant values as $s \to \infty$. This is illustrated in Fig. (VIII. 8) where the available data is plotted. Of course it is not

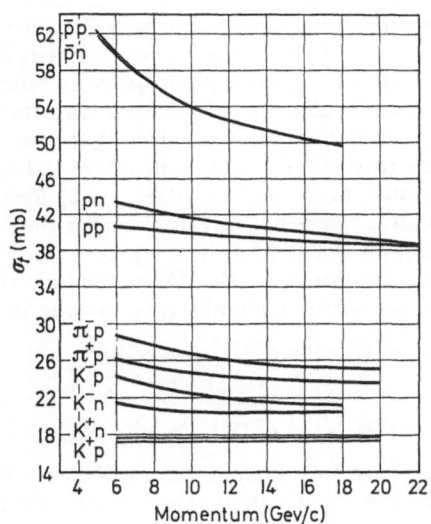

Fig. VIII.8. A plot of total cross-sections against incident laboratory momentum taken from *Lindenbaum*, 1966. Some later data at higher energies are available and confirm the trends shown in this figure.

possible to rule out a slow power variation of the high energy limit, such as s^ε where $\varepsilon \approx 0$ ($\varepsilon \lesssim 0$ from the Froissart bound), or a logarithmic variation with energy. Cosmic ray evidence, which is available for higher energies than machine energies, though with poor accuracy, lends support to a behaviour which is very close to being constant [*Perkins* (1961)], and we shall take this to be the case for the present.

An immediate consequence of such a behaviour is that the leading singularities in the J-plane must be at $\mathrm{Re}\{J\} = 1$. Thus, if we assume that the high energy limit is dominated by a Regge pole, we require the existence of a trajectory with $\alpha(0) = 1$ from (3.4). Further, total cross-sections are clearly positive, so this trajectory must contribute with the same sign to all elastic amplitudes. It must therefore have $C = P = +1$ (see below) and $B = S = I = 0$. Also, as there is no physical strongly-interacting particle with spin one and zero mass, the trajectory must have even signature. As we mentioned in section (VIII. 1) these facts were realised by *Chew* and *Frautschi* (1961b) at a time when there were no known particles which could lie on such a trajectory, so they boldly postulated one, which came to be called the "Pomeranchon", after *Pomeranchuk* (1958) who had earlier discussed certain consequences (the Pomeranchuk theorems) of a constant high energy behaviour of total cross-sections. The Pomeranchuk theorems are automatically satisfied by the assumption that the Pomeranchon trajectory dominates all elastic amplitudes at high energy. Because of the quantum numbers of the Pomeranchon trajectory it is sometimes referred to as the "vacuum" trajectory.

Since one of the objects of the Regge theory is to permit a connection to be made between particle states and cross-channel high energy behaviour, it is not a good start to have to "postulate" the leading term. However this suggestion did have some immediate successes (not all of them confirmed, as we shall see), and in particular it led to a search for spin 2^+ particles with the Pomeranchon quantum numbers. Note that $J^P = 2^+$ is the lowest J-value for which this trajectory can occur as a physical particle, since, assuming $d\alpha/dt > 0$ in the region $t < 0$ the trajectory will pass through $J = 0$ at a negative value of t. This search was quickly successful, and the f at (1254 ± 12) MeV was discovered. However, as we shall see, it is not certain whether the f and the P should be regarded as lying on the same trajectory.

In terms of the couplings of the Pomeranchuk trajectory the ten observable total cross-sections should tend to one of three values i.e.

$$\sigma^T(NN) = \sigma^T(N\overline{N}) = (\gamma^P_{N\overline{N}})^2 \qquad (3.5)$$

$$\sigma^T(\pi N) = \gamma^P_{N\overline{N}}\, \gamma^P_{\pi\pi} \qquad (3.6)$$

$$\sigma^T(KN) = \sigma^T(\overline{K}N) = \gamma^P_{N\overline{N}}\, \gamma^P_{K\overline{K}}, \qquad (3.7)$$

each relation holding for both isospin values. Here we have incorporated various factors into the γ's and used the factorization theorem.

Several alternative explanations of the constant asymptotic behaviour of total cross-sections, involving fixed singularities at $J = 1$, have been given. For example, *Oehme* (1964) suggested a fixed cut starting at $J = 1$ [this gives a $(\log s)^{-1}$ behaviour for the total cross-section], and *Oehme* (1967) has suggested a fixed pole at $J = 1$ in the even signature amplitude [such a fixed pole may be allowed if there are suitable cuts in the J-plane, as discussed in section (V. 4)]. At this stage these seem rather arbitrary suggestions, and for their motivation we refer to section (VIII. 6g).

c) The Contributions of other Trajectories

Before we consider the deviations from the Pomeranchuk limit we must analyse the restrictions on the various possible Regge pole contributions which are imposed by parity and charge-conjugation invariance. Consider first charge-conjugation invariance for the process $1 + 2 \to 1 + 2$. In the t-channel we choose eigenstates of C, i.e. the states $|1\bar{1}\rangle \pm |\bar{1}1\rangle$ and $|2\bar{2}\rangle \pm |\bar{2}2\rangle$, with $C = \pm 1$ respectively. It follows that trajectories with $C = +1$ contribute equally to the 12 and $\bar{1}2$ scattering amplitudes, whereas those with $C = -1$ give contributions which differ by a sign.

For 0^- meson-baryon scattering the two mesons will have $C = P = (-1)^J$, so only trajectories for which $C = P = \mathscr{S}$, where $\mathscr{S}$ is the signature, will contribute. It turns out [*Wagner* (1963), *Sharp* and *Wagner* (1963), *Ahmadzadeh* and *Leader* (1964)] that this property also holds for any trajectory which contributes to the spin-averaged forward nucleon-nucleon amplitude. This is an apparently accidental result which arises from the NN crossing matrix.

From section (VIII. 1) we see that from the high lying 0^-, 1^- and 2^+ nonets only the following trajectories are available for fitting the 10 total cross-sections: $f(= P?)$, f', A_2, ϱ, ω, ϕ. The contributions of these trajectories to the ten processes depend on their C values as indicated above, and on their isotopic spins. Using the isotopic spin crossing matrix [see *Barut* and *Unal* (1963) for a general discussion of isotopic spin crossing matrices], we readily obtain the entries in Table (VIII. 2). Note that this table is really three separate tables, and it must not be taken to imply that the magnitude of say the ϱ contribution to πN is in

Table VIII.2.

Process	Contributing Trajectories
$\pi^- p$	$f + f' + \varrho$
$\pi^+ p$	$f + f' - \varrho$
$K^- p$, $\bar{p}p$	$f + f' + \varrho + \omega + \phi + A_2$
$K^- n$, $\bar{p}n$	$f + f' - \varrho + \omega + \phi - A_2$
$K^- p$, pp	$f + f' - \varrho - \omega - \phi + A_2$
$K^+ n$, pn	$f + f' + \varrho - \omega - \phi - A_2$

The relative signs of the contributions of known Regge trajectories to the ten measurable total cross-sections. If the P is not identified with the f, then it occurs in this table in exactly the same way as the f and f'.

any simple way related to the magnitude of its contribution to KN. Even the relative signs between πN, KN and NN processes are not necessarily correct. With regard to the sign however, it is usually assumed (as a first guess) that the residues have the same sign as they do at the physical pole, i.e. at $t = m^2$, where m is the mass of the appropriate physical particle. This sign follows from unitarity, and the signs in Table (VIII. 2) agree with this convention.

d) Fits to the Experimental Data

To fit the experimental total cross-sections we have available 6 values of $\alpha(0)$, corresponding to the f, f', ϱ, ω and ϕ trajectories (seven if we do not identify P with f), and for each of these we have three coupling parameters, i.e. the coupling to $N\overline{N}$, $K\overline{K}$ and $\pi\pi$ states. It is not surprising, and perhaps not very significant, that a successful fit to the data is possible. The fit is not entirely trivial however, since some of the trajectories can be isolated by forming suitable combinations of cross-sections. In particular we note that only the ϱ contributes to the following three combinations

$$\Delta(\pi p) \equiv \sigma^T(\pi^- p) - \sigma^T(\pi^+ p) \tag{3.8}$$

$$\Delta(K^+ p) - \Delta(K^+ n) \equiv [\sigma^T(K^+ p) - \sigma^T(K^- p)] - [\sigma^T(K^+ n) - \sigma^T(K^- n)] \tag{3.9}$$

$$\Delta(pp) - \Delta(pn) \equiv [\sigma^T(pp) - \sigma^T(\bar{p}p)] - [\sigma^T(pn) - \sigma^T(\bar{p}n)] \tag{3.10}$$

so the same value of $\alpha_\varrho(0)$ should determine the energy dependence of these three combinations. In fact the second differences required in (3.9) and (3.10) are not very well determined experimentally, but the $(\pi^+ p)$, $(\pi^- p)$ difference is known to reasonable accuracy and provides a good test of the power behaviour predicted by Regge theory. The agreement is excellent and a least squares fit yields [*Logan* (1965)]

$$\alpha_\varrho(0) \approx 0.56 \pm 0.15 . \tag{3.11}$$

In order to reduce the number of parameters several authors have used $SU(3)$ predictions. The most complete treatment of this kind is due to *Barger* and *Olsson* (1966a) who used the 1^- and 2^+ nonet trajectories, together with a P trajectory which they did not identify with the f. They allowed for $SU(3)$ symmetry breaking by permitting $\alpha_\phi(0)$ to be different from $\alpha_\varrho(0) = \alpha_\omega(0)$. For the couplings they used exact $SU(3)$ predictions, except for the Pomeranchon couplings which they kept as free parameters. They separated the data into differences and sums by defining

$$\Delta_{AB} = \sigma^T(AB) - \sigma^T(\overline{A}B) \tag{3.12}$$

and

$$\Sigma_{AB} = \sigma^T(AB) + \sigma^T(\overline{A}B) . \tag{3.13}$$

According to our previous discussion only ϱ, ω and ϕ contribute to Δ_{AB}, so there are two values of $\alpha(0)$ and four couplings available as para-

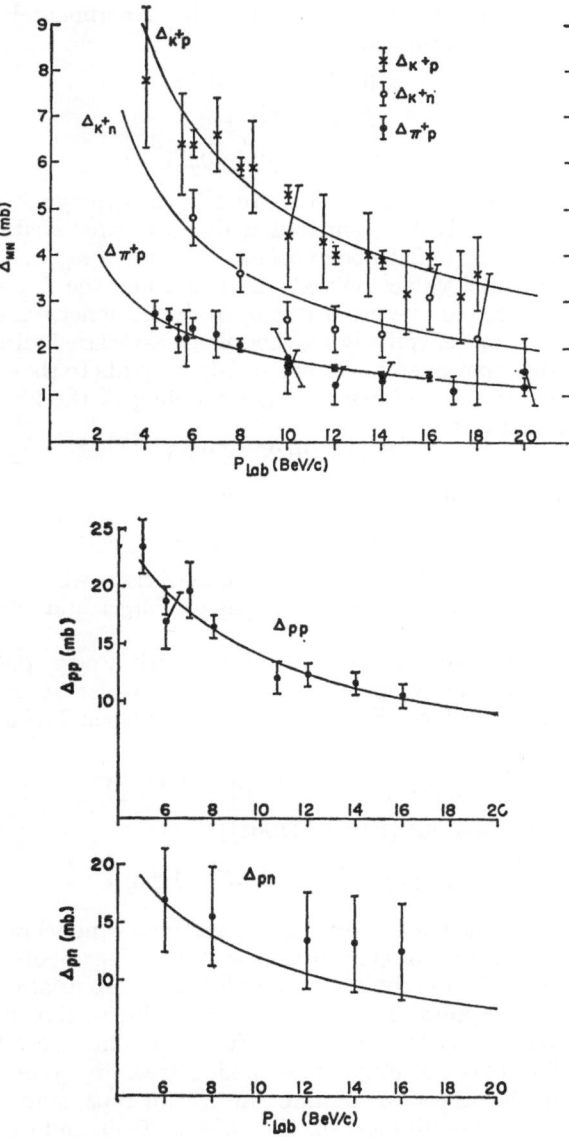

Fig. VIII.9. Experimental data and theoretical curves for the total cross-section differences $\Delta(AB) = \sigma^T(\overline{A}B)$ $-\sigma^T(AB)$ (from *Barger* and *Olsson*, 1966a).

meters. These four couplings are made up of one coupling to the mesons, a coupling of the meson octet to the baryons, with an associated f/d ratio, and a parameter to give the singlet-meson coupling to the baryons. In Fig. (VIII. 9) we quote a typical fit to the data above about 5 Gev/c.

It is clear that this fit lies well within the experimental errors. The parameter values determined were:

$$\alpha_\varrho(0) = 0.48 \pm 0.05 \tag{3.14}$$

$$f/d = -2.0 \pm 0.7 \tag{3.15}$$

$$0.4 \geq \alpha_\phi(0) \geq 0.2 . \tag{3.16}$$

The f/d ratio given here corresponds to a "spin averaged" coupling, so that it cannot readily be compared with other predicted values. The fact that $\alpha_\phi(0) < \alpha_\varrho(0)$ is to be expected since $m_\phi > m_\varrho$.

For fitting the Σ values only the 2^+ nonet and the P contribute, so there are nine free parameters (the $\alpha(0)$ of the 2^+ nonet, three Pomeranchon couplings, and effectively five couplings associated with 2^+ nonet). By varying these parameters various satisfactory fits to the experimental data are possible. The f/d ratio for the coupling of the 2^+ octet to the baryons is found to be

$$f/d = -0.20 \pm 0.06 \tag{3.17}$$

and the value of $\alpha_f(0) = \alpha_{f'}(0) = \alpha_{A_2}(0)$ is

$$\alpha_{2^+}(0) \approx 0.39 \pm 0.24 . \tag{3.18}$$

These values depend somewhat on the assumptions which are made for the couplings, and we refer to the paper of *Barger* and *Olsson* (1966a) for further details.

In general the combination of $SU(3)$ with Regge pole exchange produces equalities between total cross-sections which are not deducible from either theory alone. For example, the Johnson-Treiman relations [*Johnson* and *Treiman* (1965)]

$$\Delta(K^+ p) = 2\Delta(K^0 p) = 2\Delta(\pi p) \tag{3.19}$$

and the Freund relations [*Freund* (1965)]

$$\Delta(pp) = 5\Delta(\pi p) = \frac{5}{4}\,\Delta(pn) \tag{3.20}$$

can be obtained. The reason why a given symmetry model produces more results when it is used in conjunction with the Regge-pole theory than when it is applied directly to the amplitude is essentially that in the former case the symmetry is applied to a three-point function (i.e. a vertex), rather than to a four-point function. Thus $SU(3)$, with pure f-type coupling plus the Regge pole model, gives the Johnson-Treiman relations, which otherwise require an $SU(6)$-type symmetry to be postulated for the amplitudes [see *Carruthers* (1966), and *Gourdin* (1967) for reviews of unitary symmetries].

We shall not persue these points further here since these relations are not in fact very well satisfied, and in general there is little evidence for the validity of the $SU(3)$ couplings. The analysis of Barger and Olsson discussed above shows that $SU(3)$ plus Regge-theory is consistent with the data, but the data do not uniquely lead to $SU(3)$ values, and a wide range of values is allowed for some of the parameters.

As an example of an alternative fit to the total cross-section data we mention that due to *Högaasen* and *Frisk* [reported by *Högaasen* (1966)] who used only the ϱ, ω, A_2, $f(= P)$, and f' trajectories, and did not use any $SU(3)$ predictions. They fixed all the $\alpha(0)$ according to

$$\alpha_f(0) = 1.00 \tag{3.21}$$

$$\alpha_\varrho(0) = 0.57 \tag{3.22}$$

$$\alpha_{A_2}(0) = 0.40 \tag{3.23}$$

$$\alpha_{f'}(0) = 0.50 \tag{3.24}$$

$$\alpha_\omega(0) = 0.52 . \tag{3.25}$$

The last four values are taken from fits to other processes. The ϱ and A_2 values are well determined by $\pi^- p \to \pi^0 n$ and $\pi^- p \to \eta n$ data respectively [see section (VIII. 4)], but the f' and ω values can only be determined very approximately from other data. With these values *Högaasen* and *Frisk* obtained excellent agreement with all available total cross-section data above an incident momentum of 6 Gev/c.

Cabbibo et al. (1966a) fitted total cross-sections with a Regge pole model in which they allowed $\alpha_P(0)$ to be different from unity. The best fit yielded
$$\alpha_P(0) = 0.925 \pm 0.008 . \tag{3.26}$$

Their motivation was the desire to satisfy various relations between cross-sections deduced from a current algebra model for the universal coupling of Regge poles [*Cabibbo* et al. (1966b)] without requiring at least some of the cross-sections to be increasing functions of energy. This value of α_P ensures that all total cross-sections decrease asymptotically. The model is interesting because it obtains relations between the meson-baryon and baryon-baryon processes, but since some of the predictions, such as the Freund relations noted above, appear to be in substantial disagreement with experiment [*Barger* and *Durand* (1967)] we will not pursue it further here. [See *Ne'eman* and *Reichert* (1967) for more recent work.] It is worth remembering, however, that our present evidence, even including cosmic ray data, does not lead unambiguously to $\alpha_P(0) = 1$ [see also section (VIII. 6h)].

e) Forward Dispersion Relation Sum Rules

So far we have been concerned with fits to the total cross-section data lying in the regions of laboratory momenta from about 5 to 20 Gev/c, and it is of interest to know how well the predicted values agree with experiment outside this range. For lower energies extremely good agreement persists down to laboratory momenta of 2−3 GeV/c, provided contributions from direct channel resonances are included. We postpone discussion of this extremely interesting work to section (VIII. 8). At higher energies experimental data is not available, but something can be done using forward dispersion relations. These relate the scattering amplitude to an integral over its imaginary part in the forward direction, which in turn can be related to an integral over the total cross-section. This integral extends over all physical energies up to infinity, and can

14*

be calculated by using experimental data up to a certain point, and the Regge pole formula beyond that point.

The first application of this method was due to *Igi* (1962, 1963), who was able to show that, in addition to the Pomeranchon, there must be at least one other trajectory with $\alpha(0) > 0$, and vacuum quantum numbers (neither the f nor the f' mesons were known at this time). Igi's method used the amplitude

$$f^+(\nu) = \frac{1}{2}\frac{1}{4\pi M^2}\left[\left\langle\frac{1}{2}\Big| A(\pi^- p \to \pi^- p, s, t = 0)\Big|\frac{1}{2}\right\rangle + \right.$$
$$\left. + \left\langle\frac{1}{2}\Big| A(\pi^+ p \to \pi^+ p, s, t = 0)\Big|\frac{1}{2}\right\rangle\right], \tag{3.27}$$

where ν is the pion-energy in the s-channel laboratory system, which is related to s by
$$s = M^2 + 1 + 2M\nu,$$

M being the nucleon mass, and units being such that the pion-mass is one. Then, from the optical theorem, (3.1), we have

$$\mathrm{Im}\,\{f^+(\nu)\} = \frac{1}{2}\frac{(\nu^2 - 1)^{1/2}}{4\pi}\left[\sigma^{\mathrm{T}}(\pi^- p) + \sigma^{\mathrm{T}}(\pi^+ p)\right]. \tag{3.28}$$

We denote the Pomeranchon contribution to $f^+(\nu)$ by $f^{+\mathrm{P}}(\nu)$; then the difference
$$f'^+(\nu) = f^+(\nu) - f^{+\mathrm{P}}(\nu)$$

tends to zero as a negative power of ν, if there are no other Regge trajectories with $\alpha(0) > 0$ and vacuum quantum numbers apart from the Pomeranchon. In saying this we are ignoring the possibility of other singularities in the J-plane. Thus $f^+(\nu)$ should satisfy a single-dispersion relation with no subtractions. Putting in the nucleon pole, and using the fact that the s- and u-channels are simply related by crossing we get

$$f'^+(\nu) = \frac{1}{\pi}\int_1^\infty d\nu'\,\mathrm{Im}\,\{f'^+(\nu)\}\left[\frac{1}{\nu'-\nu} + \frac{1}{\nu'+\nu}\right] +$$
$$+ \frac{1}{4\pi}\frac{g^2}{2M}\left[\frac{1}{\nu_{\mathrm{B}}-\nu} + \frac{1}{\nu_{\mathrm{B}}+\nu}\right]. \tag{3.29}$$

Igi evaluated this sum rule at $\nu = 1$ and obtained a clear discrepancy, which he attributed to the presence of a second trajectory with vacuum quantum numbers and $\alpha(0) > 0$. In later work [*Igi* (1963)] values of t away from the forward direction were also considered, and two trajectories (denoted by P and P') with vacuum quantum numbers and $\alpha(0) > 0$ were included in the analysis. Igi found the data to be consistent with
$$\alpha_{P'}(0) \simeq 0.5. \tag{3.30}$$

Now that we know of the existence of the f and f' 2^+-mesons we can identify the P' with the f' if the identification P = f is made. Otherwise presumably Igi's P' represents the effects of both the f and the f'. The value of $\alpha_{P'}(0)$ found by *Igi* is consistent with that obtained in the direct analyses considered earlier, but neither method yields a well determined value, so it is not possible to attach very much significance to this agreement.

As a second application of this method we return to the difference between the π^-p and π^+p forward amplitudes to which, as we have seen, only the ϱ trajectory, of the trajectories we are including, contributes. *Höhler* et al. (1964) wrote the forward dispersion relation for this quantity, and used the experimental data for $\sigma^T(\pi^-p) - \sigma^T(\pi^+p)$ below 20 GeV, and the Regge-pole expression with a ϱ-trajectory above 20 GeV. The contribution of the region of integration above 20 GeV is not negligible, and determines a value

$$\alpha_\varrho(0) \backsim 0.6 , \tag{3.31}$$

which is consistent with that obtained from data below 20 GeV [see (3.11) and (3.14)].

Igi and *Matsuda* (1967) have considered the forward dispersion relation for this difference between the π^-p and π^+p forward amplitudes on the assumption that there is no contributing trajectory lying in the region $-1 < \alpha(0) < \alpha_\varrho(0)$. They proceed in an analogous way to Igi's treatment of $f^+(\nu)$ referred to above. The resulting sum rule agrees with presently available experimental data, but this is not sufficiently accurate to provide a conclusive test. In section (VIII. 7) we shall see that the presence of polarisation in π-N charge exchange scattering suggests that there are other trajectories (or singularities of some other type) in this region.

VIII.4. Inelastic Cross-Sections in the Forward Direction

a) Introduction

If high energy amplitudes are dominated by Regge poles then the forward amplitude for the process shown in Fig. (VIII. 10) should behave like $s^{\alpha_{13-24}(t_f)}$ for large s, where $\alpha_{13-24}(t)$ is the leading Regge trajectory which can couple to particles 1 and $\overline{3}$, and to particles 2 and $\overline{4}$, and t_f is the value of t in the forward direction ($z_s = 1$), i.e. from (I. 6.14)

$$t_f = -\frac{(m_3^2 - m_1^2)(m_4^2 - m_2^2)}{s} + O\left(\frac{1}{s^2}\right),$$

$$\tag{4.1}$$

$$\xrightarrow[s \to \infty]{} 0 . \tag{4.2}$$

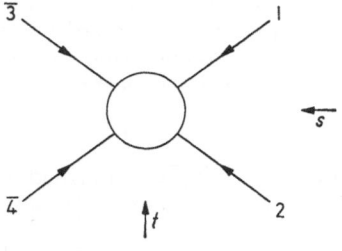

Fig. VIII.10. The process $1 + 2 \to 3 + 4$.

Similarly for the backward direction, which is the forward direction for $1 + 2 \to 4 + 3$, we expect a behaviour $s^{\alpha_{14-23}(u_f)}$ for large s, where u_f is the value of u when $z_s = -1$, i.e., from (I. 6.14),

$$u_f = -\frac{(m_4^2 - m_1^2)(m_3^2 - m_2^2)}{s} + O\left(\frac{1}{s^2}\right) . \tag{4.3}$$

Clearly we expect inelastic process to be divided according to the trajectories that are allowed, and it is convenient to group them accordingly.

b) Pomeranchon Exchange Possible

The simplest examples of this class are elastic scattering which we have discussed previously. However there are other cases for which $P(=f?)$ and f' exchange are allowed, and the forward amplitude for these processes should again go like s^1. This a remarkable prediction of Regge theory, in that "diffraction scattering" (i.e. the effect of the removal of particles from the elastic channel into the many inelastic channels), which is usually regarded as dominating elastic scattering at high energies, cannot occur for inelastic scattering, and, from this point of view, we should expect all inelastic forward amplitudes to go down relative to the elastic ones with increasing energy.

In making a comparison with experiment we are not able to use the forward amplitude, but must rather consider the total cross-section, as this is the only quantity which has been measured in the interesting cases. We have here no analogue of the optical theorem to relate this to a forward amplitude, but we note that, according to the Regge pole model the differential cross-section, expressed in terms of t rather than angle, (I. 7.12), behaves for large s, when we substitute (2.1), like

$$\frac{d\sigma}{dt} \sim \text{const. } s^{2\alpha(t)-2}. \qquad (4.4)$$

Now most of the total cross-section comes from a small region of t near $t = 0$, and this region varies at most very slowly with energy [see section (VIII. 6)], so the total cross-section for the given inelastic process is given by

$$\sigma \sim \text{const. } s^{2\alpha(0)-2}. \qquad (4.5)$$

We use σ to represent the total cross-section for a given inelastic process; in the previous section σ^T was the total cross-section for *all* the processes which are possible from the given initial state.

Thus with Pomeranchon dominance we would find

$$\sigma(s) \xrightarrow[s \to \infty]{} \text{const.} \qquad (4.6)$$

This is strikingly confirmed by the measurements of *Anderson* et al. (1966), and *Blair* et al. (1966), on the reactions

$$\text{pp} \to \text{pN*}, \qquad (4.7)$$

where the N* isobars of mass 1238, 1400, 1520, 1690 and 2190 MeV were considered. The results are shown in Fig. (VIII. 11). We see that the cross-section for N* (1238) production goes down with energy which would be expected since it has $I = 3/2$ and cannot have any contribution from P exchange, whereas the cross-sections for the other resonances, all of which have the quantum numbers of the nucleon (apart from spin), appear to be constant.

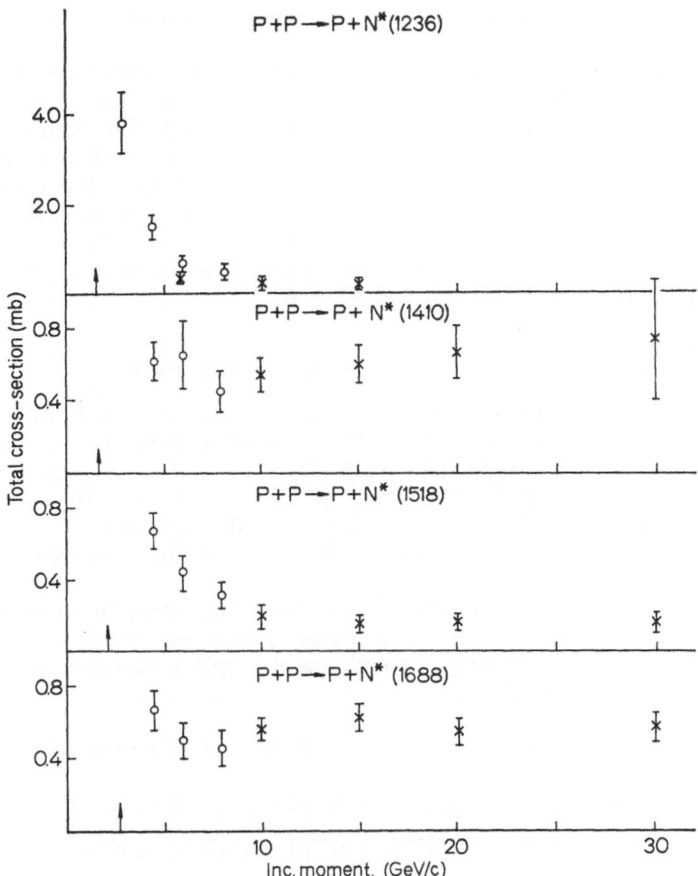

Fig. VIII.11. The cross-sections for isobar production in pp → N*p as a function of laboratory momentum (from *van Hove*, 1967).

The situation here is thus very satisfactory for the Regge model, and it is to be hoped that further experimental work will confirm it. Note that, conversely, if this behaviour did not occur we should be forced to one of two equally unsatisfactory conclusions; namely that either for some reason the P does not couple to the states $1 + 2$, though no known quantum number forbids this coupling; or else that there is no Pomeranchon trajectory, and diffraction scattering can not be explained in terms of Regge poles.

An interesting, though rather speculative, application of the ideas of this section is due to *Morrison* (1967) who has noted that the incident energy variation of the reaction $N\pi \to N + (\pi\varrho)$ in the neighbourhood of the $A_2(\pi\varrho)$ resonance suggests that there is some contribution from Pomeranchon exchange. Since this cannot occur for the A_2, he suggests that in the neighbourhood of the A_2 there is also an $I = 1$, $J^P = 1^+$ or 2^- resonance.

Gribov (1967 b) pointed out that it would be very interesting to have information on the energy variation of a process like $\pi N \to O^+ N$ where O^+ is a $J^P = 0^+$ meson (if such a meson exists). The Pomeranchon cannot couple to a $\pi + O^+$ state but the cut corresponding to double Pomeranchon exchange can contribute and must give the dominant contribution to this cross-section at high energy. In marked contrast to the other processes for which cut contributions inevitably dominate at sufficiently high energies, this process would not decrease rapidly with energy (cf. sections VIII. 4 (f) and VIII. 6 (i)).

c) Non-Strange Meson Exchange Possible, P Impossible

For this class the leading trajectories are the 1^- and 2^+ nonet trajectories, so that according to (4.5) we would expect the high energy total cross-section to behave like s^{-n} with n around unity, or rather less than unity to allow for the slope of the trajectory $[\alpha(t) < \alpha(0)$ for $t < 0]$. *Morrison* (1966 a, b) lists a large number of total cross-sections which fall into this class, and the values of n are in most cases as one would expect.

We now discuss particular examples of this class in more detail. First, we consider the charge-exchange processes. These are in fact related to elastic scattering, through isotopic spin invariance, according to the following equations:

$$A\,(\pi^- p \to \pi^0 n) = \frac{1}{\sqrt{2}}\,[A\,(\pi^+ p \to \pi^+ p) - A\,(\pi^- p \to \pi^- p)] \qquad (4.8)$$

$$A\,(K^- p \to \bar{K}^0 n) = [A\,(K^- p \to K^- p) - A\,(K^- n \to K^- n)] \qquad (4.9)$$

$$A\,(K^+ n \to K^0 p) = [A\,(K^+ p \to K^+ p) - A\,(K^+ n \to K^+ n)] \qquad (4.10)$$

$$A\,(\bar{p}p \to \bar{n}n) = [A\,(\bar{p}p \to \bar{p}p) - A\,(\bar{p}n \to \bar{p}n)] \qquad (4.11)$$

$$A\,(pn \to np) = [A\,(pp \to pp) - A\,(pn \to pn)]\,. \qquad (4.12)$$

The relations are true for each helicity amplitude, for all s and t. For meson-baryon scattering there is only the non-spin-flip amplitude at $t = 0$ [the spin flip term is zero in the forward direction by angular momentum conservation — see section (IV. 3)], so we can immediately use (4.8) to (4.10) to relate the imaginary part of the forward charge-exchange amplitude to the total elastic cross-sections. For example, from (4.7) and (3.1) we have

$$\mathrm{Im}\{A\,(\pi^- p \to \pi^0 n, s, t = 0)\} = q_s\sqrt{2s}\,[\sigma^T(\pi^+ p, s) - \sigma^T(\pi^- p, s)]. \qquad (4.13)$$

Note, however, that for nucleon-nucleon scattering the double spin flip amplitude, $\langle 1/2 - 1/2 |A| - 1/2\ 1/2 \rangle$, is not necessarily zero in the forward direction, so in this case we cannot relate the imaginary part of the forward charge-exchange amplitude to the total cross-section.

Because of (4.13) a measurement of $\operatorname{Im} A$ for the charge-exchange amplitude at $t = 0$ would not give a test of Regge theory which is independent of the one discussed in the previous section. However, experimentally it is the differential cross-section, proportional to $|\operatorname{Re} A|^2 + |\operatorname{Im} A|^2$, which is measured, and this is not simply a function of the total cross-section. The data have been analysed by *Höhler* et al. (1966) (see also the references mentioned in this paper) who fit with a single ϱ-exchange Regge trajectory (see previous section) and obtain a good fit between 4 and 18 GeV/c, with a value of $\alpha_\varrho(0)$ given by

$$\alpha_\varrho(0) = 0.58 \quad \text{or} \quad 0.57 \quad \pm 0.01 \,, \qquad (4.14)$$

the different results depending on whether the linear extrapolation to $t = 0$ is made using data in the ranges $0 < -t < 0.28$ (Gev/c)² or $0 < -t < 1.0$ (GeV/c)² respectively. This is in excellent agreement with the values [(3.11), (3.22) and (3.31)] obtained from the total cross-sections, and shows that the energy dependence of the real part of the amplitude is (within the limits of experimental error) the same as that of the imaginary part, as predicted by the Regge theory. We shall discuss this point further when we deal with the phase of amplitudes in the next section. Incidentially this result also provides confirmation of isotopic spin invariance at high energy.

As a second example of a process of this class we consider

$$\pi^- p \to \eta n \,. \qquad (4.15)$$

Here the quantum numbers are such that, of the trajectories available [section (VIII. 1)], only the A_2 can contribute. *Phillips* and *Rarita* (1965 b) have analysed the data and find a good fit to the extrapolated forward cross-section with a value of $\alpha_{A_2}(0)$ between 0.25 and 0.40. We shall discuss this, and the previous process, in more detail in section (VIII. 6).

This class of processes also includes those of the type

$$PN \to VB \qquad (4.16)$$

where P is a pseudoscalar meson, V a vector meson and B a baryon, for which a phenomenon noted at the end of section (IV. 5) occurs, and is especially significant. In the s-channel forward direction the angle z_t is not proportional to s but is equal to one, so the high energy behaviour in the forward direction $(z_s = 0)$ of t-channel amplitudes involving a change of helicity (M) is expected to be $s^{\alpha-M}$ not s^α. [See the discussion in section (IV. 3).] For trajectories with $\mathscr{S}$ (signature) $= P$ it is easy to see [L. *Jones* (1967)] that only t-channel amplitudes with $M > 0$ occur, whereas for $\mathscr{S} = - P$ there are $M = 0$ terms. Hence near the $z_s = 0$ direction the $\mathscr{S} = - P$ trajectories are enhanced relative to the $\mathscr{S} = P$ trajectories. An example is

$$\pi N \to \varrho N \,, \qquad (4.17)$$

where this mechanism allows the π trajectory to dominate the ω trajectory although it is lower. We shall see in section (VIII. 7) that there is already some slight confirmation of this behaviour from the ϱ-decay distribution.

d) Strange Meson Exchange Possible

Typical processes of this type are

$$\pi^- p \to \Lambda\ K^0 \quad \text{or} \quad \Sigma^0 K^0, \tag{4.18}$$

$$K^- p \to \Lambda\ \pi^0 \quad \text{or} \quad \Sigma^+ \pi^- \text{ etc.,} \tag{4.19}$$

$$\bar{p} p \to \bar{\Lambda}\Lambda \quad \text{etc.} \ldots \tag{4.20}$$

Since the strange mesons are heavier than the lightest non-strange mesons in the same octet their Regge trajectories are lower, and processes in this class should decrease with energy faster than processes in Class (b). This is confirmed by the data given by *Morrison* (1966b); the cross-sections varying as s^{-n} where the average value of n is about 2. The data for most of this class of processes are, however, not very accurate.

e) Baryon Exchange Possible

Examples of processes of this class are meson-baryon backward scattering for which baryon trajectories in the u-channel are expected to dominate. The question immediately arises as to whether the nucleon lies on a Regge trajectory, or whether it is an "elementary" particle (in the sense of chapter VII). One feels that if the nucleon is not "elementary" then it is unlikely that any strongly interacting particle is! If the nucleon were elementary then there would be a $\delta_{J\,1/2}$ term in the partial-wave amplitude, and the amplitude for backward πN scattering would behave like $s^{1/2}$. Otherwise the amplitude should behave like $s^{\alpha_N (0)}$, where $\alpha_N (0) < 1/2$ because of the slope of the trajectory.

Chiu and *Stack* (1967) have analysed backward $\pi^\pm p$ scattering data between 4 and 10 GeV/c. The existence of two isotopic spin states allows the separation between the nucleon and the Δ trajectories to be made, and they find

$$\alpha_N (0) \simeq -0.34. \tag{4.21}$$

This apparently confirms that the nucleon is on a Regge trajectory. Incidentally it also supports the conclusion of section (III. 8) that the Regge formula holds if $z_n \to 1$ rather than ∞ for high s. We discuss further interesting features of this process in section (VIII. 6).

f) No Known Trajectories Can be Exchanged

There are some processes which are in principle possible but for which none of the trajectories known can contribute, e.g. the process $K^- p \to K^0\ \Xi^0$, which would require a meson trajectory of strangeness equal to 2. In these processes cuts may well dominate. Thus in the example quoted the cut arising from two K*-meson exchanges, and starting at

$$\alpha_c (0) = 2\ \alpha_{K^*} (0) - 1 \tag{4.22}$$

would occur at a value of -0.4 and would probably be the leading contribution. In all such cases no events have been observed so far, which is consistent with the fact that they should decrease rapidly with energy. When better statistics become available it should be possible to obtain reasonable upper limits to the contribution of cuts. These are very much needed.

VIII.5. The Phase of the Forward Amplitude

a) The Regge Pole Prediction

A striking feature of the Regge-pole theory (which, however, is more general than this theory as we shall explain below) is that it predicts a unique relation between the phase of an amplitude and the power of its energy dependance. Thus if an amplitude is dominated by a trajectory

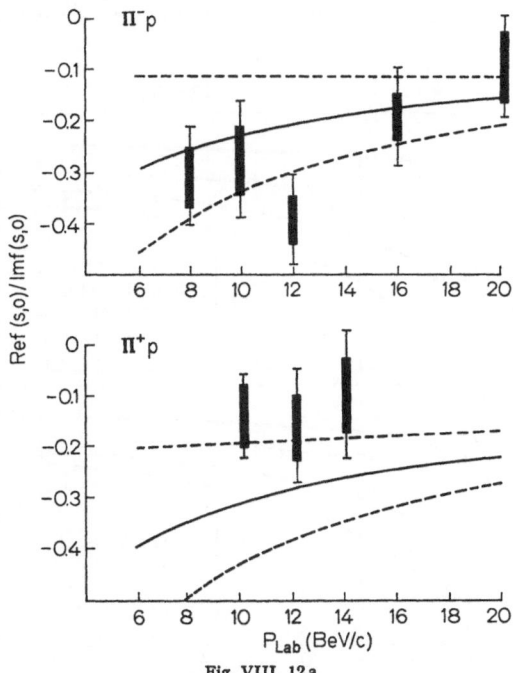

Fig. VIII. 12a

Fig. VIII.12a and b. Experimental data and theoretical curves for the ratio of the real to the imaginary parts of the forward scattering amplitudes for (a) $\pi^\pm p$, and (b) $\bar{p}p$, pp, $\bar{p}n$, pn. The dashed curves represent the error corridors of the fits to the total cross-sections (from *Barger* and *Olsson*, 1966a). (Fig. VIII. 12 b see p. 220).

$\alpha(t)$, and if $\alpha(t)$ is real for physical $t > 0$, then the high energy behaviour of the amplitude is given by $s^{\alpha(t)}$, and the phase of the amplitude by

$$\tan \Phi = \frac{-\sin \pi \alpha(t)}{\cos \pi \alpha(t) \pm 1}, \tag{5.1}$$

where the $\pm$ sign occurs when the signature is even (odd). This follows from the Regge formulae (II. 10.3) or (IV. 5.2). We noted in section (IV. 6) that kinematical cuts do not invalidate this formula in the s-channel physical region.

For the Pomeranchon trajectory $\alpha(0) = 1$, so we predict a purely imaginary forward amplitude in the high energy limit. In fact measurements for πN and NN do not give a purely imaginary amplitude, but this is not surprising since we have seen that at presently available energies other trajectories give sizable contributions [see section (VIII. 3)].

b) Some Experimental Tests

Barger and *Olsson* used the parameters of their fits to the total cross-sections, discussed in section (VIII. 3), to calculate the ratio of real part to the imaginary part of the forward elastic amplitudes. These predictions are compared with the experimental values for $\pi^{\pm} p$, pp,

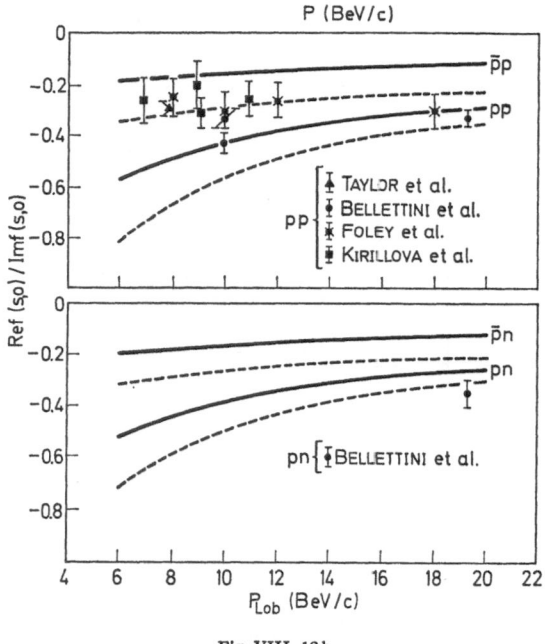

Fig. VIII. 12 b

and pn in Fig. (VIII. 12). Improved data [Foley et al. (1967)] shows a small disagreement in the πp case at higher energies, but the uncertainty in the fits makes this of minor significance. There is no experimental data for the other processes.

The $\pi^+ p$, $\pi^- p$ difference is again worth special consideration. From Table (VIII. 2) and (5.1) it follows that

$$\text{Re } A \ (\pi^+ p \to \pi^+ p, s, 0) < \text{Re } A \ (\pi^- p \to \pi^- p, s, 0) \ . \tag{5.2}$$

Since, experimentally [*Foley* et al. (1965a)], both these quantities are negative this implies

$$|\text{Re } A \ (\pi^+ p \to \pi^+ p, s, 0)| > |\text{Re } A \ (\pi^- p \to \pi^- p, s, 0)| \ . \tag{5.3}$$

Now for the total cross-sections [see Fig. (VIII. 8)] we have

$$\sigma^T \ (\pi^- p) > \sigma^T \ (\pi^+ p) \ , \tag{5.4}$$

so, combining (5.3) and (5.4), we find

$$\frac{- \text{Re} A \ (\pi^+ p \to \pi^+ p, s, 0)}{\text{Im} A \ (\pi^+ p \to \pi^+ p, s, 0)} > \frac{- \text{Re} A \ (\pi^- p \to \pi^- p, s, 0)}{\text{Im} A \ (\pi^- p \to \pi^- p, s, 0)} \ . \tag{5.5}$$

The data [Fig. (VIII. 12)] appear to be in conflict with this equation, but the errors are such that one cannot at this stage be certain. Clearly more accurate evaluations of the real parts of these amplitudes would be of great interest.

We can study essentially the same question by looking directly at the charge exchange amplitude $\pi^- p \to \pi^0 n$. Assuming that this is dominated by the ϱ trajectory we have

$$\frac{\text{Re} A \ (\pi^- p \to \pi^0 n, s, 0)}{\text{Im} A \ (\pi^- p \to \pi^0 n, s, 0)} = \tan \left(\frac{\pi \alpha_\varrho (0)}{2} \right) \tag{5.6}$$

$$\backsim 1.25 \tag{5.7}$$

using (3.11). Thus

$$\frac{d\sigma}{d\Omega} \ (\pi^- p \to \pi^0 n, s, t) \Big|_{t=0} = \frac{1}{(8\pi)^2 s} \ [1 + (1.25)^2] \ [\text{Im } A \ (\pi p \to \pi^0, n, s, 0)]. \tag{5.8}$$

We can evaluate the right-hand-side from (3.1), (4.8) and the $\pi^- p$ and $\pi^+ p$ total cross-sections, and predict the left-hand-side of (5.8). The prediction agrees well with experiment [*Barger* and *Olsson* (1966a)].

From the experimental facts that $\sigma^T \ (K^- p) > \sigma^T \ (K^- n)$ and $\sigma^T \ (K^+ p) > \sigma^T \ (K^+ n)$ the residues of the ϱ and A_2 trajectories must have the signs explicitly indicated in Table (VIII. 2) of section (VIII. 1). Then, from (4.9), (4.10) and (5.1), it follows that the contributions of ϱ and A_2 to the real parts of the charge exchange amplitudes will combine positively for $K^+ n \to K^0 p$, but tend to cancel for $K^- p \to K^0 n$. We therefore expect the real part to be larger in the former case than in the latter. Unfortunately there are no measurements for $K^+ n \to K^0 p$; the one result for $K^- p \to K^0 n$ is at 9.5 BeV/c and gives a small real part as required [see *Barger* and *Cline* (1967b)].

A case where the phase prediction causes some difficulty is np charge exchange. According to isotopic spin invariance the non-spin-flip forward amplitude is related to the difference between the pp and pn

total cross-sections (4.12). Now experimentally this difference changes sign at about 3.7 GeV/c, i.e.

$$\sigma^T \, (\mathrm{pp}) \gtrless \sigma^T \, (\mathrm{pn}) \tag{5.9}$$

for incident momenta $\lesssim 3.7$ GeV/c. This behaviour can readily be explained from Table (VIII. 2). All that is required is that the A_2 contribution should equal the ϱ contribution at an incident momentum of 3.7 GeV/c. At higher momenta the ϱ will dominate because its trajectory lies higher [c.f. (3.8) and the discussion below (4.15)]. In fact the A_2 residue is not very well determined by the higher energy fits, but they are compatible with this behaviour [see *Barger* and *Olsson* (1966a)].

Because the ϱ and A_2 have opposite signature their contributions to the real part of np charge exchange will have the same sign, but to the imaginary part they will have opposite sign. Thus the charge-exchange amplitude will be mainly real in the fairly low energy region. Since the imaginary part is already determined from the total cross-section data, this allows a lower limit to be obtained for the forward charge-exchange cross-section. It is a lower limit because double spin flip might occur in the charge-exchange process. *Högaasen* and *Frisk* (1966a) have pointed out that this lower limit is about 60 times the experimental value at 8 GeV/c. In a later paper [*Högaasen* and *Frisk* (1966b)] they include, in addition to the trajectories we have mentioned, a ϱ' trajectory with the quantum numbers of the ϱ, and are able to fit all the relevant data down to about 3 GeV/c with a value of $\alpha_{\varrho'} (0)$ given by

$$\alpha_{\varrho'} (0) \simeq - 0.63 \, . \tag{5.10}$$

They also note that an A_2' trajectory (with the quantum numbers of the A_2) would do equally well.

It is worth noting here that we expect, on the basis of simple models, that several trajectories with the same quantum numbers should exist. However, because of the signature effect, these are expected to be about two units of angular momentum apart [rather than one as required for the ϱ' with (5.10)]. For example, in a simple potential model, we would have trajectories tending to alternative integers. The ϱ' trajectory could, alternatively, be a manifestation of a cut in the J-plane. For other evidence for the ϱ' (as a Regge trajectory, or approximation to a cut contribution) we refer to section (VIII. 7).

c) Other Ways of Obtaining the Regge Prediction

As we remarked above, the Regge pole prediction of a unique relation between the phase and the index of the power behaviour in energy can be obtained without using Regge theory explicity. To see this let us assume only that (i) $A \, (s, t)$ satisfies a dispersion relation in s,

$$\text{(ii)} \; \frac{\operatorname{Im} A \, (s, t)}{|s|^\alpha} \xrightarrow[s \to \infty]{} C \, , \tag{5.11}$$

and

$$\frac{\operatorname{Im} \{ A \, (s, t) \}}{|s|^\alpha} \xrightarrow[s \to - \infty]{} \pm C \, , \tag{5.12}$$

for some real α. The number of subtractions necessary in the dispersion relation depends on α. For example, suppose $0 < \alpha < 1$, then we can write a once-subtracted dispersion relation, which we put in the form, (I. 10.8),

$$\mathrm{Re}\,\{A\,(s,t)\} = s\,\frac{\mathrm{P}}{\pi}\,\int_{\mathrm{R.H.}}\frac{\mathrm{Im}\{A\,(s',t)\}}{(s'-s)\,s'}\,\mathrm{d}s' + s\,\frac{\mathrm{P}}{\pi}\,\int_{\mathrm{L.H.}}\frac{\mathrm{Im}\{A\,(s',t)\}}{(s'-s)\,s'}\,\mathrm{d}s' +$$

$$+\ \mathrm{Re}\,\{A\,(0,t)\} \tag{5.13}$$

where P ... means the principle value of the following integral. We insert (5.11) and (5.12) into (5.13) and use the formulae [TIT, 15.2 (28) and 14.2 (5)]

$$\frac{\mathrm{P}}{\pi}\int_0^\infty \frac{\mathrm{d}s'}{s'-s}\,s'^{\alpha-1} = -\,s^{\alpha-1}\cot \pi\alpha \tag{5.14}$$

and

$$\frac{1}{\pi}\int_0^\infty \frac{\mathrm{d}s'}{s'+s}\,s'^{\alpha-1} = -\,s^{\alpha-1}\operatorname{cosec}(\alpha-1)\,\pi \tag{5.15}$$

for $s > 0$. Thus

$$\frac{\mathrm{Re}\{A\,(s,t)\}}{s^\alpha}\xrightarrow[s\to\infty]{} \to -\,C\,[\cot \pi\alpha \mp \operatorname{cosec} \pi\alpha]. \tag{5.16}$$

Combining this with (5.11) we obtain (5.1). Evidently this argument could be repeated for other ranges of α.

Thus we see that the experimental agreement with the predicted phase provides evidence in support of a power behaviour, (5.11) and (5.12), and a dispersion relation for the amplitude, but not necessarily for other aspects of the Regge theory.

VIII.6. The Angular Dependence of Differential Cross-Sections

a) Regge Poles and Differential Cross-Sections

The contribution of a single Regge pole to a scattering amplitude has the form

$$A^{\mathrm{R}}\,(s,t) = \Gamma(t)\,(s/s_0)^{\alpha\,(t)}, \tag{6.1}$$

where $\Gamma(t)$ is the product of the residue, $\beta(t)$, and kinematical factors. Now the t-dependence of $\beta(t)$ is unknown, and associated with every trajectory there are many different $\beta(t)$'s (they are not entirely independent because of the factorization theorem), so it is clear that fitting differential cross-sections is a considerably more involved process than fitting total cross-sections.

It is usual to extract various kinematical factors from the residue functions, although as we noted in section (IV. 6) there is no compelling reason for doing this. The hope here is that the reduced residue is approximately constant over the interesting region of t, but as we shall see this hope is not always fulfilled. The fact that there is no general agreement on the best way to parameterize $\beta(t)$ often makes comparison between the works of various authors difficult.

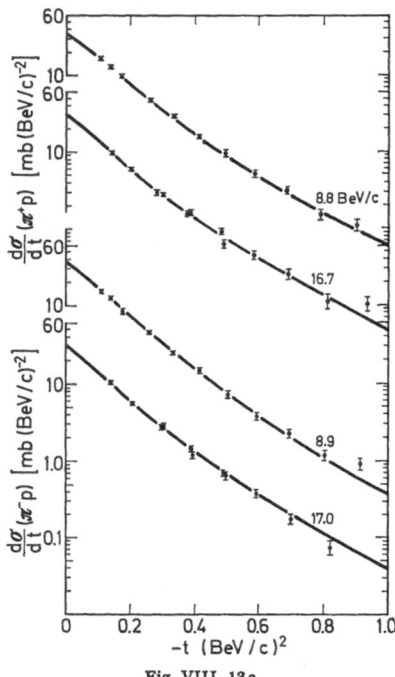

Fig. VIII. 13 a

Fig. VIII.13 a—e. The experimental differential cross-sections for (a) $\pi^\pm p$, (b) pp, (c) $\bar{p}p$, (d) K^+p, (e) K^-p at various energies. These curves are taken from *Rarita* et al., 1967, and *Phillips* and *Rarita*, 1965a, whose fits are shown as solid lines. (Fig. VIII. 13 b and c see p. 225. Fig. VIII. 13 d see p. 226 and Fig. 13 e see p. 227).

There are two particular features of (6.1) which are worth noting, namely, it predicts a "shrinking" of the angular distribution with increasing energy, and, because of zeros which occur in $\Gamma(t)$ in the physical region, it leads to dips in the angular distribution. In order to understand the first point we consider $t \backsimeq 0$ and write

$$\alpha(t) \backsimeq \alpha(0) + t\,\alpha'(0).$$ (6.2)

Then

$$A^R(s,\,t) = \Gamma(t)\,s^{\alpha\,(0)}\,e^{\alpha'\,t\,\log\,(s/s_0)}.$$ (6.3)

We note that if $\Gamma(t)$ varies slowly [of course this can only be true if we have chosen s_0 suitably in (6.1)], then the fall-off with momentum transfer squared $(= -t)$ is exponential. Further, and this is independent of $\Gamma(t)$, the rate of fall-off with momentum transfer *increases* with energy, i.e. the size of the forward diffraction peak shrinks as the energy increases. This is a striking, and rather unexpected result. It means that the "size"

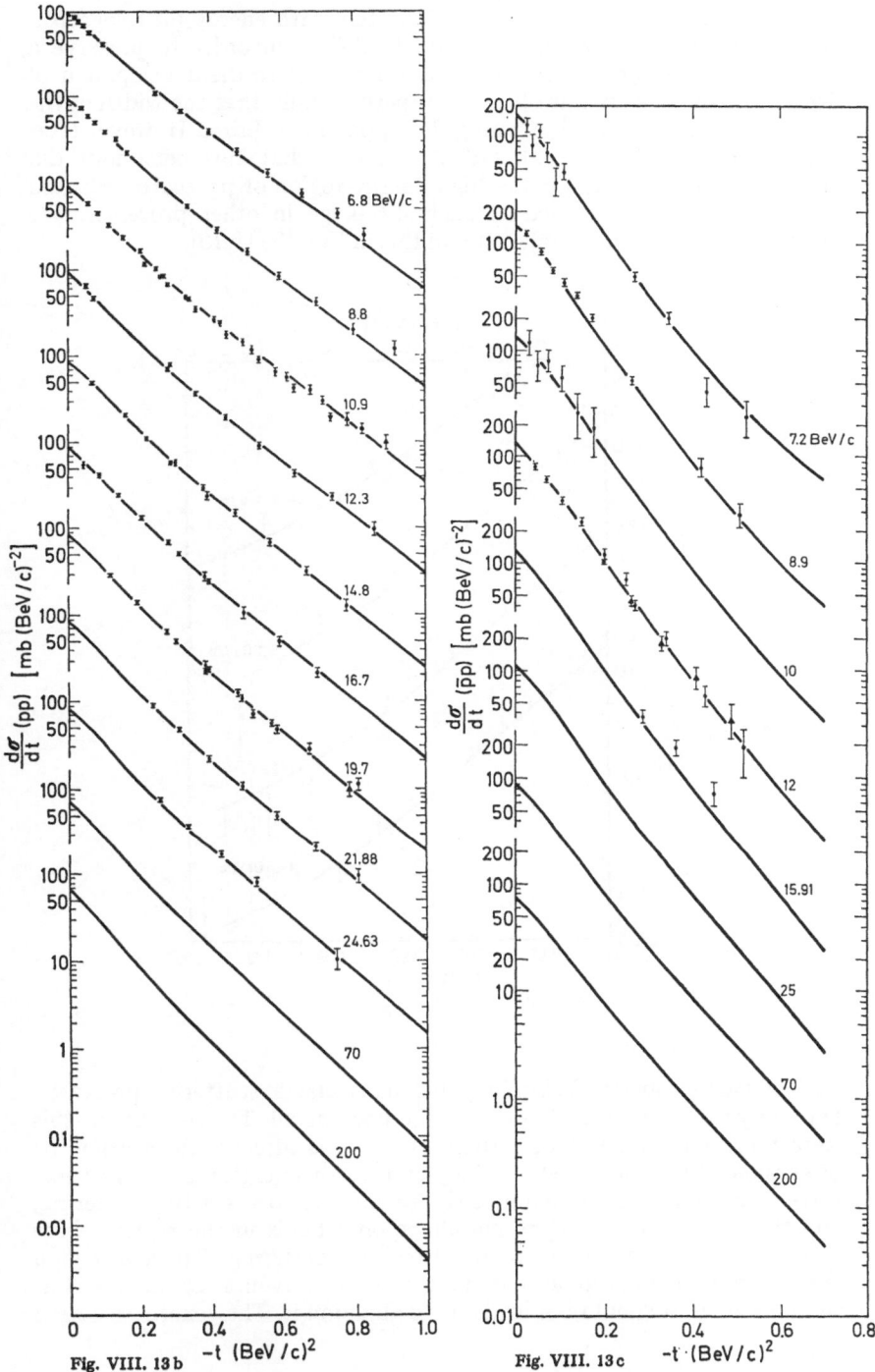

Fig. VIII. 13b

Fig. VIII. 13c

of the scattering region increases indefinitely with energy (at a logarithmic rate), but becomes correspondingly diffuse in order to preserve a
constant total cross-section. It was a great boost to the development of
Regge theory when it was observed experimentally that the width of the
diffraction peak for pp scattering did apparently shrink. However later
measurements [*Foley* et al. (1963)] showed that this behaviour did
not occur so markedly for the high energy region of pp scattering, and
that there was no evidence for such shrinkage in other processes. We
illustrate the latest available data on this in Fig. (VIII. 13).

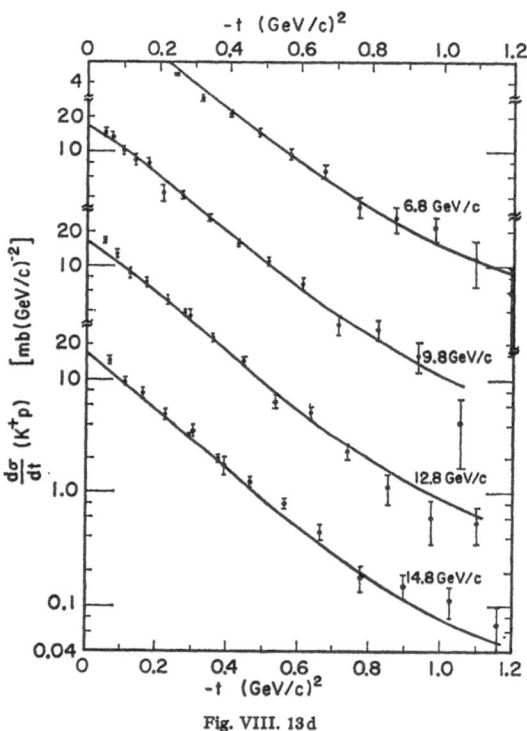

Fig. VIII. 13 d

The lack of observed shrinkage for many elastic scattering processes,
led many people to abandon the Regge pole model. The reasons for this
were more emotional than rational since it is already clear from the
discussion of section (VIII. 3) that, at present energies, several trajectories are contributing significantly to high energy elastic scattering,
and the argument leading to shrinkage only holds in the region where
one trajectory is adequate to describe the scattering. Thus, although
the lack of shrinkage somewhat spoils the attractiveness of the model at
these energies, it should not lead one to abandon it. The general situation

with regard to elastic scattering is complicated so we study first cases where only one trajectory is involved, before looking at the attempts which have been made to fit the more accurate elastic scattering data.

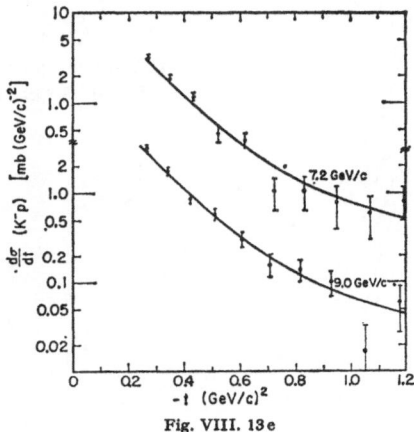

Fig. VIII. 13 e

b) π-N *Charge Exchange Scattering and the* ϱ *Trajectory*

The simplest available process is $\pi^- p \rightarrow \pi^0 n$. In the *t*-channel of this amplitude (i. e. $N\bar{N} \rightarrow \pi\pi$) there are two independent ϱ-residues, i.e. the ϱ-residues in $\langle 00|A^t| + + \rangle$ and $\langle 00|A^t| + - \rangle$. For convenience of comparison we shall use here the notation of *Höhler* et al. (1966) and write for the $\pi^- p \rightarrow \pi^0 n$ differential cross-sections

$$\frac{d\sigma}{dt} = \left(\frac{\omega}{M}\right)^{2\alpha - 2} [1 + t/4 M \omega]^{2\alpha} T(t) , \qquad (6.4)$$

with

$$T(t) = \frac{1}{2M^4} (1 - t/4M^2)^{-1} \cos^{-2} \frac{\pi\alpha}{2} \left[\frac{2^\alpha(\alpha + 1/2)!}{\alpha!} \right]^2 V(t) , \qquad (6.5)$$

and

$$V(t) = b_+^2(t) - \frac{t\alpha^2(t)}{4M^2} b_-^2(t) , \qquad (6.6)$$

where ω is the pion energy in the laboratory system, $\alpha = \alpha(t)$ is the ϱ-meson trajectory, and $b_\pm(t)$ are two independent residues which are free of kinematical singularities. The factor $\alpha^2(t)$ multiplying $b_-^2(t)$ in (6.6) arises because $J = 0$ is a nonsense value for the $\langle 00|A^t| + - \rangle$ amplitude [see section (IV. 6)]. We discuss an alternative to (6.6) below.

In Fig. (VIII. 14) we show the available data [*Sonderegger* et al. (1966)]; clearly the data contains a lot of information. In Fig. (VIII. 15) we show $d\sigma/dt$ $(\pi^- p \rightarrow \pi^0 n)$ as a function of $\log \omega$ for various values of t. According to (6.4) these plots should be straight lines with a slope given by $2\alpha(t) - 2$ (the factor in square brackets in (6.4) is negligible except for the larger *t*-values; it is included in the fits discussed below). The

15*

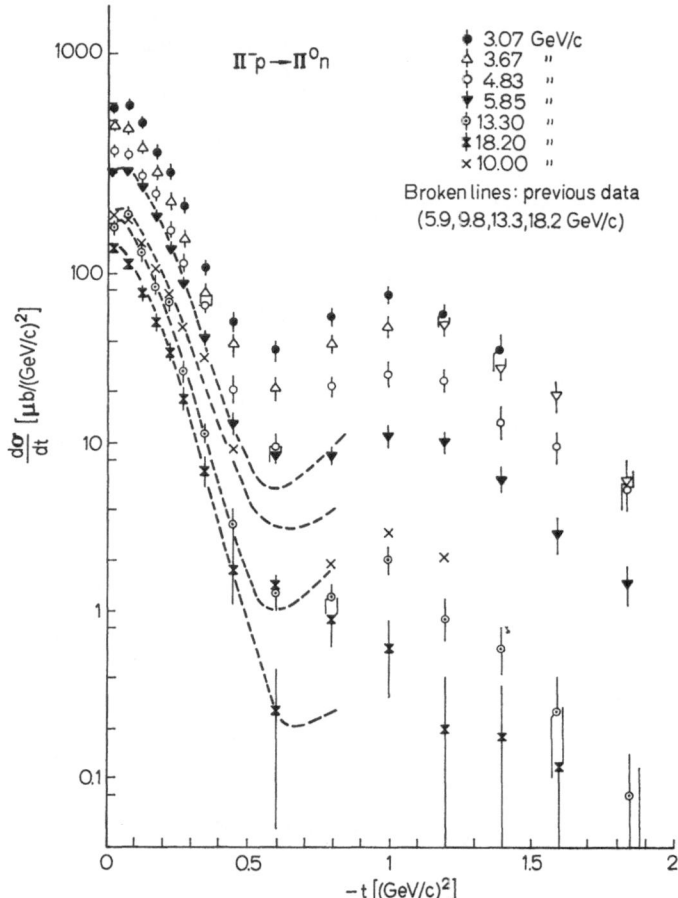

Fig. VIII.14. The differential cross-sections for $\pi^- p \rightarrow \pi^0 n$ (from *Sonderegger* et al., 1966). The broken lines are smooth curves through earlier data.

fact that the slopes of the lines in Fig. (VIII. 15) become steeper with increasing t clearly indicates shrinkage. In fact the dashed lines in this diagram give the slopes expected if we have a zero slope for the trajectory i.e. $\alpha(t) = \alpha(0) = 0.57$. A good fit to the values of $\alpha(t)$ is a linear trajectory

$$\alpha(t) = \alpha(0) + t\alpha'(0) . \tag{6.7}$$

This is illustrated in Fig. (VIII. 16a) where the values of $\alpha(t)$ computed from (6.7) are compared with the values of $\alpha(t)$ obtained directly from Fig. (VIII. 15). The best values of $\alpha(0)$ and $\alpha(t)$ are given in Table (VIII. 3); they depend a little on the energy and momentum transfer range which is included in the analysis. Also shown on Fig. (VIII. 16a)

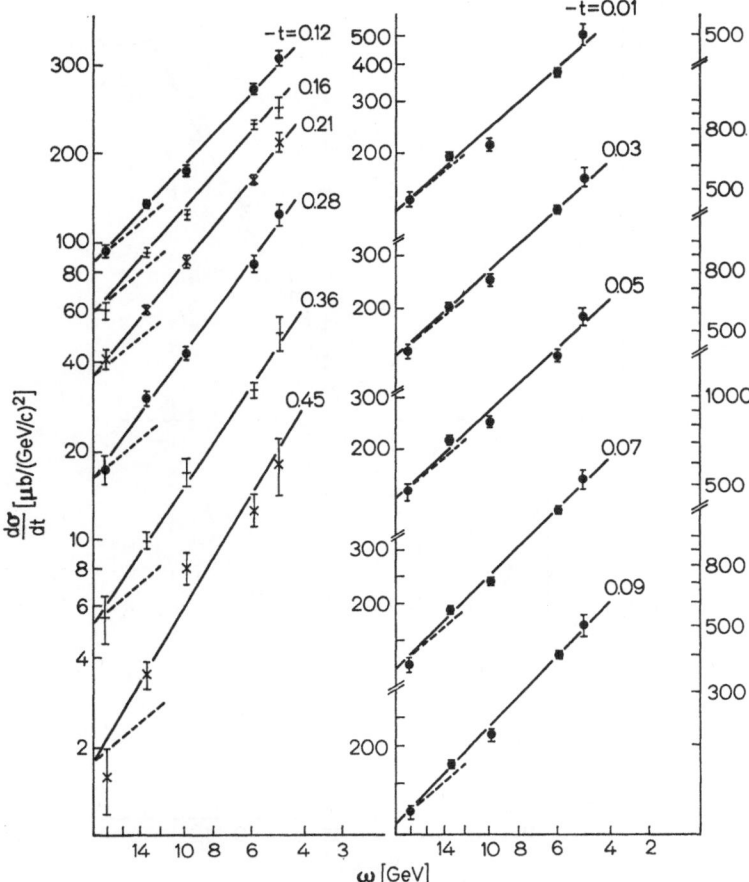

Fig. VIII.15. Log (dσ/dt) for π⁻p → π⁰n as a function of log ω. The dashed lines show the slope 2 α(0) − 2. The solid lines are the fit to (dσ/dt) with α(t) = 0.57 + 0.91 t (from *Höhler* et al., 1966).

is an earlier fit to the values of $\alpha(t)$ obtained by *Phillips* and *Rarita* (1965 c), who used a parametrization of α devised by *Pignotti*, which is constrained to pass through the physical ρ, and to have $\alpha(\infty) = -1$ [see section (V. 7)]. This tends to give the trajectory more curvature than the data warrants.

Table VIII.3.

ω-range Gev	(−t)-range (Gev/c)²	α(0)	α′ (Gev/c)⁻²	Re α(m²ρ) (linear)
5.9 — 18.2	0 — 0.28	0.58 ± 0.01	1.00 ± 0.11	1.16 ± 0.07
5.9 — 18.2	0 — 1.0	0.57 ± 0.01	0.91 ± 0.06	1.10 ± 0.04

Fit to the π⁻p → π⁰n differential cross section, from *Höhler* et al., 1966.

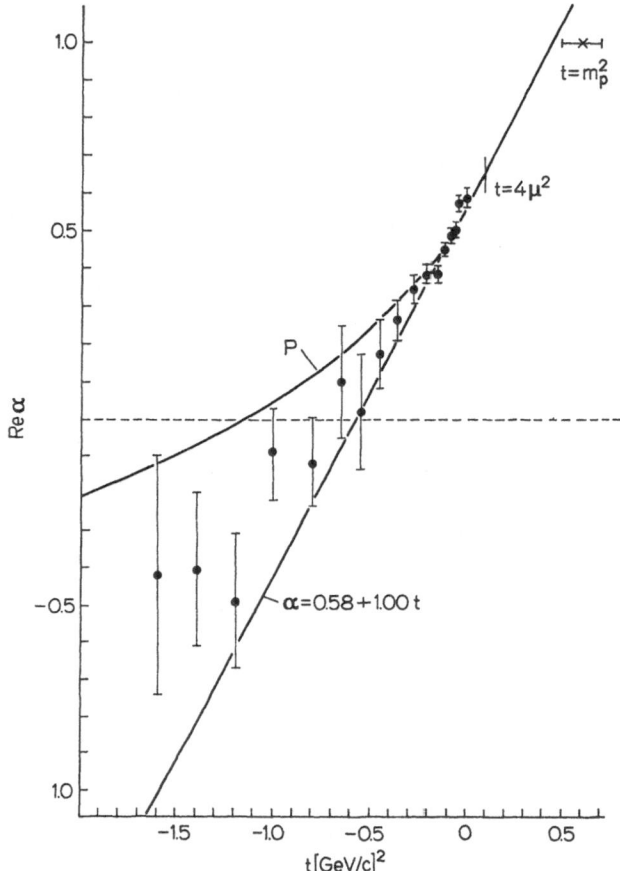

Fig. VIII.16a. The ϱ-meson trajectory from *Höhler* et al., 1966. *P* is the fit of *Phillips* and *Rarita*, 1965a, using Pignotti's formula.

It will be seen from Table (VIII. 3) and Fig. (VIII. 16a) that the straight line fit to the values of $\alpha(t)$, for $t < 0$, extrapolated to $J = 1$, passes very close to the ρ-mass. There is thus a remarkable correlation between the data presented in Fig. (VIII. 14) and the physical ρ-meson. This correlation can hardly be an accident, and provides very convincing evidence for the Regge pole model. Note that the slight evidence of curvature in Fig. (VIII. 16a), and the fact that the extrapolated straight line does not exactly pass through $J = 1$ at $t = m_\rho^2$ is not in any way a defect of the Regge theory; indeed, as we have already noted [and will discuss again in section (VIII. 9)], it is surprising that trajectories should appear to be so straight.

Turning now to the dependence on t, we show the best fit of $V(t)$ [of (6.6)] as a function of t in Fig. (VIII. 16b). It has two striking features:

a forward dip, and a second dip at $t \approx -0.6 \, \text{GeV}/c$. Both of these can be explained very readily from (6.6) without involving any t-dependance of $b_{\pm}(t)$. If we suppose that b_- is larger than b_+, the factor $t \, \alpha^2(t)$ has precisely the form required to give the dip at $t = 0$, and at $t \sim -0.6 \, \text{GeV}/c$ where, according to the above fits, $\alpha(t) \approx 0$. This is an example of the second feature mentioned at the beginning of this section, which was pointed out by *Frautschi* (1966). Note that we are here using independent information; $\alpha(t)$ above was determined entirely by the energy variation, and the absolute value at a given t was not used. In Fig. (VIII. 16b) we show the fit of *Höhler* et al. to $V(t)$ obtained with $b_{\pm}(t) = b_{\pm}(0)$ and $b_-/b_+ = 13$. These authors point out, however, that if the physical ρ plays the major part usually assigned to it in electromagnetic form

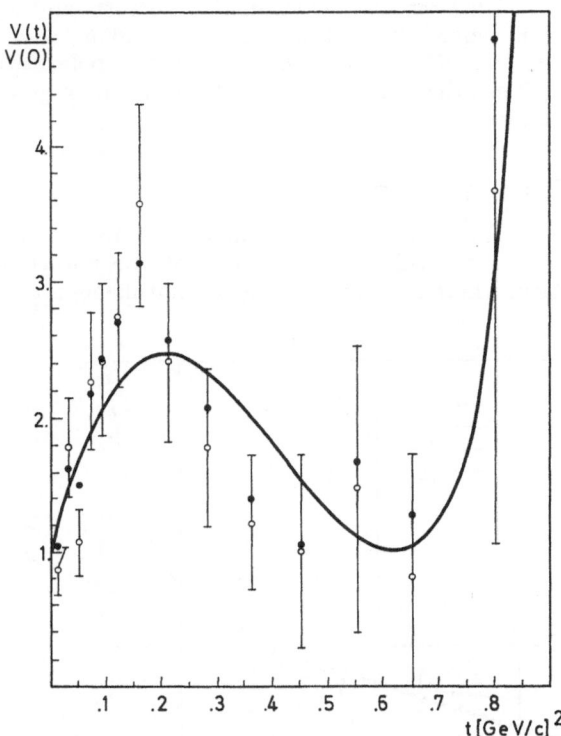

Fig. VIII.16 b. Values of $V(t)/V(0)$ obtained from the data of *Sonderegger* et al. (1966). The curve shows the fit of *Höhler* et al. (1966) referred to in the text.

factors, then the ratio b_-/b_+ cannot really be constant, since it must be approximately 3 at $t = m_\rho^2$. Similar explanations of the dip have been reported by *Arbab* and *Chiu* (1966), and *Arbab* et al. (1967).

In this discussion we have ignored the possibility [mentioned in section (V. 4)] that the fixed pole which we know occurs in the $N\bar{N} \to \pi\pi$

amplitude at the wrong-signature nonsense value, $J = 0$ [i.e. in $\langle 0 0 | A^- (N \bar{N} \to \pi \pi) | 1/2 - 1/2 \rangle$, where the superscript refers to the signature], might also occur in the residue of the ϱ-trajectory, where this passes through $\alpha_\varrho = 0$. If it did it would remove the zero due to the α_ϱ factor and hence destroy the above explanation of the dip in the differential cross-section. Nothing in the discussion of Chapter V tells us whether this pole should or should not occur in the residue, but, even if it does, it is likely to be in some sense "weak", since it is due to the presence of the third double spectral function. Thus it is likely that, in any case, the above explanation for the dip (and similar explanation of other dips, another example of which is discussed below) is basically correct.

We see then that the ϱ Regge pole seems to fit the πN charge exchange data very well, with a remarkably straight trajectory, and with approximately constant values for the functions b_+ and b_- over the region $0 > t > -0.7$ GeV/c. When we come to consider polarization in this process we shall see that this fit cannot, however, be regarded as completely satisfactory.

c) The A_2 Trajectory in $\pi^- p \to \eta^0 n$

Phillips and *Rarita* (1965b) have made a similar analysis of the process $\pi^- p \to \eta n$, to which the A_2 alone of the known mesons can contribute. Again there is clear evidence for shrinkage, and a good fit is

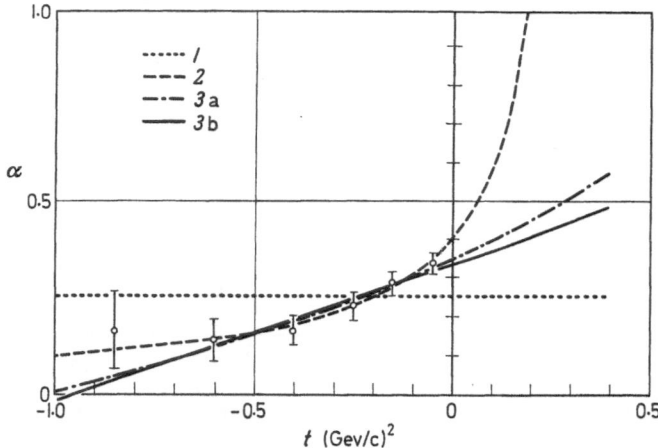

Fig. VIII.17. The A_2 trajectory $\alpha(t)$ against t, obtained from a fit to $\pi^- p \to \eta^0 n$ by *Phillips* and *Rarita*, 1965b.

obtained with the Regge formula for data from about 3 GeV/c upwards. The plot of $\alpha(t)$ against t found by *Phillips* and *Rarita* is shown in Fig. (VIII. 17) together with four possible fits:

$$\text{(i)} \quad \alpha(t) = \alpha(0) \tag{6.8}$$

$$\text{(ii)} \quad \alpha(t) = \alpha(0) + t\alpha'(0) \tag{6.9}$$

$$\text{(iii)} \quad \alpha(t) = \alpha(\infty) + \frac{(\alpha(0) - \alpha(\infty))^2}{\alpha(0) - \alpha(\infty) + t\alpha'(0)}, \tag{6.10}$$

with (iii a), $\alpha(\infty) = 0$, and (iii b), $\alpha(\infty) = -1$. Form (iii) is the one suggested by *Pignotti* (1963). The best values of the parameters, and the corresponding χ^2, are shown in Table (VIII. 4).

Table VIII.4.

Assumption	$\alpha(0)$	$\alpha'(0)$	χ^2	$\chi^2_{\text{exp.}}$
(i)	0.25 ± 0.02	—	39.9	28
(ii)	0.34 ± 0.03	0.35 ± 0.08	21.4	27
(iii a)	0.40 ± 0.04	1.19 ± 0.56	18.8	27
(iii b)	0.35 ± 0.03	0.46 ± 0.11	20.4	27

Parameters of fits to $\pi^-\text{p} \rightarrow \eta^0\text{n}$ from *Phillips* and *Rarita*, 1965 b. The difference between the 4 cases is discussed in the text. The χ^2 obtained is compared with the expected value (= no. of data — no. of parameters).

We see that, for the linear fit, the slope of the A_2 trajectory is much smaller than that of the ϱ. One reason for this appears to be that there is no dip in the differential cross section for this process analogous to that which occurs in $\pi^-\text{p} \rightarrow \pi^0\text{n}$. This means that the value of $\alpha_{A_2}(t)$ must not pass through zero, otherwise there would be a dip due to the vanishing of the crossed channel spin-flip amplitude at the value of t for which $\alpha_{A_2}(t) = 0$.

Arbab et al. (1967) have suggested how a larger slope might be obtained by allowing different properties for the residue functions. To understand their argument we recall that the t-channel spin flip amplitude $\langle + - |A^t| 0\,0 \rangle$ has a residue which behaves as $\sqrt{\alpha}$ in the neighbourhood of $\alpha = 0$, since at this value of α it is a sense-nonsense amplitude. Using the factorization theorem, we can write this residue as

$$\beta_{+-} = \gamma^N_{+-} \gamma^{\pi\eta}$$

and the sense-sense residue as

$$\beta_{++} = \gamma^N_{++} \gamma^{\pi\eta}.$$

It is natural to locate the $\sqrt{\alpha}$ factor in the γ^N_{+-} factor, so that the sense-sense amplitude $\langle ++ |A^t (\pi^-\text{p} \rightarrow \eta\text{n})| 00 \rangle$ does not become zero at $\alpha = 0$. This means that the trajectory chooses "sense" at $\alpha = 0$ (Note that it then decouples from the nonsense-nonsense $\bar{\text{N}}\text{N} \rightarrow \bar{\text{N}}\text{N}$ amplitude since this has residue $\gamma^N_{+-} \gamma^N_{+-}$). The alternative choice would be for the trajectory to choose "nonsense" at $\alpha = 0$, which would require γ^N_{++}, $\gamma^{\pi\eta}$ and $\gamma^{\pi\pi}$, etc., to have the factor $\sqrt{\alpha}$. Note that if we had made the second choice for the ϱ then both helicity amplitudes in the $\pi^-\text{p} \rightarrow \pi^0\text{n}$ would become zero at $\alpha = 0$. This would mean that, with just the ϱ trajectory, the dip would become a zero, and it would be necessary to invoke some other background contribution to explain the lack of a zero in the experimental data.

The situation is somewhat different for the A_2 since this has even signature. The residue of the A_2 at $\alpha(t) = 0$ must therefore be zero in all amplitudes, since we cannot have a bound state with $t < 0$ (the zero is required to cancel the pole coming from the $(\sin \pi \alpha)^{-1}$ in the Regge formula. A state at negative t is called a "ghost", and the removal by a zero residue is called "ghost killing". If the A_2 chooses sense at $\alpha(t) = 0$ then in order to introduce the factor α into β_{-+} we require an additional $\sqrt{\alpha}$ in each of the $\gamma(t)$ functions, so that $\beta_{+-} \sim \alpha^{2/3}$. This leads to a predicted dip at $\alpha = 0$. However if the trajectory chooses nonsense, then already $\beta_{++} \sim \alpha$ and $\beta_{+-} \sim \sqrt{\alpha}$, and no further factors are required to cancel the pole of $(\sin \pi \alpha)^{-1}$ at $\alpha = 0$ [the $d^1_{\lambda \lambda'}$ function has a factor $\sqrt{\alpha}$ in the sense-nonsense case as we have seen in section (IV. 2)]. In this case there will be no dip even if $\alpha_{A_2}(t)$ passes through zero. The removal of this constraint therefore allows considerably more freedom in fitting the data.

d) Other Meson-Nucleon Charge Exchange Processes

Apart from $\pi^- p \to \pi^0 n$ and $\pi^- p \to \eta n$ there are two other processes of this type which require $I = 1$ exchange in the t-channel, viz. $K^- p \to \overline{K}^0 n$ and $K^+ n \to K^0 p$. The latter can of course only be deduced from K-deuteron scattering, and only low energy data is available, but for the former there is good experimental information. *Arbab* et al. (1967) have analysed simultaneously the three processes $\pi^- p \to \pi^0 n$, $\pi^- p \to \eta n$, and $K^- p \to \overline{K}^0 n$ (the latter contains contributions from both ϱ and A_2 trajectories), and have considered three possible cases:

(i) ϱ and A_2 choose sense at $\alpha = 0$,
(ii) ϱ chooses sense, A_2 chooses nonsense,
(iii) ϱ and A_2 choose nonsense.

They give good fits (12 in all) for all three cases. All the fits have linear trajectories. It turns out that the ϱ parameters are not very sensitive to the choice of (i), (ii) or (iii), and in all cases are similar to those obtained above. However, the A_2 parameters, particularly the slope of the trajectory, vary a great deal. In particular with cases (ii) and (iii) a much greater slope is allowed. Even in case (i) they show that by allowing some of the residues to have non-kinematical zeros (see remarks on this later in this section), a slope

$$\alpha'_{A_2}(0) \approx 0.85 \ (\text{GeV}/c)^{-2}$$

is allowed. We refer to the original paper for further details, but it is clear from this analysis that a lot more experimental work is required before the A_2 parameters (including the question of whether it chooses sense or nonsense) can be settled.

Rarita and *Schwarzschild* (1967) have found that the ϱ and A_2 trajectories obtained from these processes are unable to explain the $K^+ n$ charge exchange data at 2.3 (GeV/c), and some other contribution (they use a hypothetical ϱ') appears to be needed.

e) Fermion Trajectories in Backward $\pi^+ p$ Scattering

An interesting process involving a fermion trajectory is $\pi^\pm p$ backward scattering. Here we expect the u-channel ($\pi N \to \pi N$) Regge poles to dominate. An excellent analysis of this is given by *Chiu* and *Stack* (1967). Two particular features of the experimental data are of note

(i) the $\pi^+ p$ backward peak is an order of magnitude larger than the $\pi^- p$ backward peak.

(ii) the $\pi^+ p$ data has a sharp dip at

$$u \simeq -0.2 \ (\text{GeV}/c)^2 \ . \tag{6.11}$$

Both these features are readily explained in terms of the N and Δ trajectories. It is assumed that the former is much more strongly coupled so, since it does not contribute to $\pi^- p$ backward scattering, the relative magnitudes are readily understood. The dip is explained as being due to the kinematical factor arising when the nucleon trajectory passes through $\alpha_N = -1/2$. This is exactly analogous to the dip mechanism mentioned earlier in πN charge exchange scattering (except that here there is only a single coupling involved), and the same remarks about the possibility that the fixed pole in the amplitude at $J = -1/2$ (this is a wrong signature value) might cancel this zero holds here equally. However we proceed on the assumption that this either does not happen, or else gives rise to a "near zero".

Because of the MacDowell symmetry [section (IV. 7)] contributions from the trajectory at both $\pm \sqrt{u}$ must be included, and, in order for the dip to occur, we must have

$$\alpha \left(+\sqrt{u}\right) \simeq \alpha \left(-\sqrt{u}\right) \simeq 0 \tag{6.12}$$

for $u = -0.2 \ (\text{GeV}/c)^2$. This suggests that the trajectory function is approximately even in $\sqrt{u}$, which appears to be confirmed by an analysis of the physical particles which lie on it [see the discussion in section (VIII. 1) and Fig. (VIII. 7)]. For further details of the fit to the data we refer to the paper by *Chiu* and *Stack*.

f) Pion Exchange in NN and N$\bar{\text{N}}$ Charge-Exchange Scattering

Cases in which the angular dependence of the differential cross section is not well understood include the NN and N$\bar{\text{N}}$ charge exchange scattering, that is

$$p + n \to n + p \tag{6.13}$$

and

$$\bar{p} + p \to \bar{n} + n \ . \tag{6.14}$$

The Regge pole model requires that these processes should be dominated, at high energy, by ϱ and A_2 exchange, and quite probably they are, but at the available energies (up to 8 GeV) it is clear that these two trajectories can not explain the data. The ϱ and A_2 give large spin flip contributions, which we have seen are required for πN charge exchange, but of course spin flip amplitudes must vanish in the forward direction. It is therefore hard to understand the sharp peak in the forward direction which is actually found for processes (6.13) and (6.14). The data at 8 GeV are shown in Fig. (VIII. 18).

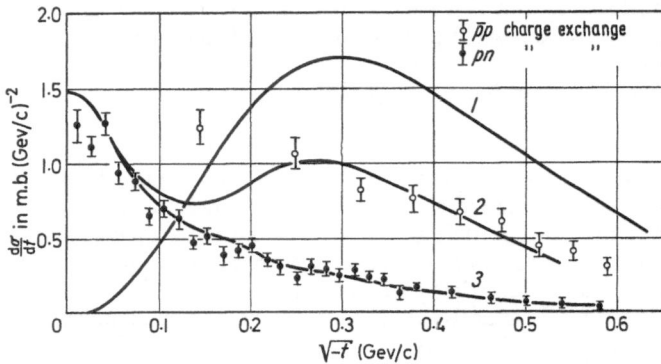

Fig. VIII.18. The differential cross-sections for $\bar{p}p$ and pn charge exchange scattering at 8 (Gev/c), plotted against $\sqrt{-t}$ to emphasize the low t region. Curve 1 is the Reggeized π-exchange contribution, while curves 2 and 3 are the fits of *Phillips*, 1967 b, using π-, conspirator- and ϱ-exchange.

The other trajectory which can contribute is the pion. Since the pion mass is very small, the extrapolation from the physical particle at $t = m_\pi^2$ to $t = 0$ is very short, and, as we know the πN coupling constant, we can predict the contribution fairly accurately. It is shown in Fig. (VIII. 18) where we see that it is much in excess of the data, except for t near zero where kinematical factors cause it to vanish. So the pion can not explain what is going on either, and there are no other trajectories known which could interfere with the π contribution in a suitable way (the B has the wrong phase). A way out of this difficulty has been suggested by *Phillips* (1967 b) who invokes the idea of a conspiracy [see sections (IV. 6) and (IV. 9)]. If there is some other trajectory which will conspire with the pion to fulfil the kinematical constraints, there is no need for the whole pion contribution to vanish at $t = 0$, and we have a good chance of fitting the data. As this is one of the few situations where the notion of conspiracy seems to be really useful it is worth examining Phillips's argument in a little more detail.

For NN and N$\bar{\text{N}}$ scattering there are five independent helicity amplitudes, which are conventionally denoted by

$$A_1 = \langle + + \mid + + \rangle$$
$$A_2 = \langle + + \mid - - \rangle$$

$$A_3 = \langle + - \mid + - \rangle$$
$$A_4 = \langle + - \mid - + \rangle$$
$$A_5 = \langle + + \mid + - \rangle \,, \tag{6.15}$$

and these are related to the differential cross section by

$$\frac{d\sigma}{dt} = \frac{\pi}{2q_s^2} \{|A_1|^2 + |A_2|^2 + |A_3|^3 + |A_4|^2 + 4\,|A_5|^2\}. \tag{6.16}$$

The pion contributes only to A_2 and A_4, and gives

$$A_2 = A_4 = \frac{\sqrt{s}}{4\pi} \left[\frac{1 + e^{-i\pi\alpha_\pi(t)}}{2\sin\pi\alpha_\pi(t)} \right] \beta_\pi^2(t) \left(\frac{s}{s_0} \right)^{\alpha_\pi(t)-1} \tag{6.17}$$

where s_0 is the usual scale factor [taken to be 2 (GeV)² as in section (VIII. 2)] and β_π^2 is the residue function. Since the πN coupling constant is $g^2 \approx 14$, we know that $\beta_\pi^2(m_\pi^2)$ must be

$$\beta_\pi^2(m_\pi^2) = \frac{4\pi^2}{s_0} m_\pi^2 g^2 \left(\frac{d\alpha_\pi(t)}{dt} \right)_{m_\pi^2} = m_\pi^2 \, b_\pi^2 \text{ (say)} \,. \tag{6.18}$$

These contributions are the same for NN and $N\bar{N}$.

Since A_4 involves helicity flip it must vanish at $t = 0$. If there is no conspiracy this can be achieved by putting

$$\beta_\pi^2(t) = t b_\pi^2 \tag{6.19}$$

with b_π^2 constant ($\neq 0$). This of course makes A_2 vanish as well and we find the contribution shown in Fig. (VIII. 18), in complete disagreement with the data. And we know that the inclusion of ϱ and A_2 could only make this disagreement worse.

Let us now try to construct a conspiracy which will ensure that the helicity flip amplitudes A_4 and A_5 vanish, but allows A_2, which has no need to be zero, to remain finite. The conspiring trajectory must have the same quantum numbers as the pion except for parity, so it must be such as to produce physical particles at $J^{PG} = 0^{+-}, 2^{+-}$, etc., if it reaches the right-half J plane, and must coincide with the pion at $t = 0$, i.e.

$$\alpha_\pi(0) = \alpha_c(0) \tag{6.20}$$

where c refers to the conspirator. In fact the full conspiracy will also require lower trajectories, but these can be neglected for our purpose. Phillips shows that, if we ignore contributions which are small near $t = 0$, the contribution of the conspirator to the amplitudes can be written

$$\begin{pmatrix} A_1 \\ A_2 \\ A_3 \\ A_4 \\ A_5 \end{pmatrix} = -\frac{\sqrt{s}}{4\pi} \left[\frac{1 + e^{-i\pi\alpha_c(t)}}{2\sin\pi\alpha_c(t)} \right] \left(\frac{s_0}{s} \right)^{\alpha_c(t)-1} \times$$

$$\times \alpha_c(t) \, b_2^2(t) \begin{pmatrix} -\tau \\ 1 \\ -\tau \\ -1 \\ -\sqrt{\tau} \end{pmatrix}, \tag{6.21}$$

where

$$\tau = - \frac{t}{4m^2} , \qquad (6.22)$$

so A_1, A_3 and A_5 vanish automatically at $t = 0$. The function $b_2^a(t)$ is a residue function. The conspiracy between (6.17) and (6.21) to make A_4 vanish at $t = 0$ requires (6.20), and

$$\alpha_c(0)\, b_2^a(0) = -\, \beta_\pi^a(0) . \qquad (6.23)$$

If we take a linear form for α,

$$\alpha_\pi(t) = \alpha_c(t) = (t - m_\pi^a)\, \alpha' \qquad (6.24)$$

with $\alpha' = m_N^{-a}$, and put

$$\beta_\pi^a(t) = \frac{4\pi^a m_\pi^a}{s_0}\, g^2\, \alpha'\, \frac{\left(1 + \dfrac{\lambda t}{m_\pi^a}\right)}{(1 + \lambda)}\, \mathrm{e}^{\nu t} , \qquad (6.25)$$

and keep $b_2^a(t)$ a constant determined by (6.23), a fairly good fit to the data can be obtained. Since both π and c contribute equally to pn and p$\bar{\mathrm{p}}$, the ϱ is also necessary if we are to produce the separation of the curves shown in Fig. (VIII. 18), but its parameters are partly determined from the πN charge exchange fits. Phillips obtains the fits shown with essentially 4 free parameters, λ and ν above, and two for the ϱ residue. The A_2 can be included, but its parameters are partly determined by the fit to $\pi^+ + \mathrm{p} \rightarrow \eta^0 + \mathrm{n}$, and its contribution is negligible. Clearly the fits could be improved with a more flexible parametrization, but with so many approximations, and such poor data, there is no point in this. The important thing is the data can be fitted by the conspiracy, but not in any obvious way without it.

There are two notable features of the fit. One is that the π residue must be rapidly varying near $t = 0$, [$\lambda \approx 1$ in (6.25)]. This has the effect of reducing the π contribution to the lower s-channel partial waves, and so acts like an absorptive correction. This is very interesting because those processes which are dominated by π exchange have been the principal successes of the absorptive peripheral model [*Jackson* (1965) and references therein], and it has never been clear how to relate these successes to the Regge-pole model because of the vanishing of the Regge-pole contribution in the forward direction. It seems more natural for absorption to be represented in terms of Regge cuts, however, as we shall see in section (VIII. 6h).

The second point to note is that the conspirator will not give rise to a light 0^+ meson, because, as is evident from (6.21), its coupling vanishes at $\alpha_c(t) = 0$, that is the trajectory "chooses nonsense" in the terminology of section (IV. 5). Alternatively a fit could be obtained without a vanishing coupling, but with a very flat trajectory, which also would not give a physical particle, or at any rate not one with a low mass. Phillips found that trajectories with neither of these features, and which therefore do contain a low mass 0^{+-} particle, do not give a satisfactory account of the data. So, even if this conspiracy is the correct way to explain the NN and N$\bar{\mathrm{N}}$ data, we can not expect to find such a particle, and of course none is known.

g) Elastic Scattering

In addition to the above fits, where we try to isolate one or two trajectories, there have been several attempts to fit the various elastic scattering processes, for which accurate data are available up to quite high energies. The advantage of better data is outweighed by the larger number of trajectories involved, and so more recent work has attempted to fit several processes at once.

There is information available about $\pi^\pm p$, $\pi^\pm n$, $K^\pm p$, $K^\pm n$, pp, pn, $\bar{p}p$ and $\bar{p}n$, but for the processes having the neutron as a target the information can only be obtained by subtracting the appropriate results with deuteron and proton targets, making suitable multiple scattering corrections [see e.g. *Galbraith* et al. (1965)], and it is not usually of the required accuracy. The main data to be fitted are thus $\pi^\pm p$, $K^\pm p$, pp and $\bar{p}p$ total cross sections and elastic differential cross sections, together with the charge exchange processes $\pi^- p \rightarrow \pi^0 n$ and $K^- p \rightarrow \bar{K}^0 n$. For most of these data is available up to at least 29 GeV/c.

The trajectories to be used are, from Table (VIII. 2),

$$\pi^\pm p; \quad f, f', \varrho$$

$$K^\pm p, pp, \bar{p}p; \quad f, f', \varrho, \pi, \varphi, A_2.$$

Evidently there is a large number of parameters available here, and many different types of fit are possible for a given elastic differential cross section, but now that the ϱ has been obtained from the πN charge exchange data, and the contribution of the A_2 pinned down to some extent by the Kp charge exchange data, there is some hope of determining the rest without too much ambiguity. Since ω and φ always contribute in the same way, and the φ coupling is expected to be small, it is usual to drop the latter and include any small effect in the ω. Similarly the ϱ and A_2 were found by *Phillips* and *Rarita* (1965a) to give very small contributions to pp and p$\bar{p}$. The main task of the elastic scattering data is thus to try and determine the f, f' and ω contributions (assuming, as we shall from now on, that f and f' are to be associated with the P and P' respectively).

For the trajectory functions the simple forms (6.9) or (6.10) are usually used, and in view of the observed straightness of the trajectories such 2 or 3 parameter forms should be adequate in the range of t usually fitted, $-1 < t < 0$ (GeV/c)^2. The same set of trajectory parameters must of course be used simultaneously for all the processes.

The residues are more complicated. Firstly, it is necessary to include various factors such as $\alpha(t)$, $\alpha(t) + 1$, etc. in order to kill ghosts, and put in sense-nonsense decoupling, at the various negative integer (and zero) values of α, through, as we have seen, there is an ambiguity as to whether a trajectory chooses sense or nonsense. Secondly, we can see from the data shown in Fig. (VIII. 13) that the elastic differential cross sections fall exponentially. In fact in the above region of t they can usually be fitted (at a given value of s) by the form

$$A(s, t) = A_0 e^{at}. \tag{6.26}$$

Now the single Regge pole expression (6.3) certainly includes an exponential factor, but this depends only on α', and is tied to the shrinkage of the diffraction peak with increasing s. Although at least the pp differential cross section does shrink a little in the energy range available, in no case can the values of α' determined in this or any other way (such as requiring the trajectory to go through the physical particles associated with it) be correlated with the values of a which are required in (6.26), so it is necessary to give the residues the desired exponential fall-off. Since this fall-off is the chief characteristic of Fig. (VIII. 13), by so doing we are obviously building the main feature of the amplitude into our parametrization arbitrarily, for there has been nothing in our previous discussion to justify such an exponential behaviour of the residue. Indeed from the dispersion relation for $\gamma(t)$ given in (III. 1.31) one would expect it to vary like $1/t$ at large negative t. The parametrization of the residues usually used is thus of the form

$$\Gamma(t) = C_0\, e^{a_1 t}\, f\,(\alpha\,(t))$$

where $f(\alpha)$ contains the various factors of α noted above.

Also, the parameters for the residues in the various processes are constrained by the factorization theorem, and the satisfaction of these constraints can cause difficulty. For example we find that the f, f' and A_2 trajectories contribute equally to pp and p$\bar{\text{p}}$ amplitudes, whereas the ϱ and ω contributions are opposite in sign. It follows that the difference between the pp and p$\bar{\text{p}}$ differential cross-sections is due to the interference between the sum of ϱ and ω and the sum of the f, f' and A_2 contributions. For details we refer to an excellent article on Regge-poles in the NN and N$\bar{\text{N}}$ system by *Leader* and *Slansky* (1966). From their results we can write

$$\left.\frac{d\sigma}{d\Omega}\right|_{\text{p}\bar{\text{p}}} - \left.\frac{d\sigma}{d\Omega}\right|_{\text{pp}} = \frac{1}{4s} \sum_{i=f,f',A_2} \sum_{j=\varrho,\omega} \times$$
$$\times \left((1 - \cot \tfrac{1}{2}\pi\alpha_i \tan \tfrac{1}{2}\pi\alpha_j) \Gamma_i\, \Gamma_j\, R_{ij}^2\, s^{\alpha_i + \alpha_j} \right. \tag{6.27}$$

Here the first bracket comes from the signature factors, and the Γ_i are the residues of the ith trajectory in a particular t-channel amplitude (various other irrelevant factors have been incorporated in Γ_i). The R_{ij}^2 term takes account of the other t-channel amplitudes and it is possible to write it in this way, with R_{ij} real (provided α_i and β_j are real for $t < 0$), because of the factorization theorem for helicity amplitudes.

Now at high energies it is found experimentally that the left-hand-side of (6.27) changes sign at about $t = -0.15$ (GeV/c)^2, being negative for $t \lesssim -0.15$ (GeV/c)^2. At such values of t the signature factor must be positive (we rule out the possibility of extremely rapidly varying trajectories which would be completely at variance with other evidence). Thus one of the residue functions must change sign for small t. In fact because this happens where the f dominates significantly over the A_2 and f', it is safe to assume that what is required is that either Γ_ϱ, Γ_ω or Γ_f must change sign. Now high energy data on $d\sigma/d\Omega|_{\text{pp}}$ and $d\sigma/d\Omega|_{\text{p}\bar{\text{p}}}$, individually, rule out the possibility that Γ_f changes sign. Further, it follows from the factorization theorem that, if Γ_ϱ passes through zero in $N\bar{N} \to N\bar{N}$, the ϱ residue in $N\bar{N} \to \pi\pi$ must pass through zero at the same value of t. This is ruled out by πN charge exchange data as discussed earlier. Hence we are forced to the conclusion that the change of sign is due to the ω contribution, and that Γ_ω passes through zero for some small value of t.

It is extremely important to try and obtain verification (or otherwise) of this fact, and *Leader* and *Slansky* (1966) suggest effects in the NN and $N\bar{N}$ system which effectively isolate the ω trajectory, and which should therefore pass through zero near $t = 0$. These tests require either measurements of spin-correlation coefficients (triple scattering parameters), or else accurate determinations of second differences, e.g. the quantity

$$\frac{d\sigma}{d\Omega}\bigg|_{pp} + \frac{d\sigma}{d\Omega}\bigg|_{pn} - \frac{d\sigma}{d\Omega}\bigg|_{\bar{p}p} - \frac{d\sigma}{d\Omega}\bigg|_{\bar{p}n}$$

depends only on the ω trajectory. Unfortunately such measurements have not been made. There has, however, been some confirmation in that a similar cross-over effect is observed in $K^\pm p$ scattering, e.g.

$$\frac{d\sigma}{d\Omega}\bigg|_{K^-p} > \frac{d\sigma}{d\Omega}\bigg|_{K^+p}, \quad t \approx 0,$$

but

$$\frac{d\sigma}{d\Omega}\bigg|_{K^-p} < \frac{d\sigma}{d\Omega}\bigg|_{K^+p} \quad t < -0.2 \text{ (GeV/}c)^2,$$

which also has to be explained by the vanishing of the ω residue for that process, and of course factorization demands that $\Gamma_{\omega,KN}$ vanish a the same points as $\Gamma_{\omega,NN}$. This could be also due to α_ω passing through zero with the couplings choosing nonsense, however. Similarly there is a crossover in $\pi^\pm p$ scattering near $t \approx 0.05$ (GeV/c)^2, which can be explained by the vanishing of the ϱ residue.

It was once thought to be undesirable to have to postulate a zero of a residue function because it is known that residues cannot pass through zero in a simple one-channel potential model [*Newton* (1964)], but in relativistic scattering this restriction does not seem to be necessary and, in any case, the introduction of many channels certainly allows residues to pass through zero even in potential models. This problem has been discussed by *Desai* (1965).

There are also constraints from factorization in relating one process to another. For example suppose we denote the residues of a given Regge pole in the t-channel amplitudes

$$\left\langle \frac{1}{2}, \frac{1}{2} \bigg| A\ (N\bar{N} \to \pi\pi)\bigg| 0, 0 \right\rangle, \left\langle \frac{1}{2}, -\frac{1}{2} \bigg| A\ (N\bar{N} \to \pi\pi)\bigg| 0, 0 \right\rangle,$$

and similarly for $N\bar{N} \to K\bar{K}$, by $\beta^{N\pi}_{++}$, $\beta^{N\pi}_{+-}$, β^{NK}_{++}, and $\beta^{N\pi}_{+-}$, respectively. According to the factorization theorem we have

$$\begin{aligned}
\beta^{N\pi}_{++} &= \gamma^N_{++}\, \gamma^\pi \\
\beta^{N\pi}_{+-} &= \gamma^N_{+-}\, \gamma^\pi \\
\beta^{NK}_{++} &= \gamma^N_{++}\, \gamma^K \\
\beta^{NK}_{+-} &= \gamma^N_{+-}\, \gamma^K,
\end{aligned} \tag{6.28}$$

so that the constraint

$$\frac{\beta_{++}^{N\pi}}{\beta_{+-}^{N\pi}} = \frac{\beta_{++}^{NK}}{\beta_{+-}^{NK}} \tag{6.29}$$

holds. This constraint has been used in the fits and does not appear to disagree with the experimental results.

After these general remarks let us look at the fits in more detail. The most comprehensive works to date, from which we shall quote, are *Phillips* and *Rarita* (1965a), who fitted $\pi^{\pm}p$ and $K^{\pm}p$ together, and *Rarita* et al. (1967) who have fitted $\pi^{\pm}p$, pp and $\bar{p}p$. These papers give a good list of references to the many earlier fits, and to the data sources which are available. They neglect NN charge-exchange scattering, however, which has been included by *Flores-Maldonado* (1967).

For $\pi^{\pm}p$ and $K^{\pm}p$ they use the amplitudes A and B of *Singh* (1963), which are related to the experimental quantities by

$$\frac{d\sigma}{dt}(s, t) = \frac{1}{\pi s}\left(\frac{M_N}{4q_s}\right)^2 \left[\left(1 - \frac{t}{4M_N^2}\right)|A|^2 - \frac{t}{4M_N^2}\left(\frac{4M_N^2 p + st}{4M - t}\right)|B|^2\right]$$

$$\sigma^T(s) = \frac{\text{Im}\,A(s, t = 0)}{p} \tag{6.30}$$

$$P(s, t) = -\frac{\sin\theta}{16\pi\sqrt{s}}\frac{\text{Im}(A\,B^*)}{\frac{d\sigma}{dt}}.$$

M_N being the mass of the nucleon, p the meson laboratory frame momentum, and P the polarization defined in Section (VIII. 7). The trajectory contributions are parametrized by

$$A_i = C_i \xi \left(\frac{E_L}{E_0}\right)^{\alpha}$$
$$B_i = D_i \xi \left(\frac{E_L}{E_0}\right)^{\alpha} \tag{6.31}$$

where ξ is the signature factor

$$\xi = \frac{e^{-i\pi\alpha} \pm 1}{\sin\pi\alpha} \tag{6.32}$$

E_L is the meson lab. energy, and E_0 is the scaling constant which is taken to be 1 GeV. The residues are written

$$C_i = \begin{cases} C_0\, e^{C_1 t}\, \alpha\,(2\,\alpha + 1)\,, & \text{for } f, f' \text{ and } A_2 \\ C_0\,[(1 + G)\, e^{C_1 t} - G\, e^{C_2 t}]\,(2\,\alpha + 1)\,, & \text{for } \varrho \text{ and } \omega, \end{cases} \tag{6.33a}$$

and

$$D_i = \begin{cases} D_0\, e^{D_1 t}\, \alpha\,, & \text{for } f, f' \text{ and } A_2 \\ D_0\,[(1 + H)\, e^{D_1 t} - H\, e^{D_2 t}]\,\alpha\,, & \text{for } \varrho \text{ and } \omega\,. \end{cases} \tag{6.33b}$$

so that for ϱ and ω it is possible to have a sign change to explain the cross-over effect. The πN and KN parameters are related by [see (6.29)]

$$\frac{A_i(KN)}{A_i(\pi N)} = \frac{B_i(KN)}{B_i(\pi N)} = F_0\, e^{F_1 t} \tag{6.34}$$

so the SU (3) prediction is $F_0 = 1$ for f and f', and 0.5 for ϱ, with $F_1 = 0$. The trajectories are parametrized in the Pignotti form (6.10). They obtained various solutions depending on the type of explanation adopted for the cross-over effect, and in Table (VIII. 5) we quote one of these which uses sign changes in A_ϱ and A_ω at the appropriate places, and Fig. (VIII. 13) demonstrates that quite a good fit is obtained. The SU (3) predictions are satisfied remarkably well for f and ϱ but not so well for f'. The slopes of the f and f' are much smaller than for the others.

Table VIII.5.

	$\alpha(0)$	$\alpha'(0)$	C_0 (mb × Gev)	C_1 (Gev^{-2})	C_2	D_0 (mb)	D_1 (Gev^{-2})	D_2	F_0	F_1 (Gev^{-2})	G	H
f	1.00	0.34	6.55	2.51	...	—7.5	0.51	...	0.901	—0.23	...	...
f'	0.50	0.34	19.6	4.04	...	—101	8.1	...	0.279	—1.61	...	...
ϱ	0.54	0.65	2.45	5.6	0.14	56.9	1.64	0.31	0.527	0.01	0.50	0.90
A_2	0.32	0.80	3.34	2.16	...	—31.2	1.76	...	...	...	...	...
ω	0.52	0.60	5.99	10.5	0.17	...	...	...	...	...	0.86	...

Parameters from a fit of $\pi^{\pm}p$ and $K^{\pm}p$ data by *Phillips* and *Rarita*, 1965 a, [we quote solution 1 of that paper].

In the more recent work of *Rarita* et al., the paramerization has been changed slightly in order to make the various contributions vanish at $\alpha = -1$ when the straight trajectory (6.9) is used, viz

$$C_i = \begin{cases} C_0 \, e^{C_1 t} \, \alpha \, (\alpha + 1) & \text{for } f \text{ and } f', \\ C_0 \, [(1 + C_2) \, e^{C_1 t} - C_2] \, (\alpha + 1) & \text{for } \varrho, \end{cases} \qquad (6.35)$$

and

$$D_i = \begin{cases} D_0 \, e^{D_1 t} \, \alpha^2 \, (\alpha + 1) & \text{for } f \text{ and } f', \\ D_0 \, e^{D_1 t} \, \alpha \, (\alpha + 1) & \text{for } \varrho, \end{cases}$$

while for pp and $p\bar{p}$ the five amplitudes (6.15) are parametrized by [following *Sharp* and *Wagner* (1963)]

$$A_k = \Gamma_k \xi \left(\frac{E_L}{E_0}\right)^\alpha, \qquad k = 1 \ldots 5, \qquad (6.36)$$

where

$$\Gamma_1 = \Gamma_3 = \frac{M_N E_0}{4\pi\sqrt{s}} \, \eta_N^2$$

$$\Gamma_2 = -\Gamma_4 = -\frac{M_N E_0}{4\pi\sqrt{s}} \, \varphi_N^2 \qquad (6.37)$$

$$\Gamma_5 = -\frac{M_N E_0}{4\pi\sqrt{s}} \, \varphi_N \, \eta_N.$$

These amplitudes are related to the physical quantities by (6.16) and

$$\sigma^T(s) = \frac{2\pi}{q_s} \, \text{Im} \, (A_1 + A_3)$$

$$P(s, t) = \frac{\pi}{2q_s^2} \, \frac{\text{Im} \, [A_5 (A_1 + A_2 + A_3 - A_4)^*]}{(d\sigma/dt)}. \qquad (6.38)$$

The residue functions Γ_k embody the required factorization properties, and are written in terms of two functions b_1 and b_2 (which assists comparison with the πN parameters),

$$\eta_N = b_1 - \left(\frac{\alpha t}{4 M_N^2}\right) b_2$$

$$\varphi_N = \left(\frac{-t}{4 M_N^2}\right)^{1/2} (b_1 - \alpha b_2) , \qquad (6.39)$$

and these in turn are parametrized by

$$b_1 = \begin{cases} F_0\, e^{F_1 t}\, [\alpha\,(\alpha + 1)]^{1/2} , & \text{for } f \text{ and } f' \\ F_0\, e^{F_1 t}\, \left[\left(1 - \frac{t}{t_0}\right) (\alpha + 1)\right]^{1/2} & \text{for } \omega \end{cases} \qquad (6.40)$$

$$b_2 = \; G_0\, e^{G_1 t}\, \left[\left(1 - \frac{t}{t_0}\right) (\alpha + 1)\right]^{1/2} \left(1 - \frac{t}{4 M_N^2}\right)^{-1} \quad \text{for } \omega .$$

The factor $(1 - t/t_0)$ is used to produce the sign change in the ω residue. Factorization then requires

$$b_2 = G_0\, e^{G_1 t}\, [\alpha\,(\alpha + 1)]^{1/2} \left(1 - \frac{t}{4 M_N^2}\right)^{-1}, \quad \text{for } f \text{ and } f', \qquad (6.41)$$

where

$$G_0 = F_0 \left(\frac{D_0}{C_0}\right) \quad \text{and} \quad G_1 = F_1 + (D_1 - C_1) .$$

Table VIII.6.

	$\alpha(0)$	$\alpha'(0)$ (Gev^{-2})	C_0 (mb × Gev)	C_1 (Gev^{-2})	C_2	D_0/C_0 (Gev^{-1})	D_1 (Gev^{-2})	F_0 (mb^2)	F_1 (Gev^{-2})	G_0/F_0	$G_1 - F_1$ (Gev^{-2})
f	1.0	0.12	7.23	2.36	...	−3.69	9.38	3.80	2.09	...	...
f'	0.73	1.50	16.35	0.44	...	−3.52	3.86	5.04	1.06	...	...
ϱ	0.58	0.94	1.47	0.20	15.2	17.6	0.34	...	...	...	...
ω	0.45	0.31	...	...	...	...	...	3.94	1.85	−16.4	4.47

Parameters from a fit to the $\pi^\pm$p, pp and $\bar{\text{p}}$p data by *Rarita* et al., 1967, [we quote from solution 1 of that paper]. $t_0 = -0.13$.

We quote a solution in Table (VIII. 6). The discrepancy between these values and those of Table (VIII. 5) gives some idea of the large uncertainties in the parameters. The f slope is very much reduced, while the f' has a large slope so that the dip and second maximum in the $\pi^\pm$ p differential cross sections can be associated with the zero of $\alpha_{f'}$. The factorization constraints appear to be compatible with such polarization data as exists, but this is not a very stringent test. The results are also in reasonable agreement with the sum rules of *Igi* and *Scanio* discussed in section (VIII. 4). The fact that the f' intercept is so high means that if this fit is correct the cross sections found in the present energy range are very far from their asymptotic values. In fact *Rarita* et al. predict

$$\sigma_{NN}^T = 28.8 \text{ m.b.}$$

$$\sigma_{\pi N}^T = 14.5 \text{ m.b.}$$

$$\sigma_{\pi\pi}^T = \; 7.3 \text{ m.b.}$$

compared with the values of Fig. (VIII. 8). This also explains why there is so little shrinkage. The f will not dominate until very much higher energies than are presently available. The small shrinkage observed in pp is due to a partial cancellation of the f' and ω contributions. As we noted in section (VIII. 3b) it was this lack of shrinkage which at one time led to suggestions that the Pomeranchon trajectory is flat, or that diffraction scattering can not be explained in terms of Regge poles.

Obviously these fits must be treated with caution since until we have really good polarization data there will remain many ambiguities, and we know that there are difficulties in situations where only one pole can be exchanged, but none-the-less it is encouraging that so much experimental information can be summarized so economically by a few Regge poles.

h) Production Processes

In general the available data on the angular distributions of resonance production processes is inadequate to test the Regge pole model or to test the possibility of conspiracies. However some work on this subject has been done and there will doubtless be much more in the future.

Wang (1967) has given a useful tabulation of Regge-pole contributions to a variety of production processes (though as we noted in section (IV. 6) it may be that the explicit insertion of all the kinematical singularities into the residues, as is done in this paper, is liable to be misleading; and some kinematical constraints are not given). *Frautschi* and *Jones* (1967b) have studied the processes $\pi N \to \rho N$, $KN \to K^*\Delta$, $\pi N \to f\Delta$ and $\pi N \to fN$, paying particular attention to the nature of the kinematical constraints on the residue of the pion trajectory which is expected to dominate these processes. They obtain good fits to the data, and find that the variation of the residue can be explained entirely in terms of "kinematical" effects. This latter result is in contrast to the earlier work of *Thews* (1967) who did not make use of all the threshold constraints.

Frautschi and *Jones* (1967c), *Högaasen* and *Salin* (1967) and *Le Bellac* (1967) have discussed the form of possible conspiracies in vector meson production processes. In particular Le Bellac notes that if there is a pion conspiracy in NN scattering, then factorization requires that there be a forward dip in processes like $\pi N \to \rho\Delta$, $f\Delta$ and $KN \to K^*\Delta$. The data do not allow this to be tested directly, but the parametrization used by *Frautschi* and *Jones* in fitting the these processes (see above) seems to support the existence of such a dip.

Several inelastic processes have been satisfactorily fitted with just ϱ exchange, using the ϱ-trajectory obtained earlier, and constant reduced residues. These are listed in Table (VIII. 7). In each case only one parameter is available with which to fit the energy and momentum transfer variation. *Roy* (1966) has also fitted rather poor data on $p\bar{p} \to \Lambda\bar{\Lambda}$, $\Lambda\bar{\Sigma}$ and $\Sigma\bar{\Sigma}$, with K^* exchange and $\alpha_{K^*}(t) \approx 0.4 + 0.7\,t$.

Table VIII.7.

Reference	Process	Energy	$-t$
Krammer and *Maor*, 1967	$\pi^+ p \to \pi^0 \Delta^{++}$ (1238)	4, 8 Gev/c	$0 < -t < 0.8$ (Gev/c)²
Caprasse and *Stremnitzer*, 1966	$\pi^+ p \to \pi^0 \Delta^{++}$ (1238)	4, 8 Gev/c	$0 < -t < 0.8$ (Gev/c)²
Sundarm and *Sridar*, 1967	$\pi^+ p \to A_2^+ p$	4, 6 Gev/c	$0 < -t < 1.0$ (Gev/c)²
Roy, 1965	$K^+ p \to K^0 \Delta^{++}$ (1238)	3, 3.5, 5 Gev/c	
	$\pi^+ p \to \pi^0 \Delta^{++}$ (1238)	2.8, 3.5, 4 Gev/c	$0 < -t < 1.0$ (Gev/c))
	$\pi^+ p \to \omega \Delta^{++}$ (1238)	2.4, 2.9, 4 Gev/c	

Some recent Regge pole fits to inelastic cross-sections.

i) Regge Cuts

Although there is every reason to believe that there are cuts in the angular-momentum plane, as we have seen in Chapters V and VII, there has been success so far in fitting most of the experimental data with just poles. But there is no reason to suppose that the poles must always dominate, and there has been a temptation to invoke cuts wherever the simple Regge pole fits have seemed unsatisfactory. Unfortunately the contribution of a cut permits the introduction of many free parameters, and of course a cut can always be represented in any finite region by a superposition of poles, so the effect of adding a cut is difficult to distinguish from that of adding further poles.

From (II. 9.10), the contribution of a cut to an amplitude takes the form

$$A^{\pm}(s,t) = \int_{\text{(background)}}^{\alpha_0(t)} g(l,s) \left(\frac{s}{s_0}\right)^l dl \tag{6.42}$$

where $g(l,s)$ is obtained from the discontinuity across the cut. If $g(l,s)$ is taken to be regular at α_c the asymptotic contribution of the cut is

$$A^{\pm}(s,t) \to \text{const.} \left(\frac{s}{s_0}\right)^{\alpha_c(t)} \frac{1}{\log\left(\frac{s}{s_0}\right)} \tag{6.43}$$

which only differs from a pole by the logarithmic factor, but of course different models for $g(l,s)$ can give different sorts of behaviour [see e.g. *Oehme* (1964)]. We have found [see sections (V. 3) and (VI. 4)] that the cut resulting from the exchange of two trajectories $\alpha_1(t)$ and $\alpha_2(t)$ is [see (V. 3.4)]

$$\alpha_c(t) = \text{Max}\,[\alpha_1(t_1) + \alpha_2(t_2) - 1]\,, \tag{6.44}$$

and with n identical trajectories $\alpha(t)$ we get [see (V. 3.9)]

$$\alpha_c^n(t) = n\,\alpha\left(\frac{t}{n^2}\right) - n + 1. \tag{6.45}$$

Thus if α is the Pomeranchon trajectory (f), and $\alpha_f(0) = 1$ as is usually assumed, we find

$$\alpha_c^n(0) = 1 \qquad (6.46)$$

for all n, and as $n \to \infty$ the slope of the cut contribution $\alpha_c^{n'}(t) \to 0$. These results lead us to expect that whereas the Pomeranchon pole may still dominate over α_c^n at $t = 0$, because of the logarithmic factor, this will not be true for $t < 0$, and of course with an infinite number of cuts it is not obvious that the f will dominate even at $t = 0$. Similarly in cases where the f can not be exchanged, such as in πN charge exchange scattering, discussed above, which the ϱ is expected to dominate, we can also expect cuts from the exchange of $\varrho + nf$, which will result in the same sort of accumulation of cuts at $\alpha_\varrho(0)$. It is possible therefore that the cross section fits which we have discussed in sections (VIII. 3 and 4) have really been using the $s^{\alpha_c(0)}$ dependence of such cuts, and that the difficulties in fitting the differential cross sections, and in particular the anomalously small slope of the P in Table (VIII. 1), indicates that we are observing the effects of cuts. However, it would then be very hard to understand why we find such a large slope for the ϱ and A_2 trajectories, so this argument must be regarded with some suspicion.

Pursey and *Sertorio* (1967) have tried to discover whether a good fit to the $\pi^\pm p$ total cross sections can be obtained by including cuts. By looking at

$$\Sigma(\pi p) = \sigma^T(\pi^+ p) + \sigma^T(\pi^- p)$$

they eliminate the contribution of the ϱ [see Table (VIII. 2)], and according to the models of section (VIII. 3) Σ can be fitted by just the f and f'. They find, though this is not very surprising, that an acceptable fit can also be obtained by combining a cut with a f' trajectory. They allow $\alpha_c(0)$ to be an arbitrary parameter, however, and it is not possible to associate their various values with the exchange of any particular sets of trajectories. Note that if there is no f trajectory, and such cuts do dominate, then the total cross section will vanish asymptotically.

One way in which the difficulty of an infinite number of cuts occuring at $\alpha_r(0)$ (for any trajectory r) could be avoided would be for $\alpha_f(0)$ to be less than 1. We have noted in section (VIII. 3) that $\alpha_f(0) = 0.93$ seems to be equally acceptable. Then, as *Srivastava* (1967) has noted, if

$$\alpha_f(0) = 1 - \varepsilon, \qquad (6.47)$$

the cut resulting from the exchange of $r + nf$ is at

$$\alpha_c^n(t) = \alpha_r(0) - n\varepsilon + \frac{\alpha_f' \alpha_r'}{\alpha_f' + n\alpha_r'} t \qquad (6.48)$$

in a linear approximation, and as n is increased we get a series of cut trajectories of smaller slope but also lower intercept as in Fig. (VIII. 19). So the pole will always dominate for a range of t near zero, but the cuts will be important for sufferently large $|t|$. This might explain why the slopes determined for $t < 0$ are smaller than those required by the Regge recurrances of Figs. (VIII. 2–6) [see Table (VIII. 1)], since really for $t < 0$ we would be seeing the effect of the curve in Fig. (VIII. 19), called a "twisting" trajectory by Srivastava.

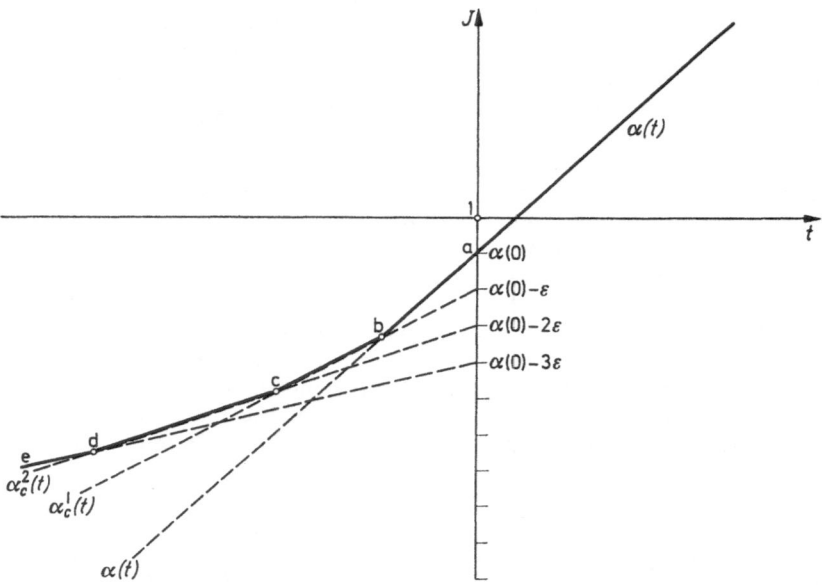

Fig. VIII.19. Pole and cut trajectories with the Pomeranchon intercept $\alpha_P(0) = 1 - \varepsilon$. The leading singularity in the negative-t region is the curve abcde . . ., while for positive t the pole still dominates (from *Srivastava*, 1967).

Another possible use for cuts has been found by *Huang* et al. (1967). They note that the differential cross section for p-p scattering at 90° exhibits the discontinuous behaviour [*Akerlof* et al. (1966)]

$$\left(\frac{d\sigma}{dt}\right)_{90°} = \begin{cases} C_1 \exp\left(-3.29\, q_s^2\right), & q_s^2 < 3.40 \\ C_2 \exp\left(-1.51\, q_s^2\right), & q_s^2 > 3.40 \end{cases} \tag{6.49}$$

C_1 and C_2 being constants, and q_s the centre of mass momentum in Gev/c, and suggest that the low q_s^2 region may be controlled by the f trajectory while the behaviour in the $q_s^2 > 3.40$ (Gev/c)^2 region is due to the cut from the exchange of $2f$'s. Note that, since

$$\cos\theta_s = 1 + \frac{t}{2q_s^2}, \tag{6.50}$$

$\theta = 90°$ implies

$$q_s^2 = -\frac{t}{2}, \tag{6.51}$$

and hence the larger q_s^2 values correspond to the more negative values of t. They suppose the discontinuous behaviour to result from the vanishing of the f contribution due to the signature factor (II. 10.3) at the value of t where $\alpha_f(t) = -1$, which from (6.49) and (6.51) must be at $t \approx -6.8$ (Gev/c)². This is of course only true if there is no pole in the residue at this wrong-signature nonesense value. Taking

$$\alpha_f = 1 + \alpha' t \tag{6.52}$$

this requires

$$\alpha' = 0.29 \; (\text{Gev}/c)^{-2}$$

which is reasonably compatible with other values [see Table (VIII. 1)]. *Huang* et al. were able to obtain a good fit to the data in the neighbourhood of the discontinuity, but complications due to spin, etc., were ignored, and further work is promised.

This discontinuous behaviour of the differential cross sections would appear to be part of a more general phenomenon. Although the form

$$\frac{\mathrm{d}\sigma}{\mathrm{d}t} = A_0^{at} \tag{6.53}$$

appears to hold for both πp and pp elastic scattering [*Harting* et al. (1965), *Foley* et al. (1965b)] as well as many production processes [*Anderson* et al. (1966)] out to about $t = -1$ (Gev/c), at larger momentum transfers the fall-off is less rapid [*Orear* (1964), *Orear* et al. (1965)] being more like

$$\frac{\mathrm{d}\sigma}{\mathrm{d}t} = B_0 \, e^{-b(-t)^{1/2}} . \tag{6.54}$$

This behaviour is however the fastest permitted by the analytic properties of the scattering amplitude and the constancy, or near constancy of the total cross sections [*Martin* (1965)], and one would like to be able to understand its connection with the J-plane singularities. The concentration of the amplitude near to the forward and backward directions is much stronger than would be expected from simple peripheralism, i.e. the dominance of crossed channel poles. Further discussion of the t-dependance of cut contributions and the way in which they may reproduce (6.54) is given by *Anselm* and *Dyatlov* (1967).

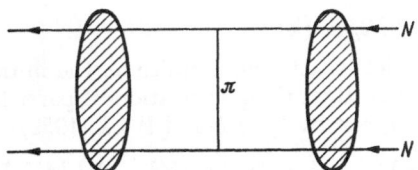

Fig. VIII.20. Absorptive corrections to π-exchange in NN scattering.

Cuts have also been invoked by *De Laney* et al. (1967) to explain the πN charge-exchange polarization, which we shall discuss in the next section. It is also natural to think of invoking cuts wherever absorptive corrections to single particle exchange have proved particularly successful in explaining a process [see *Jackson* (1965)]. An example is the NN charge exchange amplitude, which, as we noted above, might be explained by an absorbed pion, but which Regge poles can account for only if there is a pion conspiracy. The usual picture of absorptive corrections, such as Fig. (VIII. 20) is most naturally interpreted as $\pi + nf$ exchange etc., the f (and of course other trajectories) giving the initial and final state interactions. It has not so far proved possible to put this idea in a more concrete form, however [see *Gribov* (1967a)].

It is unlikely that the data will be good enough for a long time to permit a convincing separation in situations where cuts and poles can both contribute, and *Phillips* (1967a) has suggested that the best place to look for cuts is probably in amplitudes where no known single particle can be exchanged [as we noted in section (VII. 4f)]. Examples are double charge exchange processes such as

$$\pi^- p \to K^+ \Sigma^-$$

and

$$K^- p \to K^+ \Xi^-$$

for which $\varrho + K^*$ and $K^* + K^*$ cuts (respectively) are likely to be the highest J-plane singularities. Using the intercepts $\alpha_\varrho(0) = 0.58$ and $\alpha_{K^*}(0) \approx 0.4$ we get

$$\frac{d\sigma}{dt}(\pi^+ p \to K^+ \Sigma^-) \sim \frac{1}{s^2}$$

$$\frac{d\sigma}{dt}(K^- p \to K^+ \Xi^-) \sim \frac{1}{s^{2 \cdot 4}}$$

apart from logarithmic factors. It will be interesting to see if such predictions can be verified.

However the present situation can be summarized by saying that, even through we know cuts must be present, there is no convincing experimental evidence for them.

VIII.7. Polarization and Spin Density Matrices

a) Polarization and Regge Poles

If, in a two-particle scattering amplitude, the initial particles, 1 and 2 say, are unpolarized, then the polarization of particle 3, normal to the scattering plane, is given by [*Jacob* and *Wick* (1959)]

$$P = \frac{\sum_{\lambda_1 \lambda_2 \lambda_3 \lambda_4} [(\sigma_3 + \lambda_3)(\sigma_3 - \lambda_3 + 1)]^{1/2} \, \text{Im} \, [\langle \lambda_3 - 1 \, \lambda_4 | A | \lambda_1 \lambda_2 \rangle \langle \lambda_3 \lambda_4 | A | \lambda_1 \lambda_2 \rangle^*]}{\sum_{\lambda_1 \lambda_2 \lambda_3 \lambda_4} |\langle \lambda_3 \lambda_4 | A | \lambda_1 \lambda_2 \rangle|^2} .$$

$$(7.1)$$

It follows from this equation that polarization can only occur if there is a phase difference between the helicity amplitudes. Now the phase of the contribution of a single Regge pole in the physical region of the crossed channel is determined solely by the signature factor, and is therefore the same for all helicity amplitudes [see section (IV. 6)]. It follows that polarization can only arise from interference between the contributions of two Regge-poles. An immediate consequence of this is that all polarizations tend to zero at high energy in the Regge-pole model. Even if cut contributions are included this statement remains true, although the rate at which the polarization tends to zero may then be much slower.

b) Polarization in $\pi^- p \to \pi^0 n$

A sensitive test of the polarization prediction is in $\pi^- p \to \pi^0 n$, where we have seen that a good fit to the differential cross-section can be obtained from the ϱ trajectory alone. From the above discussion we see that the ϱ trajectory gives zero polarization. The measured polarization however, is certainly not zero, and does not show any marked tendency to decrease with energy (although the errors are large). The best available data is shown in Fig. (VIII. 21) [taken from the paper of *Bonamy* et

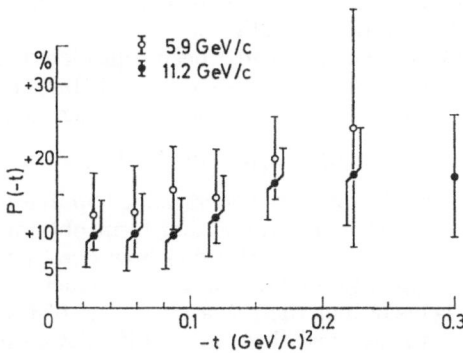

Fig. VIII.21. The polarization parameter $P_0(t)$ at 5.9 and 11.2 Gev/c, as measured by *Bonamy* et al., 1966.

al. (1966)]. If the experimental values are averaged over the range $0.04 \leq -t \leq 0.24$ $(\text{GeV}/c)^2$ then

$$\langle P(t) \rangle_{\text{AV}} = (16 \pm 3.5) \% \quad \text{at } 5.9 \text{ GeV}/c \tag{7.2}$$

and

$$\langle P(t) \rangle_{\text{AV}} = (14 \pm 4.5) \% \quad \text{at } 11.2 \text{ GeV}/c . \tag{7.3}$$

It is not clear at present to what this polarization should be ascribed. In a pure Regge pole model it is necessary to include, in addition to the ϱ, at least one other trajectory. An obvious candidate would be a trajectory with the same quantum numbers as the ϱ, say a ϱ'. We have already noted the possibility of such a trajectory in section (VIII. 5b). *Högassen* and *Frisk* (1966b) show that such a trajectory — if its parameters are chosen so as to disturb the ϱ fit to the differential cross-section as little as possible — will give a polarization of the observed sign. The residue of the ϱ' trajectory can be adjusted to agree with experiment at one energy, but the value of $\alpha_{\varrho'}(0)$ given in (5.10) would result in an energy dependence of the polarization given by

$$P(s) \backsim \text{const. } s^{-\alpha_\varrho + \alpha_{\varrho'}} \tag{7.4}$$

$$\backsim \text{const. } s^{-1 \cdot 2} \tag{7.5}$$

if we assume that the trajectories are approximately parallel over the interesting region. The experiments do not seem to support such a rapid decrease with energy, though it should be noted that, because of the large errors, they do not completely rule it out. Note that the ϱ' here cannot be the daughter of the ϱ since this would require that it had opposite signature, and hence at $t = 0$ (where we would have $\alpha_{\varrho'} = \alpha_\varrho - 1$) exactly the same phase.

Although the prediction (7.5) is based on the doubtful value of $\alpha_{\varrho'}(0)$ given by (5.10), it is clear that there cannot be any trajectory, having the quantum numbers of the ϱ, much higher than this, otherwise presumably it would have been seen as a physical particle; so the lack of a rapid decrease in the polarization with energy, if it is confirmed, will require the introduction of other effects.

Several possibilities have already been considered. *De Lany* et al. (1967) have suggested that the $\varrho + f$ cut contribution is responsible. Since, for $t < 0$, the branch point $\alpha_c(t)$ is higher than $\alpha_\varrho(t)$, i.e.

$$\alpha_c(t) > \alpha_\varrho(t) \quad \text{for } t < 0 \tag{7.6}$$

see section (V. 3), the factor $s^{-\alpha_\varrho + \alpha_c}$ contained in the polarization will increase with energy. Of course the actual form obtained in any given region of energy depends very much upon how one "weights" the contribution of the cut, i.e. upon its discontinuity, which is a function of J as well as of t. It is clear that there is a great deal of freedom, and that if a cut contribution is allowed there is no difficulty in fitting the data. For example, *Chiu* and *Finkelstein* (1967) give fits to all the πN scattering data and to the charge exchange polarization.

A third possible origin of the observed polarization is that $\alpha(t)$ and/or $\beta(t)$ become complex somewhere below threshold. As we saw in section (III. 1) this would require two trajectories to coincide. Again, once this is allowed, there is enough freedom to allow a good fit to the presently available data [see *Bali* et al. (1967a)]. Note that any polarization obtained by this method is predicted to be independent of energy.

Lastly we mention that several authors, following *Phillips* (1966), have considered the possibility that the polarization in this process can be explained as an interference between direct channel resonances and the ϱ contribution. We discuss this further in section (VIII. 8). We can summarize the situation by saying that the presence of polarization in πN charge exchange scattering can be explained in a variety of ways, without essentially disturbing the fits to the other πN data. It will require accurate measurements at higher energies to distinguish between the various possibilities.

c) Polarization in Other π-N Elastic Scattering Processes

The simple Regge pole model gives much better agreement with the other available π-N data. As we have seen in section (VIII. 6b), the ϱ trajectory gives a large spin-flip term, but the absence of forward dips in the $\pi^{\pm}p$ differential cross-sections precludes a large spin-flip contribution from the f and f'. So polarization is expected to come mainly from interference between the ϱ spin flip term and the $f + f'$ non-spin-flip term. This predicts opposite signs for the polarzation in $\pi^{+}p$ and $\pi^{-}p$, and a change of sign of the polarization at a value of t near to where $\alpha_{\varrho}(t) = 0$ (assuming that this is the origin of the dip in the charge exchange differential cross-section as discussed in section (VIII. 6b)). Both these features are observed. In Figs. (VIII. 22) and (VIII. 23) we compare the

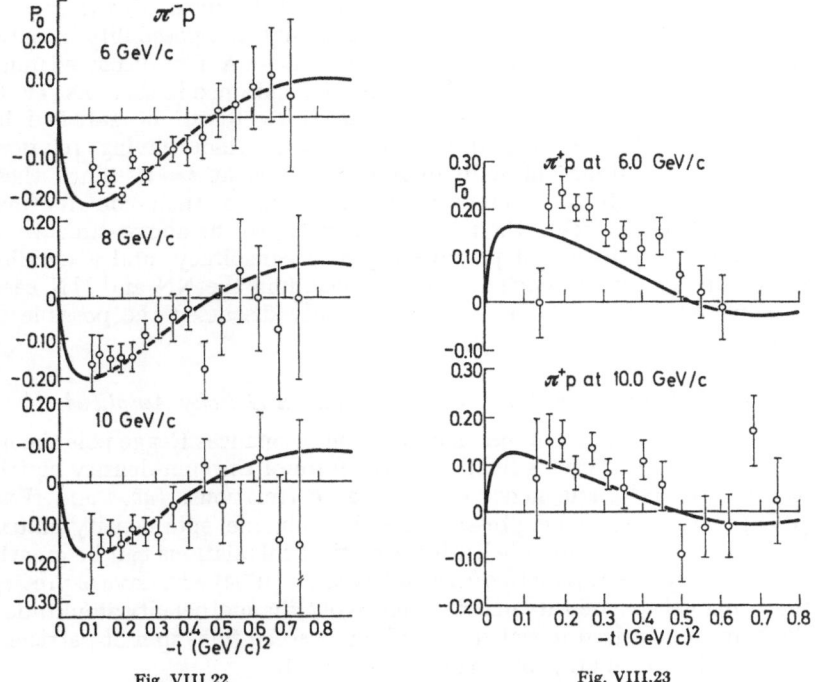

Fig. VIII.22
Fig. VIII.23

Fig. VIII.22. Polarization in elastic π⁻p scattering at various laboratory momenta, compared with the fit of *Chiu* et al., 1967, (from *Durand*, 1966).

Fig. VIII.23. Polarization in elastic π⁺p scattering at various laboratory momenta, compared with the fit of *Chiu* et al., 1967, (from *Durand*, 1966).

calculated results of *Chiu* et al. (1967) with the best available data [*Borghini* et al. (1967)]. Some of the present data were not available when the fit was made so a better fit is presumably possible. In view of the extra effects which are needed to explain the charge exchange polarization, however, this is hardly a worthwhile task at this stage.

d) *Spin Structure in* NN *Scattering*

The NN system offers a rich variety of polarization and spin-structure properties, but as yet there is little high energy data available for analysis. An excellent discussion, including a variety of proposed tests of Regge theory is given by *Leader* and *Slansky* (1966). In particular these authors point out that the factorization theorem could be tested if the relevant experimental measurements were made.

In addition *Leader* noticed that when the kinematical constraints on the amplitudes — in particular the constraints that the spin-flip amplitudes are zero in the forward direction — are combined with the factorization theorem, then all possible Regge-pole contributions to the double-spin-flip amplitudes [A_5 of (6.15)] are zero in the forward direction. Since there is no direct kinematical reason for this amplitude to be zero we have here a curious and rather unsatisfactory result, and this motivated the recent suggestion of "conspiring" trajectories [although the possibility of satisfying kinematical constraints through conspiracy was first discussed much earlier by *Volkov* and *Gribov* (1963)]. As we mentioned in sections (IV. 6) and (IV. 9) the kinematical constraints at $t = 0$ can be satisfied by having several cancelling trajectories, rather than having relations between the residues of individual trajectories at $t = 0$. Since these trajectories can have different quantum numbers, their contributions to A_5 need not cancel in the forward direction, so the above difficulty is removed. A discussion of possible types of conspiracy, and a detailed treatment for several particular cases, including the NN and N$\bar{\text{N}}$ case, is given by *Leader* (1967). We have already discussed the possible π-conspiracy in section (VIII. 6f).

e) *Density Matrices in Terms of Cross-Channel Helicity Amplitudes*

For the production of meson and bayon resonances Regge pole theory makes strong predictions relating to the form of the spin-density matrix of the final state, and hence to its decay correlations. *Jacob* and *Wick* (1959) give the formal expressions for the s-channel spin density matrix in terms of the s-channel helicities. Practical calculations can be greatly facilitated by the work of *Gottfried* and *Jackson* (1964) who have expressed the s-channel spin density matrix in terms of t-channel helicity amplitudes. With an unpolarized initial state and unobserved final spin of particle 3 the spin density matrix of particle 4 is given by

$$\langle m|\varrho_4| m\rangle = N \sum_{\bar{\lambda}_2 \bar{\lambda}_3 \lambda_1} \langle m\bar{\lambda}_2|A^t(s_2 t)| \bar{\lambda}_3 \lambda_1\rangle \langle m\bar{\lambda}_2|A^t(s_2 t)| \bar{\lambda}_3 \lambda_1\rangle^* , \quad (7.7)$$

where $\langle|A^t(s, t)|\rangle$ is the t-channel helicity amplitude and N is defined so that

$$\text{Tr}\,\varrho_4 = 1 . \quad (7.8)$$

With (7.7) it is possible to go directly from the Regge expression for the t-channel helicity amplitudes to the density matrix, and hence to the angular correlations of the decaying particles. We refer the reader to the article by *Gottfried* and *Jackson* for further details. The results depend sensitively upon the parities and signatures of the contributing trajectories, and will constitute a good test of the theory. Results at appropriate energies are just beginning to appear.

f) Density Matrices in $\pi N \to \rho N$ and $\pi N \to \rho \Delta$

As an illustrative example [*L. Jones* (1967)], and one which has several interesting features, we consider the processes

$$\pi N \to \rho N \tag{7.9}$$

and

$$\pi N \to \rho \Delta \tag{7.10}$$

for which measurements have been made in the pion momentum range 2.5 to 8 GeV/c [*Miller* et al. (1967)].

The ω, A_2 and π trajectories can contribute to these processes, and since the ω and A_2 are higher they should dominate over the π. The measured density matrix, however, in the region near $t = 0$, has the properties to be expected from the π-trajectory ($\varrho_{00} = \varrho_{1-1} = 0$ and $\mathrm{Re}\,\varrho_{10} = 0$), and only at larger t does it have the form expected from the ω and A_2 trajectories ($\varrho_{00} = \mathrm{Re}\,\varrho_{10} = 0$). To understand why this might be so we note first that the parity operator on the $\pi \varrho$ initial state (in the t-channel) gives

$$P|JM \lambda 0\rangle = (-1)^{J+1}|JM - \lambda 0\rangle \tag{7.11}$$

where λ is the ϱ helicity. It follows that the $\lambda = 0$ state is an eigenstate of parity with parity $(-1)^{J+1}$. It therefore does not couple to the ω and A_2 trajectories, so these only contribute to amplitudes with $\lambda = \pm 1$. Now in the s-channel forward direction, $t = 0$, we have the effect discussed in sections (III. 6) and IV. 3), i.e. that z_t does not tend to infinity with s, but becomes unity. Hence, as noted in section (IV. 3), the high energy behaviour at $t = 0$ of a given Regge trajectory will be $s^{\alpha(0)-\lambda}$. Thus at $t = 0$ the ω and A_2 contributions behave approximately like $s^{-0.5}$ whereas the pion contribution behaves approximately as s^0. For any given value of the energy there is a transition region of t from the $s^{\alpha(t)-\lambda}$ to the $s^{\alpha(t)}$ behaviour [an estimate of the size of this region, which shrinks with increasing s, is given by *L. Jones* (1967)], so we have a natural explanation of the observed behaviour of the ϱ density matrix.

Further experimental data will be necessary, particularly at higher energy, before this can be considered as a convincing explanation, and as definite evidence in favour of the $s^{\alpha-\lambda}$ behaviour of the forward amplitude. The situation is complicated by the fact that the π contribution is likely to be enhanced near the forward direction in any case, because at $t = 0$ it is very close to being a real physical particle, i.e. the value $\alpha_\pi(0)$ is very close to zero, and there is a pole due to the zero of $\sin \pi \alpha_\pi$ at this point (this is the Regge-pole way of saying that we are near the physical pole at $t = m_\pi^2$). For the ϱN final state there is a kinematical factor t in the residue which effectively kills this, but *Rogers* (1967) has pointed out that this factor t would not be present if there were a suitable pion conspiracy. In this case the closeness of the pion-pole could well explain the dominance of the π-trajectory near the forward direction at present energies.

Paciello and *Pugliese* (1967) have analysed the $\pi^- p \rightarrow \rho^- p$ data at 8 GeV/c using the π and ω trajectories. They give parametrized fits for the residue functions but do not explicitly extract any of the kinematic factors, so their fits cannot be easily interpreted in the above form.

It is clear that this field is a large one and that a great deal of experimental and theoretical work remains to be done.

VIII.8. Lower Energies and Interference with Direct Channel Resonances

It was found in sections (VIII. 3—7) that crossed channel Regge poles can provide a good description of scattering amplitudes near the

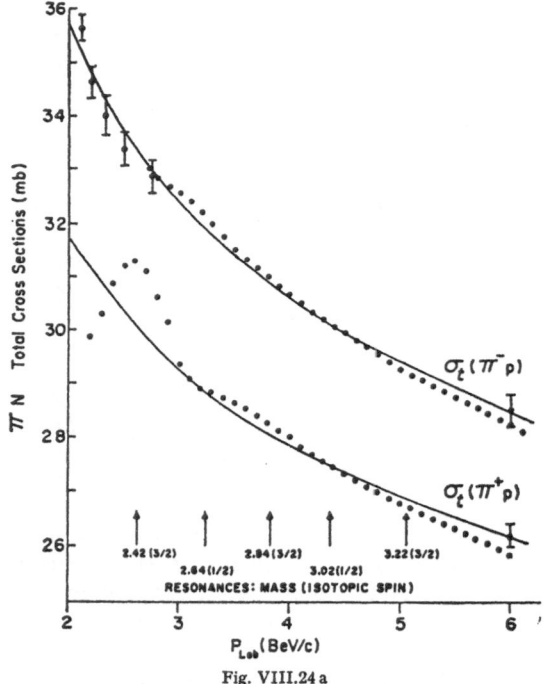

Fig. VIII.24 a

Fig. VIII.24 a—c. A comparison between the total cross-section data on (a) π^+p, (b) K$^\pm$p, K$^\pm$n, and (c) pp, pn $\bar{p}$p and $\bar{p}$n, at low energies, and predictions of the high energy fit of *Barger* and *Olsson*, 1966 a, with SU (3) couplings (from *Barger* and *Olsson*, 1966 b). (Fig. VIII. 24 b see p. 257 and Fig. VIII. 24 c see p. 258).

forward and backward directions at high energies. In accordance with our discussion of section (VIII. 2) "high energy" has been taken to mean above a few Gev, and typically the Regge pole fits have used data from about 5 to 20 Gev. But what happens when we look at lower energies? Obviously there is no longer any reason why the background and low-lying trajectories should be negligible compared with the high-lying trajectories. In fact we know that the low energy region in any given channel tends to be dominated by its low mass resonances, and these can not be represented by the sum of a few crossed-channel Regge poles.

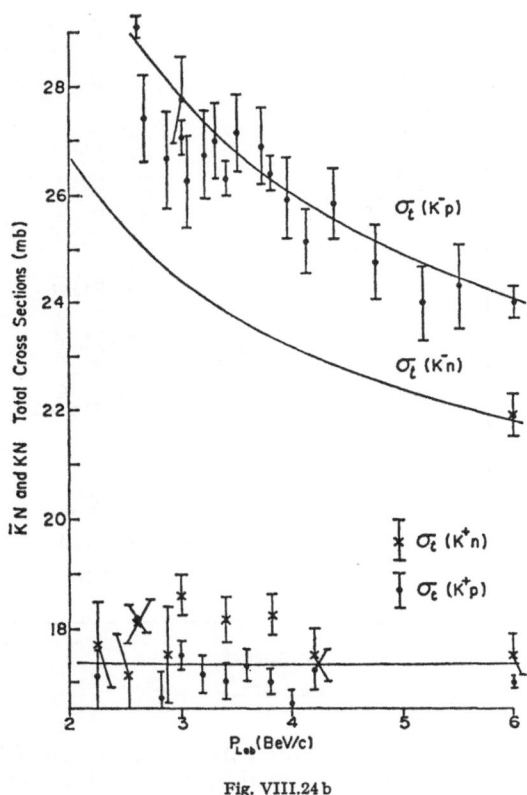

Fig. VIII.24 b

Barger and *Olsson* (1966b) used the parameters which they had obtained from fitting cross sections in the $p_{lab} = 6-20$ (Gev/c) range [described in section (VIII. 3); *Barger* and *Olsson* (1966a)] to extrapolate the cross sections down to about 1 (Gev/c). Their results are shown in Fig. (VIII. 24), and evidently quite a good representation of the data is obtained. In πN and NN scattering, for which the cross sections are known with high accuracy, some oscillation of the experimental points about the predicted curve is to be seen, and these oscillations can be correlated with the masses of the various direct channel resonances.

This suggests that we should write the amplitude as a sum of the leading crossed channel Regge poles, plus a background consisting of just the direct channel resonances. This representation corresponds to the one which we used for dynamical calculations in section (VI. 6),

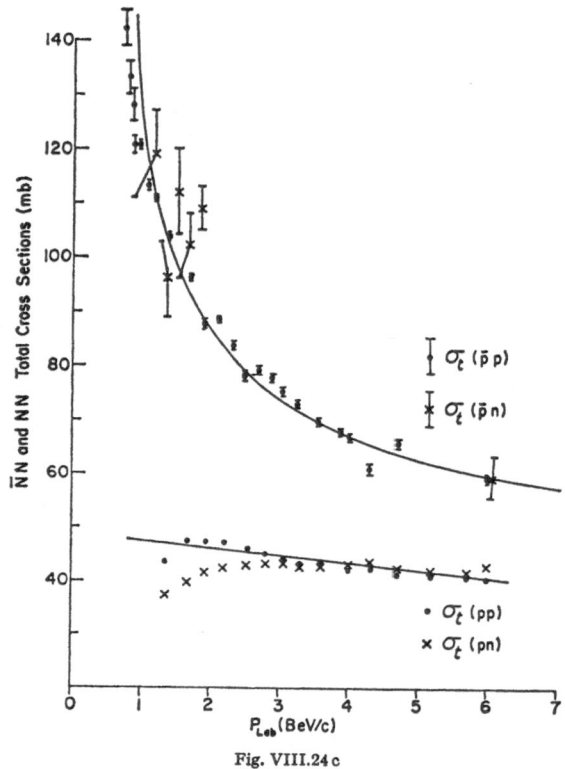

Fig. VIII.24 c

where the amplitude was represented by a sum of the Regge poles in each channel. But care is needed in making this combination, because we are really using two equivalent ways of representing the amplitude, and there is a danger of double counting. The one method involves making an s-channel partial-wave expansion which is then expressed in terms of the resonances plus a non-resonant background. In the other method we represent the amplitude by a Sommerfeld-Watson transform of the t-channel partial waves, i.e. in terms of the leading J_t-plane Regge poles, and a Regge background. We are then supposing that we can add the leading Regge singularities to the resonances and get a good fit to the amplitude, which will be true if the resonances give almost all of the Regge background, or equivalently — if the non-resonant part of the amplitude is almost all due to the Regge poles. Note however, that if we were to go into the left-half J_t-plane we could expect to pick up singularities which stem from the s-channel resonances. But these can not contribute to the right-half J_t-plane, so provided we only invoke the highest Regge poles there will be no double counting.

The contributions of the resonance poles will interfere with the Regge term, but of course this interference will only be important in those cases where the two types of contribution are of comparable magnitude. *Heinz* and *Ross* (1965) proposed that the interference between direct- and crossed-channel effects might prove useful as a method for determining the spins of resonances, and they applied it to backward π^+p elastic scattering, which they represented by the various isospin 3/2 resonances interfering with a fixed spin (not Regge pole), crossed channel, nucleon pole.

Later the method was exploited by *Barger* and *Cline* (1966, 1967a) who tried to analyse backward π^-p scattering into the contributions of the N_α, N_γ and Δ_δ Regge recurrences in the direct channel, and the

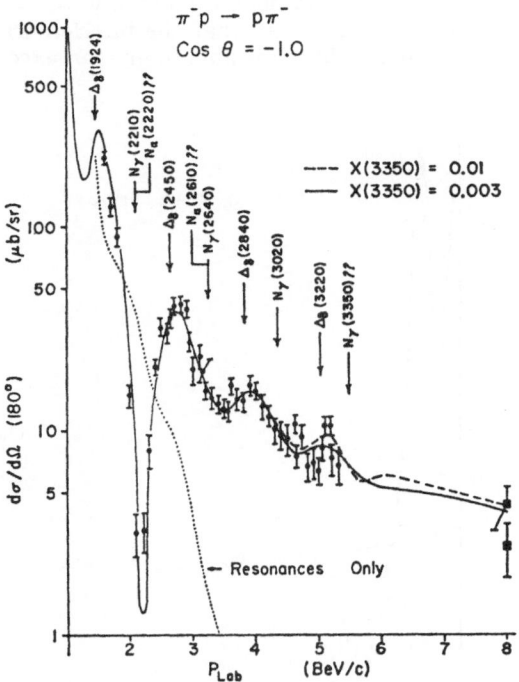

Fig. VIII.25. Theoretical fit to the 180° π^-p elastic scattering data. The arrows indicate the positions of the direct channel Δ_δ, N_α and N_γ recurrences, (from *Barger* and *Cline*, 1967a).

Δ_δ Regge trajectory in the crossed channel. Their success in fitting all the resonances on straight Regge trajectories has already been noted in section (VIII. 1). For the Δ_δ they used the trajectory

$$\mathrm{Re}\{\alpha(\sqrt{u})\} = 0.15 + 0.9\,u\,, \tag{8.1}$$

17*

which they obtained by fitting the established recurrences. They then wrote, (I. 7.10),

$$\frac{d\sigma}{d\Omega}\,(s, \theta_s = 180°) = \frac{1}{(8\pi)^2 s}\,|A^{\text{res}}(s, u = 0) + A^{\text{Regge}}(s, u = 0)|^2 \quad (8.2)$$

the resonances giving a contribution of the form

$$A^{\text{res}}(s, u = 0) = \sum_i \frac{x_i M_i \Gamma_i (-1)^{l_i} \left(J_i + \frac{1}{2}\right)}{M_i^2 - s - i M_i \Gamma_i} \times \left(\begin{array}{l} \frac{1}{3} \text{ for } \Delta \\ \frac{2}{3} \text{ for } N \end{array}\right) \quad (8.3)$$

summed over all the resonances of mass M_i, width Γ_i, spin J_i and parity $(-1)^{l_i}$, x_i being the elasticity of the resonance, which varies between 1, for a purely elastic resonance, and 0 for a wholly inelastic one. They included all the known resonances together with some higher mass peaks in the cross sections, to which they gave the assignments of Figs. (VIII. 3—5). Experimental elasticities were used where known, but were left as adjustable parameters if not known. The results are shown in Fig. (VIII. 25), and we see that the interference between the two contributions to the amplitude is able to give an excellent explana-

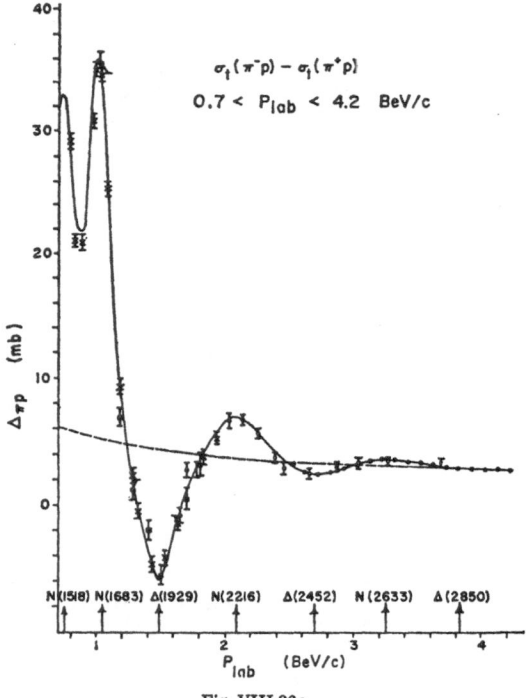

Fig. VIII.26a

Fig. VIII.26. The total cross-section difference $\Delta(\pi p)$ as a function of laboratory momentum. The solid curve represents the theoretical fit to the data. The dashed curves represent the ϱ-exchange contribution. The three plots show the structure in different energy ranges (from *Barger* and *Olsson*, 1966c). (Fig. VIII. 26b see p. 261 and Fig. VIII. 26c see p. 262).

tion of the oscillations of the data. The factor $(-1)^4$ in (8.3) means that the sign of the interference depends on the parity of the states, and so in most cases the parities of the resonances are determined unambiguously.

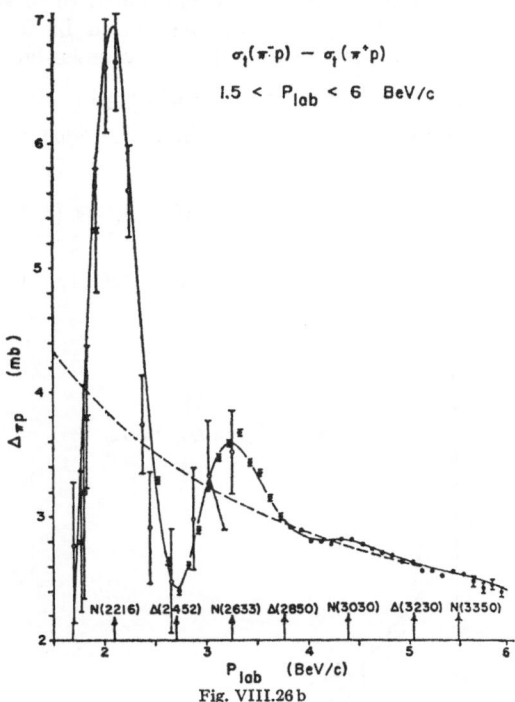

Fig. VIII.26 b

The actual J_i values are not really determined though, as only the factor $x_i(J_i + 1/2)$ is fixed by the fit, and the elasticities of the higher resonances are small. There is of course no guarantee that the sum of the amplitudes in (8.2) will satisfy unitarity, but in fact, because of the small elasticity, the resonance terms only reach a small fraction of the unitarity limit, and there is no problem.

Barger and *Cline* (1967a) also fitted $\pi^+ p$ data, but in this case isospin factors cause the direct channel resonances to dominate over the crossed channel contribution, at any rate up to about 4 (Gev/c), though N_α, N_γ, and Δ_δ exchanges are all possible.

More recently *Dikmen* (1967) has shown that a fit as good as that of *Barger* and *Cline* to the backward $\pi^- p$ differential cross section can be obtained using only the resonances and completely neglecting the crossed channel Regge trajectory, if slightly different values for some of the widths and elasticities are used. So the work of *Barger* and *Cline* is to be regarded as a help in establishing the nature of the resonances, rather than as a demonstration that there is an interference with the Regge term, but of course we know that the crossed channel contribution must exist. There is probably a danger of double counting in that the Regge pole fits to the high energy data, even in the 6—20 (Gev/c) range, must inadvertently be including some direct channel effects.

The interference method has also been applied to forward πp amplitudes by *Barger* and *Olsson* (1966c). We found in sections (VIII. 3) and (VIII. 4) that the ϱ is the only known crossed-channel pole contributing to

$$\Delta\,(\pi\mathrm{p}) = \sigma^{\mathrm{T}}\,(\pi^-\mathrm{p}) - \sigma^{\mathrm{T}}\,(\pi^+\mathrm{p})\,, \qquad (8.4)$$

and the $\pi^-\mathrm{p} \to \pi^0\mathrm{n}$ differential cross section, $\mathrm{d}\sigma^{\mathrm{c.\,e.}}/\mathrm{d}t\,(s,\,t)$. Their model is just the same; to represent

$$\Delta\,(\pi\mathrm{p}) = \frac{1}{2\,q_s\,\sqrt{s}}\,\mathrm{Im}\,[A^{\mathrm{res}}\,(s,\,0) + A^{\mathrm{Regge}}\,(s,\,0)]\,, \qquad (8.5)$$

and

$$\frac{\mathrm{d}\sigma^{\mathrm{c.\,e.}}}{\mathrm{d}t}\,(s,\,0) = \frac{1}{(8\pi)^2 s}\,|A^{\mathrm{res}}\,(s,\,0) + A^{\mathrm{Regge}}\,(s,\,0)|^2\,, \qquad (8.6)$$

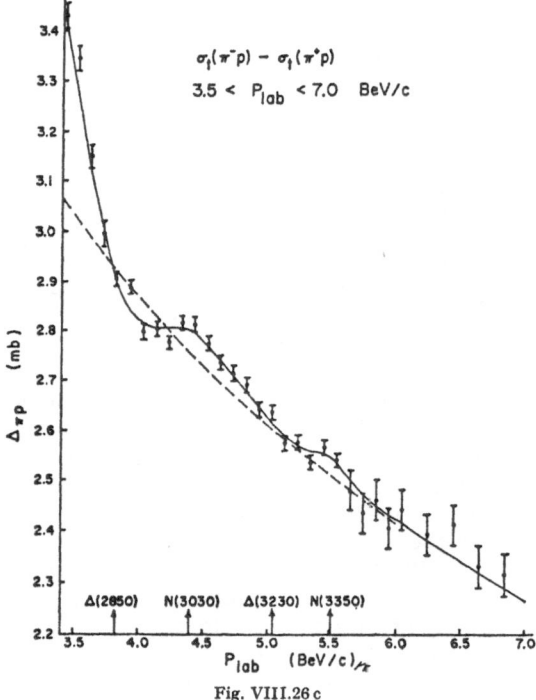

Fig. VIII.26 c

where A^{Regge} is just the ϱ trajectory contribution, and A^{res} contains the N_α, N_γ, and Δ_δ resonances of Figs. (VIII. 3—5). Curves of the results with

$$\alpha_\varrho(0) = 0.58 \pm 0.01$$

are shown in Fig. (VIII. 26), and they account well for the departures from pure ϱ exchange. However, *Hoff* (1967) has shown that at these low energies $d\sigma^{c.e.}/dt$ can equally well be fitted by just the sum of direct channel resonances. In fact this works for all t out to and beyond the dip at $t \approx -0.6$ (Gev/c)^2 discussed in section (VIII. 6), and explained by *Frautschi* (1966) in terms of the vanishing of $\alpha_\varrho(t)$. Whether this alternative explanation could work at the much higher energies fitted by *Arbab* et al. (1967), where the dip still persits, is doubtful, but we shall see in section (VIII. 10) that there are sum rules which restrict the non-Regge contribution.

It has also been noted, in section (VIII. 7), that ϱ exchange alone is unable to account for the polarization of π^-p charge-exchange scattering, and *Logan* and *Sertorio* (1966) have tried to explain the polarization in terms of this interference between the ϱ trajectory and the resonances. By including just the N_α, N_γ, and $\varDelta_\delta$ states of Figs. (VIII. 3—5) they were able to fit the 6 Gev polarization data, but they predicted a rapid decrease of the polarization at higher energies. This is because the ϱ can interfere only with the tails of the resonances at such energies. In fact, because of the small elasticities of the higher resonances, it is found that it is the resonances of medium energy which make the most important contribution. *Phillips* (1966) has also used this sort of model to fit the 6 Gev data successfully.

The difficulty with these models is that experimentally the polarization does not seem to be decreasing with energy, though the errors are too large to be sure about this, and among the various explanations, some of which we noted in section (VIII. 7), *Desai* et al. (1967) have proposed that the trajectories of Fig. (VIII. 3—5) continue to higher values of J, giving higher mass recurrences, so that even at very high energies the ϱ can interfere with a nearby resonance to produce the polarization. The variation of the polarization with energy then depends

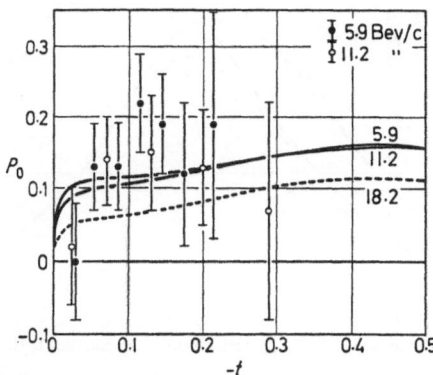

Fig. VIII.27. The polarization, P_0, versus momentum transfer, — (Gev/c)^2. The solid curves are the best fits at 5.9 and 11.2 (Gev/c), and the dashed curve is the prediction at 18.2 (Gev/c) obtained by *Desai* et al., 1967

very much on how the elasticities of the resonances change as we go up a trajectory. *Desai* et al. assumed a modified form of exponential decrease of x_i with energy, and obtained the results shown in Fig. (VIII. 27), but a suitable modification could probably account for almost any sort of variation of the polarization.

The process $\pi^- p \to \eta n$ is controlled at high energies by only the A_2 Regge pole, and so will also have no polarization in the absence of interference with some other contribution. *Logan* and *Sertorio* (1967) have given a prediction of the polarization (which is to be measured soon) based on their model mentioned above, and it will be interesting to see how this turns out.

VIII.9. Straight-Line Trajectories and Exchange Degeneracy

a) The Dynamics of Straight-Line Trajectories

We have noted several times that Regge trajectories, for both $t > 0$ and $t < 0$, appear to be straighter than would be expected on a simple bootstrap model, or indeed on any dynamical scheme where the scale of mass is about 1 GeV.

A suggested explanation of this behaviour in the $t > 0$ region is to link it with the onset of inelastic states with increasing energy, the idea being that the higher resonances are more properly thought of as being bound or resonant states of the inelastic channels. To see how this might work we imagine a one-channel problem, with threshold at $s = s_1$ and zero-spin particles, in which the force gives rise to the type of trajectory,

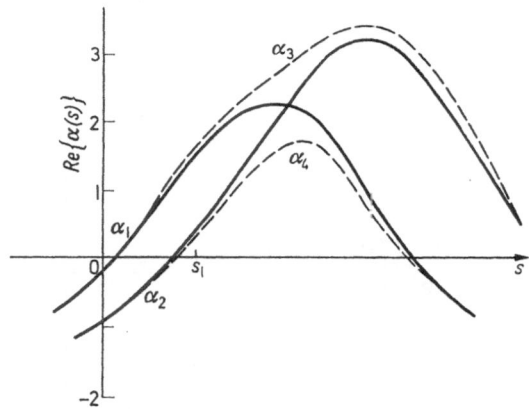

Fig. VIII.28. Illustrating the behaviour of Regge trajectories when coupling between two channels is introduced.

α_1, shown in Fig. (VIII. 28). This has one $J = 0$ bound state and two resonant states at $J = 1$ and 2 (we ignore the signature effect which is an irrelevant complication here). In addition we imagine a second one-channel problem, involving a particle of spin 3 and one of spin 0, and having its threshold at $s = s_2 > s_1$. We suppose the force to be such that this has a trajectory α_2 containing an $l = 0$, i.e. $J = 3$, resonance, as shown in Fig. (VIII. 28). The states with $J = 0$, 1 and 2 on α_2 are nonsense states and play no role in the discussion. Now we couple these two one-channel systems together to form a single two-channel system. For weak coupling there will be little effect on the trajectories, but as the coupling is increased a point will come when the trajectories do not cross but exchange tails [see *Newton* (1964), p. 88] and the state with $J = 3$ will lie on the same trajectory as the states with $J = 0$, 1 and 2, e.g. the dotted trajectory α_3 in Fig. (VIII. 28). The original α_2 trajectory will then become something like that shown as α_4. In drawing α_3 and α_4 we have used the result that the coupling in of extra states must reduce the energy, at least of a bound state, as follows from the variational principle. Note that "crossing" of Regge trajectories above threshold is not strictly forbidden since in fact they do not actually meet in the complex J-plane, as they will have different imaginary parts, $\operatorname{Im}\alpha_1 \neq \operatorname{Im}\alpha_2$. Thus for weak coupling the trajectories will cross, and it is only when the coupling becomes sufficiently strong that the flip-over of the trajectories will occur.

It is clear that we could imagine a situation like the above repeated indefinitely, with still higher thresholds keeping the trajectory from falling. For example, we could imagine the first resonance on the N trajectory to be essentially a $\pi - \Delta_{3/2}$ state, the next mainly a $\pi - \Delta_{7/2}$ state, etc. A model of this type is discussed by *Carruthers* (1966).

The apparent lack of trajectory curvature in the $t < 0$ region must be a related phenomenon, although dynamically the connection is obscure [note that it is not in any sense "established" experimentally, but the data is fitted by approximately straight line trajectories much better than those contrained to pass through -1, e.g. see section (VIII. 6b)]. *Mandelstam* (1966) has suggested that this can be understood as a natural consequence of the non-existence of elementary particles. To see this we recall that the Born approximation for a two-particle amplitude, e.g.

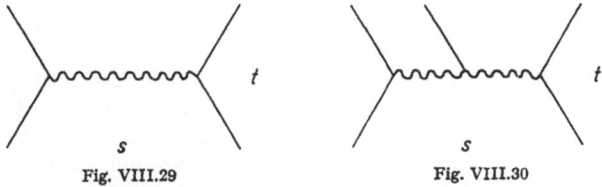

Fig. VIII.29   Fig. VIII.30

Fig. VIII.29. A Born diagram behaving as t^{-1} for large t.

Fig. VIII.30. A diagram behaving like t^{-2} for large t.

Fig. (VIII. 29), behaves as $1/t$ for large t, which is the origin of the fact that, in the absence of any "cancellations", the leading Regge trajectory tends to -1 as $t \to \infty$. However, if we replace one of the external lines by a two particle state, the Born approximation (Fig. VIII. 30) behaves like t^{-2} and so we find $\alpha(-\infty) = -2$ for the leading trajectory. The continuance of this process naturally leads to the notion of a leading trajectory which tends to $-\infty$ as $t \to -\infty$.

It is not easy to see how the above argument can be put in a more acceptable form, but *Stack* (1967) has recently produced an extremely interesting model in which he endeavours to make a nucleon out of a π-N pair, and to take account of the fact that the external nucleon is also a π-N bound state. This leads to a nucleon which is a $\pi\pi$N state, and this in turn is regarded as the external nucleon, and so on. It is hoped that iteration of this process may lead to a final bound-state, N_∞, which closely resembles the real nucleon. The exact meaning of all this is rather obscure, but Stack shows that the form factor of N_∞ goes down exponentially, i.e. obeys an infinite number of superconvergence relations. Clearly this is related to trajectory functions which obey $\alpha(t) \to -\infty$ as $t \to -\infty$.

Some of the implications of trajectories which go to infinity as $t \to \pm\infty$ have been discussed by *Khuri* (1967) and by *Jones* and *Teplitz* (1967a). It is clear that this behaviour is possible, but none of the mechanisms which might give such a behaviour offer any understanding of why the trajectories appear to be linear functions of t. *Brower* and *Harte* (1967) have pointed out that states on linear trajectories find it increasingly hard to decay to other states as t increases owing to the orbital angular momentum barrier; their dynamics has to be dominated by channels in which they are bound states, rather than resonances.

b) The Quark Model for Trajectories

Since any curve is approximately linear if we observe a sufficiently small part of it, a natural explanation of this apparent linearity is that the scale of the dynamics is much larger than the few GeV over which trajectories have been observed. An obvious, if highly speculative, example of such a theory is the quark-model, in which hadrons are assumed to be composed of "quarks" [*Gell-Mann* (1964), *Zweig* (1964); see *Dalitz* (1966) and *Squires* (1967) for general discussion of quark models and further references]. For our purpose here the essential property of quarks is that they are heavy ($M_q > 5$ GeV), and that mesons are regarded in the model as quark-antiquark bound states and baryons as bound states of three quarks. The "potentials" between the quarks required to produce the necessary binding energies must be very deep (we use the language of potential scattering; it has been suggested [*Morpurgo* (1965)] that this might be a reasonable approximation for the bound states, but in any case it will do for our purpose here), and such a potential, of suitable range ($\sim 10^{-13}$ cm), could well produce a trajectory of the form given by the Focacci plot, for example [see section VIII. 1)]. The trajectories would turn over eventually, but would be expected to pass through a large number of bound states (i.e. bound with respect to the $q\bar{q}$ threshold) first.

An additional feature of this model is that it explains the lack of any "signature" effect, at least for the mesons. To see this we note that the direct force between a $q-\bar{q}$ pair can be caused by the exchange of mesons (note that these give a potential of the correct range). However the exchange force arises from u-channel exchanges and these have the quantum numbers of a qq pair (Baryon number 2/3). It is known that there cannot be such a particle with mass less than about 5 GeV, so the exchange force is of extremely short range, and its small effect is therefore easily understood. The lack of exchange force, leading to overlapping trajectories of opposite signature, is sometimes refered to as "exchange degeneracy".

We note that this behaviour in the quark model is in striking contrast to the situation in conventional bootstrap theories of the type described in Chapter VI. In these theories the ϱ is assumed to be predominantly a $\pi\pi$ state, whereas the A_2 cannot contain any $\pi\pi$ contribution ($A_2 \to \pi\pi$ is forbidden by G-parity) so the A_2 and ϱ trajectories have completely different origins, and their degeneracy is apparently an accident.

However, the quark model also has its difficulties. Besides the obvious embarrassment due to the fact that quarks have not been observed (actually if they are much heavier than 5 GeV it is hard to see how they could be observed, even in cosmic rays) there are problems with the baryon states and with the lack of apparent 4 or 5 quark states which ought, at least on simple models, to be even more strongly bound than 3 quark states. At a deeper level it is very hard to see why the admixture of lower threshold states does not destroy completely the simple quark model of the hadrons [see *Squires* and *Watson* (1967) for a discussion of this problem]. It has, however, had some success in explaining mass differences, weak and electromagnetic properties of hadrons, and high energy scattering data [reviews and references are given by *Squires* (1967) and *Dalitz* (1966)].

We note that the quark model of high energy scattering is not inconsistent with the Regge-pole model, and it might be profitable to combine them by Reggeizing the qq and $q\bar{q}$ amplitudes. Some work along these lines has been done by *Ahmadzadeh* (1966), and by *James* and *Logan* (1967).

The exchange degeneracy assumption has been extended by several authors to include the couplings, i.e. it is assumed that the ϱ and A_2 trajectories have the same residue functions [see *Arnold* (1966) for application of this idea and further references]. This equality of the residues — which is a worthwhile starting assumption — does not,

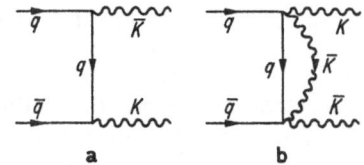

Fig. VIII. 31 a and b. The longest range forces for (a) $q\bar{q} \to \bar{K}K$ and (b) $q\bar{q} \to K\bar{K}$.

however, follow in any simple way from the quark model discussed above. To see this we consider, for example, the evaluation of the ϱ (and A_2) coupling to $K\bar{K}$ on this model. This could be done by introducing

$q\bar{q} \to K\bar{K}$ and $q\bar{q} \to K\bar{K}$ forces [see Fig. (VIII. 31a and b)], performing a two channel ($q\bar{q}$ and $K\bar{K}$) calculation and extracting the residue of the ϱ-pole in the $q\bar{q} \to K\bar{K}$ amplitude [see *Squires* and *Watson* (1967) for details of this type of calculation]. The largest range t-channel coupling is due to exchange of a quark (Fig. 31a) whereas there is no known single particle which can be exchanged in the u-channel (Fig. VIII. 31b). However the range of the force due to (quark + meson) which can be exchanged is not very different from that due to a quark alone, so it is unlikely that t-channel forces dominate u-channel ones to any significant extent. Now it may be that the forces in Fig. (VIII. 31a and b) do not greatly alter the bound state masses, and so do not affect the exchange degeneracy of the trajectories, but the residues of the two trajectories would only be approximately equal if one of Fig. (VIII. 31a or b) was dominant. There is some slight suggestion from $\pi^- p \to \pi^0 n$, $n\eta$ that the exchange degeneracy assumption may not in fact apply to residues [see section (VIII. 6d)].

VIII.10. Applications of Superconvergence Relations

In section (V. 5) we saw that integrals over certain single spectral functions have to be zero. *De Alfaro* et al. (1966) suggested that one might approximate these single spectral functions by the contributions of low lying particles and resonances, and hence obtain relations between the couplings and masses of these states. There are no convincing reasons why this should be a good approximation, and indeed there is plenty of evidence to demonstrate its inadequacy, but the method is easy, and produces quick results, so it has naturally attracted a great deal of attention. With suitable modifications, which we mention below, it may produce reasonable results, and indeed form the base of a dynamical model, so we discuss it briefly here.

First we note that the superconvergence relations we have obtained are integrals over the s- and u-discontinuities of t-channel amplitudes. In order to fit these with s- and u-channel bound states and resonances we must cross to the $s(u)$ channel. A direct method of doing the helicity crossing, which may in some cases be more convenient than the method using the Trueman-Wick crossing matrix, is given by *Odorico* (1967). We then insert a set of bound states and resonances into the discontinuities and assume that all other contributions can be neglected. If, for simplicity, we replace the resonance contributions by δ-functions (many authors use the Breit-Wigner form which is presumably preferable), we obtain relations of the form

$$\sum_i C_i R_i d_{\lambda\lambda'}^{\sigma_i}(z_t(t, s = m_i^2)) = 0 \qquad (10.1)$$

for all $t <$ (some t_0), where the sum is over a given set of states of spin σ_i and mass m_i. The coefficient R_i is the pole residue (a product of two coupling constants), and C_i is a coefficient that arises from the crossing matrix. Although not explicitly indicated the sum in (10.1) is understood to include u-channel states as well as s-channel states.

We note immediately that (10.1) cannot be satisfied over a range of t by a finite number of states, unless it is also satisfied by the states of each spin and mass separately (except that the states with $\sigma = 0$ need not have the same mass since $d^0_{00}(z)$ is a constant independent of z). These points were made by *Sakita* and *Wali* (1967), who postulated that pairs of states with the same mass and spin should each obey the superconvergence relations. These pairs however do not seem to occur in nature.

Fubini (1967) and *Klein* (1967) have discussed the possibility of saturating the superconvergence relations with an infinite number of states. Since superconvergence relations will exist in general for all possible scattering processes between these states, this is clearly a complicated problem which we shall not discuss further here.

Many workers have concentrated on superconvergence relations near $t = 0$, and have not hesitated to obtain cancellation between states of different spins and masses. *Frampton* and *Taylor* (1967) have offered a possible justification for this by noting that if instead of considering relations of the form

$$\int\limits_{s_0}^{\infty} \operatorname{Im} A\,(s,\,t)\,\mathrm{d}s = 0 \tag{10.2}$$

for all t, we consider the infinite set

$$\int\limits_{s_0}^{\infty} \operatorname{Im} A\,(s,\,0)\,\mathrm{d}s = 0 \tag{10.3}$$

$$\int\limits_{s_0}^{\infty} \operatorname{Im} \frac{\mathrm{d}A}{\mathrm{d}t}\,(s,\,0)\,\mathrm{d}s = 0 \tag{10.4}$$

$$\int\limits_{s_0}^{\infty} \operatorname{Im} \frac{\mathrm{d}^2 A}{\mathrm{d}t^2}\,(s,\,0)\,\mathrm{d}s = 0 \tag{10.5}$$

etc., which are equivalent to (10.2), then, whereas the first equation gets contributions from zero spin resonances, the second requires $\sigma \gtreqqless 1$, the next $\sigma \gtreqqless 2$, and so on. They argue that whereas the lower mass states (which tend to have low spin) may perhaps be adequate for saturating (10.3), the other relations are likely to require the addition of higher mass states. Note that in the above equations we have for convenience just written the integral over the s cut; in general we have a sum of contributions from s and u cuts [see section (V. 5)].

A simple example, first treated by *De Alfaro* et al. (1966), is $\rho\pi$ scattering in the s-channel. In the t-channel the helicity amplitudes have $\lambda = 0$, 1 and 2, and there are three isotopic spin states $I = 0$, 1 and 2. For $I = 0$, the Pomeranchon dominates so at $t = 0$ we have no superconvergence relations (though we will have them for $t < 0$). For $I = 1$ the ϱ trajectory, with $\alpha_\varrho(0) \approx 0.5$ should dominate so there will be a superconvergence relation for the $\lambda = 2$ amplitude (since $\alpha_\varrho(0) -$ $- \lambda \approx - 1.5 < - 1$). For $I = 2$ we know of no trajectory with $\alpha(0) > 0$, but the combination of two ϱ-poles gives a cut starting at $\alpha_c(t) \approx 0$ at $t = 0$. Assuming this to be below zero we will have three more superconvergence relations, i.e. for $\lambda = 2$ there will be the zeroth and first moment relation (since $\alpha_c(0) - \lambda < - 2$) and for $\lambda = 1$ there will be just the zeroth moment relation. Even if $\alpha_c(0)$ is not below zero these relations should hold reasonably well since the contribution of the cut above $J = 0$ is presumably small.

The u-channel for this problem is identical to the s-channel, and the crossing matrix shows that one of the above four superconvergence relations is identically satisfied by direct cancellation between the s and u cuts. For the other three the contribution of the u-cut is exactly the same as that of the s-cut. Note that if we ignore the fixed pole caused by the third double spectral function in the wrong signature amplitude [see the discussion in section (V. 5)] then we must take the s and u cuts separately, and hence we will have four superconvergence relations for this problem.

A discussion of the saturation of the three right-signature superconvergence relations with the π, ω, A_1 and A_2 mesons is given by *Frampton* and *Taylor* (1967), who give references to earlier work. The results are satisfactory for $\operatorname{Im} A(s, 0)$, i.e. the equations corresponding to (10.3), but not for the higher derivatives.

Superconvergence relations which arise in the (baryon octet)-(meson octet) scattering amplitudes because of the absence of a high Regge trajectory corresponding to the 27 representation of $SU(3)$ have been studied by *Sakita* and *Wali* (1967), *Babu* et al. (1967) and *Altarelli* et al. (1967). *Jones* and *Scadron* (1967) have also studied the (baryon octet)-(meson octet) → (baryon decouplet)-(meson octet), and (baryon decouplet)-(meson octet) → (baryon decouplet)-(meson octet) processes.

The inadequacy of the fits mentioned above (and there are many others in the literature) have led to various suggestions for obtaining more reasonable results. For example, *Ademollo* et al. (1967) have tried fitting the $\pi\varrho$ relations, mentioned above, by splitting the region of integration into two, and approximating $\operatorname{Im} A$ by direct channel states in $s < s_N$, and by a Regge pole form in $s > s_N$. The superconvergence relations then have the form

$$\int_{s_0}^{s_N} \operatorname{Im} A^{\text{Res}}(s, t)\, \mathrm{d}s + \int_{s_N}^{\infty} \operatorname{Im} A^{\text{Reg}}(s, t)\, \mathrm{d}s = 0 \qquad (10.6)$$

(again we do not explicitly put in the u-cut contributions). *Ademollo* et al. use just the ϱ-trajectory for A^{Reg}, and, with a reasonable choise of s_N and the ϱ Regge pole parameters, obtain a good fit to the three superconvergence relations over a wide range of values of t using the π, ϱ, A_1 and A_2 states. They are similarly able to fit the superconvergence relations in the $SU(3)$ related processes

$$\pi K^* \to \pi K^* \qquad (10.7)$$

and
$$K\rho \to K\rho .\tag{10.8}$$

Rubinstein (private communication) reports that this method also permits a reasonable fit to the wrong-signature superconvergence relation for these processes.

Similar in spirit to the above is the work of *Dolan* et al. (1967 a and b), who consider amplitudes which do not superconverge, but which can be made to do so by subtracting their leading Regge pole contributions [following the work of *Igi* (1962, 1963), and *Igi* and *Matsuda* (1967), referred to in section (VIII. 3e), and also that of *Logunov* et al. (1967)]. That is, they consider the quantity $[A(s, t) - A^{\text{Reg}}(s, t)]$, where $A^{\text{Reg}}(s, t)$ represents the sum of the leading Regge pole contributions, such that, with suitable assumptions about other J-plane singularities, $A - A^{\text{Reg}}$ is superconvergent, i.e.

$$\int_{s_0}^{\infty} [\text{Im}\,A(s, t) - \text{Im}\,A^{\text{Reg}}(s, t)]\,\mathrm{d}s = 0 .\tag{10.9}$$

Next they choose a of value $s = s_N$ such that, hopefully, the approximation $A = A^{\text{Reg}}$ holds for $s > s_N$, so that

$$\int_{s_N}^{\infty} \text{Im}\,A(s, t)\,\mathrm{d}s = \int_{s_N}^{\infty} \text{Im}\,A^{\text{Reg}}(s, t)\,\mathrm{d}s ,\tag{10.10}$$

and also such that A for $s < s_N$ can be represented by a few direct channel states, i.e.

$$\int_{s_0}^{s_N} \text{Im}\,A(s, t)\,\mathrm{d}s = \int_{s_0}^{s_N} \text{Im}\,A^{\text{Res}}(s, t)\,\mathrm{d}s .\tag{10.11}$$

Combining (10.11), (10.10) and (10.9) we obtain

$$\int_{s_0}^{s_N} \text{Im}\,A^{\text{Res}}(s, t)\,\mathrm{d}s = \int_{s_0}^{s_N} \text{Im}\,A^{\text{Reg}}(s, t)\,\mathrm{d}s .\tag{10.12}$$

Note that if A is superconvergent itself then A^{Reg} will also be superconvergent, and (10.12) becomes equivalent to (10.6).

Dolan et al. (1967 a and b) apply this method to πN charge exchange scattering, and by calculating $\int_{s_0}^{s_N} \text{Im}\,A^{\text{Res}}(s, t)\,\mathrm{d}s$ for various values of t, and putting the ϱ-trajectory for $A^{\text{Reg}}(s, t)$, they are able to predict the parameters of the ϱ trajectory, in excellent agreement with those found by conventional high energy Regge pole analysis. They apply this method to the amplitudes of both signatures and are able to show explicitly the failure of the wrong-signature relations, a failure which can be rescued by the expected fixed pole at $J = 0$ in the wrong signature amplitude. They find that this fixed pole does not occur in the ϱ residue and therefore does not spoil the explanation of the dip in the differential cross section as being due to the ϱ trajectory passing through $\alpha_\varrho = 0$. The fact that they need the fixed pole here is not obviously consistent with Rubinstein's result mentioned above, but since different processes are being considered this is not a serious contradiction.

It is important to note that in the above two methods the direct channel resonances and the Regge pole approximations are made in different regions and no question of double counting arises. This is different from a procedure which has been used in the literature (and which is wrong) which consists of writing

$$A(s, t) = A^{\text{Res}}(s, t) + A^{\text{Reg}}(s, t),\qquad(10.13)$$

assuming $A(s, t) - A^{\text{Reg}}(s, t)$ superconverges and hence obtaining

$$\int\limits_{s_0}^{\infty} \text{Im}\, A^{\text{Res}}\, ds = 0.\qquad(10.14)$$

It is clear that this equation is not compatible with (10.12). The chief cause of error here lies in the fact that if sufficient Regge terms have been subtracted from $A(s, t)$ to enable it to superconverge then the Regge terms already contain large parts of $A^{\text{Res}}(s, t)$ (and vice-versa) so that the separation in (10.13) is not valid. These points have been discussed by *Barger* and *Phillips* (1967).

Mandelstam (1967) has suggested that a dynamical scheme for the calculation of Regge-trajectories might be developed by using the same class of trajectories to describe both the low energy resonance region (in the direct channel) and the high energy behaviour (in the crossed channel). Then by demanding that the appropriate superconvergence relations of the type considered by *Dolan* et al. (as discussed above) are satisfied, self-consistency conditions on the trajectories and residues can be obtained. At present this work is in a preliminary stage.

VIII.11. Electromagnetic and Weak Processes

Our discussion so far has been entirely devoted to hadrons, but the ideas can also be applied to particles with only weak interactions (we include electromagnetic interactions in this expression). Although there are as yet few suitable experiments, there are some interesting features of these processes which have recently received attention in the literature.

First we consider Compton scattering

$$\gamma A \to \gamma A\qquad(11.1)$$

in the s-channel, where A is any hadron. In order to find the high s (small t) behaviour of this process we must consider the Regge expansion in the t-channel, i.e.

$$\gamma\gamma \to A\bar{A}.\qquad(11.2)$$

Now the photon has only two helicity states ± 1, so there are effectively two values of λ, the difference of the photon helicities, namely $\lambda = 2$ and $\lambda = 0$.

At $J = 1$ the $\lambda = 2$, two-photon, state is a nonsense state so, independent of the helicities of A and $\overline{A}$, there will be a superconvergence relation for the right signature amplitude (i.e. odd signature) to remove the fixed pole [or (pole)$^{1/2}$] at $J = 1$. An example of such a superconvergence relation is the Drell-Hearn sum rule [*Drell* and *Hearn* (1966)] for Compton scattering off protons.

It is important to note that these superconvergence relations will not necessarily hold in approximate calculations. For example, the necessity for the fixed pole to be removed follows from unitarity, so if the amplitude is calculated to lowest order in e^2, and hence does not satisfy unitarity, the fixed pole could be present. Such a fixed pole was suggested by *Bronzan* et al. (1967a and b), and by *Singh* (1967), in the case where we replace the γ-ray by a particle coupled to the isovector current, e.g. an intermediate vector boson W (assuming such a particle to exist). If we treat this vector boson as an elementary particle, then its exchange (see Fig. VIII. 32) will give rise to a Kronecker-delta, δ_{J1}, term in the amplitude. The assumptions of current algebra then imply the exis-

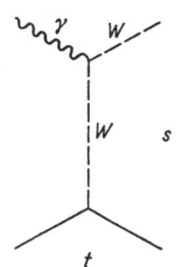

Fig. VIII.32. W-meson exchange in $\gamma N \rightarrow WN$.

tence of a fixed pole at $J = 1$ in the amplitude [*Fubini* (1966), *Dashen* and *Gell-Mann* (1966)]. It is clear that if we regard all the particles as elementary, then the arguments of chapter VII show that a fixed pole at $J = 1$ will occur in the Born approximation for the amplitude. *Bronzan* et al. (1967a and b) have shown that, at least in certain models, the fixed pole remains when the strong interactions are treated exactly, provided that the weak interaction is calculated only in lowest order.

If we replace the elementary W meson by a compound state (e.g. a ϱ particle) then the Kronecker-delta disappears and so does the fixed pole [*Rubinstein* et al. (1967)]. The effect on the fixed pole of keeping the W elementary but taking account of the non-elementary nature of the hadrons is not so clear, but it is unlikely that a fixed pole actually occurs in scattering amplitudes (even in lowest order).

If we consider, not the lowest order (in the weak interaction) amplitude, but the exact amplitude, then any fixed poles must become moving (Regge) poles. From the Froissart-bound these must lie in

$$\text{Re}\,\alpha \leq 1 \qquad (11.3)$$

for $t \leq 0$, so the appropriate superconvergence relations (e.g. the Drell-Hearn sum rule mentioned above) will certainly hold for the exact amplitudes at least for $t < 0$. The trajectory of such a moving pole would have the form

$$\alpha(t) = 1 + f(t) \cdot O(g^2) \qquad (11.4)$$

where g is the weak coupling constant, so to any finite order of g^2 the amplitude will have a fixed pole at $J = 1$. These "weak trajectories" do not apparently manifest themselves as particles, so if present they are rather obscure objects which deserve further study.

An additional consequence of the fixed poles in the sense-nonsense amplitudes is that, in second and higher order, they would lead, by the mechanism discussed in section VII. 6 to Kronecker delta singularities in the sense-sense amplitudes.

It will be noted that in the above discussion we have ignored fixed poles arising from the third double spectral function. These are at wrong signature values, and do not affect the asymptotic behaviour of the amplitude.

Returning to our physical problem, i.e. Compton scattering of ordinary photons, we expect the Pomeranchon trajectory to dominate. There is some doubt, however, about whether this is so in the forward direction ($t = 0$). To see this we consider, for simplicity,

$$\gamma \pi \to \gamma \pi \,, \tag{11.5}$$

for which there are two s-channel helicity amplitudes, $\langle 1|A^s| 1\rangle$ and $\langle 1|A^s| -1\rangle$. These are related to the t-channel helicity amplitudes by the helicity crossing matrix, which is here very simple,

$$\langle 1|A^s| 1\rangle = \langle |A^t| 1-1\rangle \tag{11.6}$$

$$\langle 1|A^s| -1\rangle = \langle |A^t| 1\ 1\rangle \,. \tag{11.7}$$

Now in the s-channel forward direction $\langle 1|A^s| -1\rangle$ is zero (by angular momentum conservation as explained in section IV. 3) so only $\langle |A^t| 1-1\rangle$ contributes. However, this has $\lambda = 2$, and at $J = 1$ is a sense-nonsense amplitude, so, assuming sense-nonsense decoupling, which must hold if we ignore the third double spectral function, it must be zero at $J = 1$. It follows that the residue of the Pomeranchon trajectory is zero at $\alpha_I(t) = 1$, i.e. at $t = 0$ (with the usual assumption). Thus, if we ignore the third double spectral function, the Pomeranchon trajectory does not contribute in the forward direction, and the $\gamma \pi$ total cross-section (related to the forward amplitude by the optical theorem) therefore goes to zero with increasing energy. This result also holds for Compton scattering off spin 1/2 targets [*Shepherd* (1967)].

As we have seen the third double spectral function implies the presence of a fixed pole at $J = 1$ in the wrong-signature amplitude (which in itself does not contribute to the asymptotic behaviour because of the signature factor), and this pole may occur in the residue of the Pomeranchon trajectory, thereby allowing it to dominate the forward amplitude, and hence allowing the total cross-section to tend to a constant. It will be curious however if such an apparently simple process as Compton scattering should be dominated at high energy by third-double spectral function effects. Further discussion of these points is given by *Mueller* and *Trueman* (1967b).

Photoproduction of π and K mesons has been discussed by *Ader* et al. (1967) and by *Frautschi* and *Jones* (1967a). There is obviously a resemblance between this problem and the $\pi N \to VB$ problem discussed in section (VIII. 6), but the fact that the γ-ray cannot have zero helicity means that the suppression in the forward direction due to unequal mass kinematics is present for all photoproduction amplitudes. When combined with the factors of t which arise from the s-channel kinematical factors, and the factorisation theorem (see section IV. 6), this fact gives rise to sharp forward dips in the predicted cross-sections in the absence of any conspiracy [this was first noticed by *Drell* and *Sulivan* (1967)]. It is clear that good data on these processes — at rather higher energy than is available at present — may provide convincing evidence in favour of at least some conspiracies. The paper by *Frautschi* and *Jones* (1967a), referred to above, discusses the effect of such a conspiracy in detail. See also *Halpern* (1967), and *Adler* and *Gilman* (1966), in this connection. The best available data on $\gamma p \rightarrow \pi^+ n$ does not seem to be compatible with the Regge pole model unless there is a pion conspiracy.

Photoproduction of ω mesons is discussed by *Drechsler* (1967b), and of ω, ϕ and ϱ^0 by *Buccella* and *Colocci* (1967).

Postscript

A considerable amount of work on Regge theory has come to our attention since the manuscript for this book was completed, and in this section we shall refer, very briefly, to some of it. Good reviews of recent developements were given by *L. Bertocchi* and *R. Omnes* at the 1967 Heidelberg International Conference on High Energy Physics; these have been published in the Proceedings of the Conference (*Filthuth*, 1968). The references for this section are on page 287.

(i) Daughters, Conspiracies and $O(4)$

Some evidence that the $\delta(965 \text{ MeV})$ resonance is not a 0^+ particle (as would be required for it to be the first daughter of the ρ; see section VIII. 1b) is given by *Allison* et al. (1967).

Calculations by *Cutkosky* and *Deo* (1967) show that the nature of the daughter trajectories obtained from the Bethe-Salpeter equation is very dependent on the potential used, and that they exhibit peculiar properties. The deviation from the symmetry as t moves away from zero can be large and irregular.

Confirmation of the forward peak in $\gamma p \to \pi^+ n$, which appears to require a pion conspiracy [see section VIII. 11, and also *Mitter* (1967)] is given by *Buschhorn* et al. (1967). The pion conspiracy in many such processes (including $np \to pn$ discussed in section VIII. 6f) is reviewed by *Durand* (1967). The $O(4)$ classification of such conspiracies has been discussed by *Sawyer* (1967). *Sertorio* and *Toller* (1967) have shown that the ϱ' trajectory introduced to explain the πN charge exchange polarization (see section VIII. 7b) needs a vanishing (or very small) residue in the non-spin-flip amplitude at $t = 0$ if the forward non-flip sum rule of Sertorio and Toller (section VIII. 3e) is to remain true. This would occur naturally if the ϱ' were part of a conspiracy (involving the B but not the ϱ). Unfortunately the data is really not yet good enough to give real confidence in this conclusion.

Further work on the use of Lorentz poles to fit high-energy amplitudes (see the end of section IV. 9) is reported by *Nwachuku* et al. (1967a and b), *Komy* and *Samiullah* (1967), and *Rashid* and *Samiullah* (1967). The conclusion that more than one Lorentz pole is needed to explain effects for which a single Regge pole is satisfactory remains. A useful review of the "little group" decomposition is given by *Feldman* and *Matthews* (1967), and the complex rotation group is discussed by *Iverson* (1967).

A clarifying discussion of kinematic singularities and threshold conditions is given by *Jackson* and *Hite* (1967).

Mandelstam (1967) has shown that many of the results of *PCAC* follow from the kinematic constraints on the pion residue at $t = 0$.

(ii) Zeros of Residues

The nature of the zeros of residue functions, depending on whether trajectories choose sense or nonesense, and the corresponding dips of the cross sections (see sections IV. 6 and VIII. 6), have been reviewed by *Chiu* et al. (1967), and by *Drechsler* (1967). *Matsuda* and *Igi* (1967) show how finite-energy sum rules can help to determine whether trajectories choose sense or nonsense; in particular the A_2 in π-p $\rightarrow$ ηn (see section VIII. 6c).

Contigorous and *Tran Thanh Van* (1967) have analysed $\pi^{\pm}$N $\rightarrow$ $\rho^{\pm}$N with particular reference to combinations of amplitudes for which the ω is the only known trajectory which can contribute. They find good agreement, but the fit does not support the suggestion that the "cross-over" effect in pp $\rightarrow$ pp and p$\bar{\text{p}}$ $\rightarrow$ p$\bar{\text{p}}$ [see the discussion below (VIII. 6.27)] is due to a change of sign of the ω residue. A similar conclusion is reached by *Barger* and *Durand* (1967) in their analysis of γp $\rightarrow$ π^0p. They explain the cross-over in terms of lower-lying trajectories.

(iii) Cuts

A general account of the properties of Regge cuts is given by *Rothe* (1967). *Dunne* (1967) discussed the evaluation of cut contributions, and gives references to other work. The manner in which cuts shield fixed J-plane poles so that these do not violate unitarity (see section V. 4) has been studied by *Bronzan* and *Jones* (1967), who argue that it is essential for the cut discontinuities to be singular and vanish at their end points. *Schwarz* (1967) distinguishes two types of two-particle Regge cuts; type 1 occuring when the two exchanged particles are Reggeized, and type 2 when one is a Regge pole and the other a normal t-plane pole. He shows that for unequal mass problems it is the latter which are required to shield the fixed J-plane poles by covering the two-particle unitary cut. *Hwa* (1967) makes the remarkable claim that, if the usual unitarity formula, like (V. 1.3), holds for amplitudes with Regge poles in the intermediate states, then all Regge trajectories must have the same slope. *Finkelstein* and *Tan* (1967) point out that the shielding mechanism would not work for the Pomeranchon if it were regarded as a fixed pole (see section VIII. 3b). This makes it very unlikely that the P is in fact a fixed pole.

Squires (1968) has given arguments suggesting that cuts might dominate elastic scattering, but be small in inelastic scattering where their effect is equivalent to applying absorptive corrections to the Regge pole model.

(iv) Infinitely Rising Trajectories

Goldberg (1967) has shown that the observed behaviour of Regge residues is approximately consistent with the exponential form required by *Jones* and *Teplitz* (see section VIII. 9a) for trajectories which tend to infinity. The implications of the lack of power boundedness of the amplitudes if $\alpha \xrightarrow[s \to \infty]{} \infty$ for *Martin's* theorem, on the maximum rate of fall of the scattering amplitude with momentum transfer at fixed energy (section VIII. 6i), are considered by *Chiu* and *Tan* (1967). Further work using new data on backward $\pi^{\pm}$p scattering, similar to that discussed in section VIII. 6c, is reported by *Barger* and *Cline* (1967).

(v) Sum Rules

Many authors have used finite energy sum rules to determine various Regge parameters. In particular *Borgese* et al. (1967) have examined the f, f', ϱ, ω and A_2 trajectories in the KN system. *Olsson* (1967) has shown that the use of low energy data on $\Delta(\pi p)$ to determine the ϱ parameters gives results consistent with those obtained from high energy fits to this cross-section difference.

(vi) Bootstraps

Finite energy sum rules in semi-bootstrap schemes have been discussed by *Gross* (1967) and by *Ademollo* et al. (1967).

The difficulty of reconciling straight Regge trajectories with a bootstrap dynamics of the type discussed in Chapter VI has been emphasized by *Collins* et al. (1968). They point out that the higher resonances in the *Focacci* et al. plot of the mesons [see Fig. (VIII. 2)] appear to be of too small a width to lie on the $\varrho(A_2)$ trajectory if the bootstrap hypothesis is correct.

January, 1968   *P. D. B. Collins*
                                                     E. J. Squires

References

Abbe, W. J., P. Kaus, P. Nath, and *Y. N. Srivastava:* Phys. Rev. **140**, B 1595 (1965).
— — — — Phys. Rev. **141**, 1513 (1966).
— — — — Phys. Rev. **154**, 1515 (1967).
Abers, E., and *V. L. Teplitz:* Nuovo Cimento **39**, 739 (1965).
— — Phys. Rev. **158**, 1365 (1967).
—, and *C. Zemach:* Phys. Rev. **131**, 2305 (1963).
Ademollo, M., H. R. Rubinstein, G. Veneziano, and *M. A. Virasoro:* Nuovo Cimento **51**, 227 (1967).
Ader, J. P., M. Capdeville, and *Ph. Salin:* Nuclear Physics B3, 407 (1967).
Adler, S. L., and *F. J. Gilman:* Phys. Rev. **152**, 1460 (1966).
Ahmadzadeh, A.: Ph. D. Thesis, University of California UCRL-11096, 1963 (unpublished).
— Phys. Letters **22**, 96 (1966).
— *P. G. Burke,* and *C. Tate:* Phys. Rev. **131**, 1315 (1963).
—, and *E. Leader:* Phys. Rev. **134**, B 1058 (1964).
Akerlof, C. W. et al.: Phys. Rev. Letters **17**, 1105 (1966).
Akyeampong, D. A., J. F. Boyce, and *M. A. Rashid:* Trieste preprint IC/67/61 (1967 a).
— — — Trieste preprint IC/67/59 (1967b).
de Alfaro, V., S. Fubini, C. Rossetti, and *G. Furlan:* Phys. Letters **21**, 576 (1966).
—, and *T. Regge: Potential scattering.* Amsterdam: North-Holland 1965.
Altarelli, G., F. Buccella, and *R. Gatto:* Phys. Letters **24B**, 57 (1967).
Amati, D., S. Fubini, and *A. Stanghellini:* Phys. Letters **1**, 29 (1962).
Anderson, E. W. et al.: Phys. Rev. Letters **16**, 855 (1966).
Andrews, M., and *J. Gunson:* J. Math. Phys. **5**, 1391 (1964).
Anselm, A. A., and *I. T. Dyatlov:* Phys. Letters **24B**, 479 (1967).
Arbab, F., and *C. B. Chiu:* Phys. Rev. **147**, 1045 (1966).
— *N. F. Bali,* and *J. W. Dash:* Phys. Rev. **158**, 1515 (1967).
Arnold, R. C.: Talk at the symposium on Regge poles held at Argonne National Laboratory (Dec. 1966), Argonne report 1966.
Atkinson, D., K. Dietz, and *D. Morgan:* Unpublished CERN report Th 531, 1965.
— — — Ann. Phys. **37**, 77 (1966).
—, and *A. P. Contogouris:* Nuovo Cimento **39**, 1082 and 1102 (1965).
—, and *M. B. Halpern:* Phys. Rev. **150**, 1377 (1966).
Azimov, I. A., A. A. Anselm, and *V. M. Shekhter:* JETP **17**, 726 (1963).
Azimov, Ya. I.: Phys. Letters **3**, 195 (1963).
Babu, P., F. T. Gilman, and *M. Suzuki:* Phys. Letters **24B**, 65 (1967).
Baker, M.: Ann. Phys. **4**, 271 (1958).
Balazs, L. A. P.: Phys. Rev. **128**, 1939 (1962).
— Phys. Rev. **129**, 872 (1963).
— Phys. Rev. **137**, B 1510 (1965a).
— Phys. Rev. **139**, B 1646 (1965b).
Bali, N.: Phys. Rev. **150**, 1358 (1966).
Bali, N. F., G. F. Chew, and *S.-Y- Chiu:* Phys. Rev. **150**, 1352 (1966).
—, and *S.-Y. Chiu:* Phys. Rev. **153**, 1579 (1967).
— — *R. W. Haymaker,* and *C.-I. Tan:* Phys. Rev. **161**, 1450 (1967a).
— *J. S. Ball, G. F. Chew,* and *A. Pignotti:* Phys. Rev. **161**, 1459 (1967b).
— *G. F. Chew,* and *A. Pignotti:* Phys. Rev. **163**, 1572 (1967c).
— — — Phys. Rev. Letters **19**, 614 (1967d).
Bander, M., P. W. Coulter, and *G. L. Shaw:* Phys. Rev. Letters **14**, 230 (1965).
Bardakci, K.: Phys. Rev. **127**, 1832 (1962).
Barger, V., and *D. Cline:* Phys. Rev. Letters **16**, 913 (1966).
— — Phys. Rev. **155**, 1792 (1967a).
— — Phys. Rev. **156**, 1522 (1967b).
—, and *M. Olsson:* Phys. Rev. **146**, 1080 (1966a).
— — Phys. Rev. **148**, 1428 (1966b).

Barger, V., and *M. Olsson:* Phys. Rev. 151, 11 (1966c).
—, and *R. J. N. Phillips:* Phys. Letters 25 B, 351 (1967).
—, and *L. Durand:* Phys. Rev. 156, 1525 (1967).
Bargmann, V.: Ann. Math. 48, 568 (1947).
Barut, A. O., and *B. C. Unal:* Nuovo Cimento 28, 112 (1963).
—, and *D. E.Zwanziger:* Phys. Rev. 127, 974 (1962).
Le Bellac, M.: Phys. Letters 25 B, 524 (1967).
Bjorken, J. D.: Phys. Rev. Letters 4, 473 (1960).
—, and *T. T. Wu:* Phys. Rev. 130, 2566 (1963).
Blair, I. M. et al.: Phys. Rev. Letters 17, 789 (1966).
Blankenbeckler, R., *M. L. Goldberger, N. N. Khuri*, and *S. B. Trieman:* Ann. Phys.
    10, 62 (1960).
— — Phys. Rev. 126, 766 (1962).
Bogoliubov, N. N., and *D. V. Shirkov:* Introduction to the theory of quantized
    fields. New York: Interscience 1959.
Bonamy, P. et al.: Phys. Letters 23, 501 (1966).
Borghini, M. et al.: Phys. Letters 24 B, 77 (1967).
Bottino, A., A. M. Longoni, and *T. Regge:* Nuovo Cimento 23, 954 (1962).
Boyce, J. F.: J. Math. Phys. 8, 675 (1967).
— *R. Delbourgo, A. Salam*, and *J. Strathdee:* Trieste preprint IC/67/9, 1967.
Bransden, B. H., P. G. Burke, J. W. Moffat, R. G. Moorhouse, and *D. Morgan:*
    Nuovo Cimento 30, 207 (1963).
Brittin, W. E., and *A. O. Barut* (Eds.): Lectures in theoretical physics, VII A;
    Lorentz group.: University of Colorado Press 1965.
Bronzan, J. B., I. S. Gerstein, B. W. Lee, and *F. F. Low:* Phys. Rev. Letters, 18,
    32 (1967 a).
— — — — Phys. Rev. 157, 1448 (1967 b).
Brower, R. C., and *J. Harte:* Phys. Rev. 164, 1841 (1967).
Buccella, F., and *M. Colocci:* Phys. Letters 24 B, 61 (1967).
Cabibbo, N., J. J. J. Kokkedee, L. Horwitz, and *Y. Ne'eman:* Nuovo Cimento 45,
    275 (1966 a).
— *L. Horwitz*, and *Y. Ne'eman:* Phys. Letters 22, 336 (1966 b).
Calogero, F., and *J. M. Charap:* Ann. Phys. 26, 44 (1964).
— —, and *E. J. Squires:* Ann. Phys. 25, 325 (1963 a).
— — — Proceedings of the Sienna Conference, 1963 b.
Capps, R. H.: Phys. Rev. 150, 1263 (1966).
Caprasse, H., and *H. Stremnitzer:* Nuovo Cimento 44 A, 1245 (1966).
Carruthers, P.: Introduction to unitary symmetry. New York: Interscience 1966.
— Phys. Rev. 154, 1399 (1967).
Castillejo, L., R. H. Dalitz, and *F. J. Dyson:* Phys. Rev. 101, 453 (1956).
Challifour, J. L., and *R. J. Eden:* Nuovo Cimento 27, 1104 (1963 a).
— — Phys. Rev. 129, 2349 (1963 b).
Chan, Hong-Mo, P. C. De Celles, and *J. E. Paton:* Nuovo Cimento 33, 70 (1964).
Charap, J. M.: Nuovo Cimento 31, 452 (1963).
—, and *E. J. Squires:* Phys. Rev. 127, 1387 (1962 a).
— — Ann. Phys. 20, 145 (1962 b).
— — Ann. Phys. 21, 8 (1963 a).
— — Ann. Phys. 25, 143 (1963 b).
Cheng, H.: Phys. Rev. 130, 1283 (1963).
—, and *D. Sharp:* Ann. Phys. 22, 481 (1963 a).
— — Phys. Rev. 132, 1854 (1963 b).
Chew, G. F.: Rev. Mod. Phys. 33, 467 (1961).
— S-matrix theory of strong interactions. New York: Benjamin 1962 a.
— Rev. Mod. Phys. 34, 394 (1962 b).
— Phys. Rev. Letters 9, 233 (1962 c).
— Phys. Rev. 129, 2363 (1963 a).
— Phys. Rev. 130, 1264 (1963 b).

Chew, G. F.: In: *Jacob, M.,* and *G. F. Chew:* Strong interaction physics. New York: Benjamin 1964.
— In: *de Witt, B.,* and *M. Jacob* (Eds.): High energy physics; the 1965 Les Houches Lectures. New York: Gordon, and Breach 1965 a.
— Progress in theoretical physics. Supplement (extra number. Commemoration issue for 30 th anniversary of Meson Theory by Dr. H. YUKAWA) 1965 b.
— Phys. Rev. **140**, B 1427 (1965 c).
— The dynamical S-matrix. New York: Benjamin 1966.
—, and *S. C. Frautschi:* Phys. Rev. Letters **5**, 580 (1960).
— — Phys. Rev. **123**, 1478 (1961 a).
— — Phys. Rev. Letters **7**, 394 (1961 b).
— — Phys. Rev. Letters **8**, 41 (1962).
— —, and *S. Mandelstam:* Phys. Rev. **126**, 1202 (1962).
—, and *C. E. Jones:* Phys. Rev. **135**, B 208 (1964).
—, and *S. Mandelstam:* Phys. Rev. **119**, 476 (1960).
— — Nuovo Cimento **19**, 752 (1961).
—, and *V. L. Teplitz:* Phys. Rev. **136**, B 1154 (1964).
— — Phys. Rev. **137**, B 139 (1965).
Chisholm, J. S. R.: Proc. Cambrigde Phil. Soc. **48**, 300 (1952).
Chiu, C. B., and *J. Finkelstein:* Nuovo Cimento **48** A, 820 (1967).
— *R. J. N. Phillips,* and *W. Rarita:* Phys. Rev. **153**, 1485 (1967).
—, and *J. D. Stack:* Phys. Rev. **153**, 1575 (1967).
Chung, V., and *D. R. Snider:* Phys. Rev. **162**, 1639 (1967).
—, and *J. Wright:* Phys. Rev. **162**, 1716 (1967).
Cohen-Tannoudji, G., A. Morel, and *H. Navelet:* Saclay Preprint. Kinematical singularities, crossing matrix and kinematical constraints for two-body helicity amplitudes. 1967.
Collins, P. D. B.: Phys. Rev. **136**, B 710 (1964).
— Phys. Rev. **139**, B 696 (1965).
— Phys. Rev. **142**, 1163 (1966).
— Phys. Rev. **157**, 1432 (1967).
—, and *V. L. Teplitz:* Phys. Rev. **140**, B 663 (1965).
Coulter, P. W., A. Scotti, and *G. L. Shaw:* Phys. Rev. **136**, B 1399 (1964).
Cutkosky, R. E.: Phys. Rev. **96**, 1135 (1954).
— J. Math. Phys. **1**, 429 (1960).
— Rev. Mod. Phys. **33**, 446 (1961).
— Phys. Rev. **131**, 1888 (1963).
— Proceedings of the Rochester Conference on Theoretical Physics, 1967; see also *Cutkosky, R. E.* and *B. B. Deo,* Phys. Rev. Letters **19**, 1256 (1967).
Dalitz, R. H.: In: *De Witt, B. S.,* and *M. Jacob* (Ed.): High energy physics, the 1965 Les Houches lectures. New York: Gordon and Breach 1966.
Dashen, R., and *M. Gell-Mann:* Phys. Rev. Letters **17**, 340 (1966).
De Laney, V. M., D. J. Gross, I. J. Muzinich, and *V. L. Teplitz:* Phys. Rev. Letters **18**, 149 (1967).
Delbourgo, R., A. Salam, and *J. Strathdee:* Phys. Rev. **164**, 1981 (1967 a).
— — — Phys. Letters **25** B, 230 (1967 b).
Desai, B. R.: Phys. Rev. **138**, B 1174 (1965).
— Phys. Rev. Letters **17**, 498 (1966).
— *D. T. Gregorich,* and *R. Ramachandran:* Phys. Rev. Letters **18**, 565 (1967).
—, and *R. G. Newton:* Phys. Rev. **129**, 1437 (1963 a).
— — Phys. Rev. **129**, 1445 (1963 b).
—, and *B. Sakita:* Phys. Rev. **136**, B 226 (1964).
Dikmen, F. N.: Phys. Rev. Letters **18**, 798 (1967).
Dolan, R., D. Horn, and *C. Schmidt:* Phys. Rev. Letters **19**, 402 (1967 a).
— — — Caltech. preprint. Finite energy sum rules and their application to πN charge exchange scattering. 1967 b. (Phys. Rev., to be published)
Domokos, G.: Phys. Rev. **159**, 1387 (1967).
—, and *P. Suranyi:* Nucl. Phys. **54**, 529 (1964).

Drechsler, W.: CERN preprint Th 773, 1967a. [Now published: Nuovo Cimento **53A**, 115 (1968)].
— Phys. Letters **23**, 272 (1967b).
Drell, S. D., and *A. C. Hearn:* Phys. Rev. Letters **16**, 908 (1966).
—, and *J. D. Sullivan:* Phys. Rev. Letters **19**, 268 (1967).
Drummond, I. T.: Nuovo Cimento **29**, 720 (1963).
— Phys. Rev. **140**, 1368 (1965).
Durand, L.: Talk at the Symposium on Regge poles held at Argonne National Laboratory (Dec. 1966), Argonne report 1966.
— Phys. Rev. Letters **18**, 58 (1967).
Eden, R. J., P. V. Landshoff, D. I. Olive, and *J. C. Polkinghorne:* The analytic S-matrix. London: Cambridge University Press 1966.
Edmonds, A. R.: Angular momentum in quantum mechanics.: Princeton University Press 1957.
Erdelyi, A. (Ed.): The Bateman manuscript project, higher transcendental functions, 3 Volumes; and tables of integral transforms, 2 Volumes. New York: McGraw-Hill 1953.
Fearing, H. W.: Stanford University preprint. Daughter trajectories and the Regge formula for the scattering of unequal mass particles, 1967.
Federbush, P. G., and *M. T. Grisaru:* Ann. Phys. **22**, 263 (1963a).
— — Ann. Phys. **22**, 299 (1963b).
Ferro-Luzzi, M.: Rapporteur's talk in Garnjost (Ed.). Proceedings of the XIIIth International Conference on High Energy Physics (University of California Press) 1967.
Finkelstein, J.: Phys. Rev. **145**, 1185 (1966).
— Phys. Rev. **154**, 1596 (1967).
—, and *J. M. Wang:* Lawrence Radiation Laboratory report, UCRL-17500, 1967.
Flores-Maldonado, V.: Phys. Rev. **155**, 1773 (1967).
Focacci, M. N. et al.: Phys. Rev. Letters **17**, 890 (1966).
Foley, K. J. et al.: Phys. Rev. Letters **10**, 376 (1963).
— Phys. Rev. Letters **14**, 862 (1965a).
— Phys. Rev. Letters **15**, 45 (1965b).
Foley, K. J., et al.: Phys. Rev. Letters **19**, 193 (1967).
Fox, G. C.: Phys. Rev. **157**, 1493 (1967).
—, and *E. Leader:* Phys. Rev. Letters **18**, 628 (1967).
Frampton, P. H., and *J. C. Taylor:* Superconvergence sum rules in $\varrho\pi$ scattering. Oxford preprint 1967.
Frautschi, S. C.: Regge poles and S-matrix theory. New York: Benjamin 1963.
— Phys. Rev. Letters **17**, 722 (1966).
— *M. Gell-Mann,* and *F. Zachariasen:* Phys. Rev. **125**, 2204 (1962).
—, *L. Jones:* Phys. Rev. **163**, 1820 (1967a)
— — Phys. Rev. **164**, 1918 (1967b)
— — Caltech. preprint 68—142 (1967c). (Phys. Rev., to be published)
— *P. Kaus,* and *F. Zachariesen:* Phys. Rev. **133**, B 1607 (1964).
Freedman, D. Z., and *J. M. Wang:* Phys. Rev. Letters **17**, 569 (1966).
— — Phys. Rev. **153**, 1596 (1967a).
— — Phys. Rev. Letters **18**, 863 (1967b).
— — Phys. Rev. **160**, 1560 (1967c)
— *C. E. Jones,* and *J. M. Wang:* Phys. Rev. **155**, 1645 (1967).
Freund, P. G. O.: Phys. Letters **3**, 123 (1962).
— Phys. Rev. Letters **15**, 929 (1965).
Froissart, M.: Invited paper at the La Jolla Conference on Weak and Strong Interactions (unpublished) 1961a.
— Nuovo Cimento **22**, 191 (1961b).
— Phys. Rev. **123**, 1053 (1961c).
Frye, G., and *R. L. Warnock:* Phys. Rev. **130**, 478 (1963).
Fubini, S.: Nuovo Cimento **43**, 475 (1966).
— Lectures on "Current algebra and sum rules", also private communication 1967.
Fulco, J., G. L. Shaw, and *D. Wong:* Phys. Rev. **137**, B 1242 (1965).

Galbraith, W. et al.: Phys. Rev. **138**, B 913 (1965).
Gell-Mann, M.: Proceedings of the 1962 CERN Conference on High Energy Physics 1962a.
— Phys. Rev. Letters **8**, 263 (1962b).
— Phys. Letters **8**, 214 (1964).
—, and *M. L. Goldberger:* Phys. Rev. Letters **9**, 275; sell also the erratum in Phys. Rev. Letters **10**, 39 (1962).
— — *F. E. Low, E. Marx,* and *F. Zachariasen:* Phys. Rev. **133**, B 145 (1964a).
— — — *V. Singh,* and *F. Zachariasen:* Phys. Rev. **133**, B 161 (1964b).
Goldberger, M. L., M. T. Grisaru, S. W. MacDowell, and *D. Y. Wong:* Phys. Rev. **120**, 2250 (1960).
—, and *C. E. Jones:* Phys. Rev. Letters **17**, 105 (1966a).
— — Phys. Rev. **150**, 1269 (1966b).
Goldhaber, G.: Rapporteur's talk in Garnjost (Ed.). Proceedings of XIIIth International Conference on high energy physics.: University of California Press 1967.
Gottfried, K., and *J. D. Jackson:* Nuovo Cimento **33**, 309 (1964).
Gourdin, M.: Unitary symmetries. Amsterdam: North-Holland 1967.
Gribov, V. N.: JETP **41**, 677 (1961). [Translation, Soviet Physics JETP **14**, 478 (1962)].
— Soviet J. Nucl. Phys. **5**, 138 (1967a).
— Proceedings of the Rochester Conference on Theoretical Physics 1967b.
—, and *I. Ya. Pomeranchuk:* Phys. Rev. Letters **8**, 343 (1962a).
— — Proceedings of the 1962 CERN Conference on High Energy Physics, p. 522, 1962b.
— — Phys. Rev. Letters **9**, 238 (1962c).
Gunson, J.: J. Math. Phys. **6**, 827, 845 and 852 (1965). (The original version of these papers was available in 1962).
Hadjioannou, F. T.: Nuovo Cimento **44**, 185 (1966).
Hadjioannou, F., R. J. N. Phillips, and *W. Rarita:* Phys. Rev. Letters **9**, 183 (1962).
Hall, D., and *A. S. Wightman:* Kgl. Danske Vidensk. Selsk. Mat.-Fys. Medd. **31**, No. 5 (1957).
Halliday, I. G.: Nuovo Cimento **30**, 177 (1963).
—, and *J. C. Polkinghorne:* Phys. Rev. **132**, 2741 (1963).
Halpern, M. B.: Phys. Rev. **160**, 1441 (1967).
Hamilton, J.: In: *Moorehouse, R. G.* (Ed.): Strong interactions and high energy physics. London: Oliver and Boyd 1964.
Hankins, D., P. Kaus, and *C. J. Pearson:* Phys. Rev. **137**, B 1034 (1965).
Hara, Y.: Phys. Rev. **136**, 507 (1964).
— Phys. Letters **23**, 696 (1966).
Harting, D. et al.: Nuovo Cimento **38**, 60 (1965).
Heinz, R. M., and *M. H. Ross:* Phys. Rev. Letters **14**, 1091 (1965).
Herglotz, A.: Ber. Verh. Sächs. Akad. Wiss. Leipzig, Math. Naturw. Kl. **63**, (1911).
Högaason, H.: Herceg Novi Lectures 1966 (unpublished).
—, and *A. Frisk:* Phys. Letters **22**, 90 (1966a).
— — Phys. Letters **22**, 516 (1966b).
—, and *Ph. Salin:* Nucl. Phys. **B2**, 657 (1967).
Höhler, G., J. Baacke, J. Giesecke, and *N. Zovko:* Proc. Roy. Soc. A **289**, 500 (1965).
— — *H. Schaile,* and *P. Sonderegger:* Phys. Letters **20**, 79 (1966).
— *G. Ebel,* and *J. Giesecke:* Z. Physik **180**, 430 (1964).
Hoff, G. T.: Phys. Rev. Letters **18**, 816 (1967).
Van Hove, L.: Rapporteur's talk in: Garnjost (Ed.); Proceedings of the XIIIth International Conference on High Energy Physics.: University of California Press 1967.
Huang, K., C. E. Jones, and *V. L. Teplitz:* Phys. Rev. Letters **18**, 146 (1967).
Igi, K.: Phys. Rev. Letters **9**, 76 (1962).
— Phys. Rev. **130**, 820 (1963).
—, and *S. Matsuda:* Phys. Rev. Letters **18**, 625; 822 (E) (1967).
Inönu, E., and *E. P. Wigner:* Nuovo Cimento **9**, 707 (1952).
Jackson, J. D.: Rev. Mod. Phys. **37**, 484 (1965).

Jacob, M., and *G. C. Wick:* Ann. Phys. **7**, 404 (1959).

James, P. B., and *R. K. Logan:* Phys. Letters **25** B, 38 (1967).

Johnson, K., and *S. B. Trieman:* Phys. Rev. Letters **14**, 189 (1965).

Jones, C. E.: Lawrence Radiation Laboratory report UCRL-10700 1962 (unpublished).

— Phys. Rev. **135**, B 214 (1964).

— Nuovo Cimento **40**, 761 (1965).

—, and *J. B. Hartle:* Phys. Rev. **140**, B 90 (1965).

—, and *V. L. Teplitz:* Phys. Rev. Letters **19**, 135 (1967 a).

— — Phys. Rev. **159**, 1271 (1967 b).

—, and *G. Tiktopoulos:* J. Math. Phys. **7**, 311 (1966).

Jones, H. F.: Nuovo Cimento **50**, 814 (1967).

—, and *M. D. Scadron:* Nuclear Physics B **4**, 267 (1967).

Jones, L.: Phys. Rev. **163**, 1523 (1967).

Joos, H.: In: Lectures in theoretical physics, Vol. 7 A. (Ed.) *Britten, W. E.*, and *A. O. Barut* (University of Colarado Press).

Källén, G.: Elementary particle physics. Reading, Mass: Addison Wesley 1964.

Kawai, T.: A generalised factorisation theorem and the asymptotic behaviour of the scattering amplitude. Osaka preprint 1967.

Khuri, N. N.: Phys. Rev. **130**, 429 (1963 a).

— Phys. Rev. Letters **10**, 420 (1963 b).

— Phys. Rev. **132**, 914 (1963 c).

— Phys. Rev. Letters **18**, 1094 (1967).

Kibble, T. W. B.: Phys. Rev. **117**, 1159 (1960).

— Phys. Rev. **131**, 2282 (1963).

Klein, S.: Phys. Rev. Letters **18**, 1074 (1967).

Krammer, M., and *U. Maor:* Nuovo Cimento **50** A, 963 (1967).

Kretzschmar, M.: Nuovo Cimento **32**, 1405 (1964).

Landau, L. D.: Nucl. Phys. **13**, 181 (1959).

Leader, E.: Conspiracy and evasion; a property of Regge poles. Cambridge preprint 1967. (Phys. Rev., to be published).

—, and *R. C. Slansky:* Phys. Rev. **148**, 1491 (1966).

Lee, B. W., and *R. F. Sawyer:* Phys. Rev. **127**, 2266 (1962).

Lehman, H.: Nuovo Cimento **10**, 579 (1958).

Leitner, J.: An experimental review of SU (3), Syracuse preprint 1966.

Levinson, N.: Kgl. Danske Vidensk. Selsk., Mat.-Fys. Medd. **25**, No. 9, 1949.

Lévy-Leblond, J.-M.: Nuovo Cimento **45**, 772 (1966).

Lin, K. Y.: Phys. Rev. **155**, 1515 (1967).

Lindenbaum, S. J.: In the Proceedings of the 1965 Oxford International Conference on Elementary Particles (Rutherford Laboratory) 1966.

Logan, R. K.: Phys. Rev. Letters **14**, 414 (1965).

—, and *L. Sertorio:* Phys. Rev. Letters **17**, 834 (1966).

— — Urbana Preprint ,,A Regge pole model of $\pi^- + p \rightarrow \eta + n$ Polarization" (1967).

Logunov, A., *L. D. Soloviev*, and *A. N. Tabkhelidze:* Phys. Letters **24** B, 181 (1967).

Lovelace, C., and *D. Masson:* Nuovo Cimento **26**, 472 (1962).

Luming, M.: Phys. Rev. **136**, B 1120 (1964).

MacDowell, S.: Phys. Rev. **116**, 774 (1959).

Mandelstam, S.: Phys. Rev. **112**, 1344 (1958).

— Phys. Rev. **115**, 1741 (1959 a).

— Phys. Rev. **115**, 1752 (1959 b).

— Ann. Phys. **19**, 254 (1962).

— Nuovo Cimento **30**, 113 (1963 a).

— Nuovo Cimento **30**, 1127 (1963 b).

— Nuovo Cimento **30**, 1148 (1963 c).

— Ann. Phys. **21**, 302 (1963 d).

— Phys. Rev. **137**, B 949 (1965).

Mandelstam, S.: In: *Takeda, G.,* and *A. Fujii (Ed.):* 1966 Tokyo summer lectures in theoretical physics, Part II. New York: Benjamin 1966.
— Dynamics based on rising Regge trajectories. University of California, Berkeley, preprint 1967. (Phys. Rev., to be published).
—, and *L.-L. Wang:* Phys. Rev. **160**, 1490 (1967).
Martin, A.: Phys. Rev. Letters **9**, 410 (1962).
— Nuovo Cimento **37**, 671 (1965).
Miller, D. H., et al.: Phys. Rev. **153**, 1423 (1967).
Morrison, D. R. O.: Phys. Letters **22**, 528 (1966a).
— Review of inelastic two-body reactions, paper given to the Stony Brook conference, April 1966, CERN preprint TC/Physics 66-20. 1966b.
— Phys. Letters **25**B, 238 (1967).
Morpurgo, G.: Physics **2**, 95 (1965).
Mueller, A. H., and *T. L. Trueman:* Phys. Rev. **160**, 1296 (1967a).
— — Phys. Rev. **160**, 1306 (1967b).
Murphy, P. G.: Rapporteurs talk in: Garnjost (Ed.): Proceedings of the XIIIth International Conference on High Energy Physics.: University of California Press 1967.
Muzinich, I.: J. Math. Phys. **5**, 1481 (1964).
Nakanishi, N.: Phys. Rev. **138**, B 1182 (1965).
— Prog. Theor. Phys. **37**, 618 (1967).
Nath, L. M.: Nuovo Cimento **52**, 944 (1967)
Ne'eman, Y., and *J. D. Reichert:* Phys. Rev. Letters **18**, 1226 (1967).
Newton, R. G.: The complex *j*-plane. New York: Benjamin 1964.
Oakes, R. J.: Phys. Letters **24**B, 154 (1967).
Odorico, R.: Nuovo Cimento **51**A, 1021 (1967).
Oehme, R.: In: *Moorhouse, R. G. (Ed.):* Strong interactions and high energy physics. London: Oliver and Boyd 1964.
— Phys. Rev. Letters **18**, 1222 (1967).
—, and *G. Tiktopoulos:* Phys. Letters **2**, 86 (1962).
Olive, D. I.: Phys. Rev. **135**, B 745 (1964).
Omnès, R., and *M. Froissart:* Mandelstam theory and Regge poles. New York: Benjamin 1963.
—, and *E. Leader:* On the removal of singularities in the Regge Asymptotic behaviour. Cambridge preprint 1967.
Orear, J.: Phys. Rev. Letters **12**, 113 (1964).
— et al.: Phys. Rev. Letters **15**, 309 (1965).
Paciello, M. L., and *A. Pugliese:* Phys. Letters **24**B, 431 (1967).
Peierls, R. F., and *T. L. Trueman:* Phys. Rev. **134**, B 1365 (1964).
Perkins, D. H.: Proceedings of the International Conference of Theoretical Aspects of Very High Energy Phenomena (CERN, Geneva). (1961).
Phillips, R. J. N.: Nuovo Cimento **45**, 245 (1966).
— Phys. Letters **24**B, 342 (1967a).
— Nucl. Phys. B**2**, 657 (1967b).
—, and *W. Rarita:* Phys. Rev. **139**, B 1336 (1965a).
— — Phys. Letters **19**, 598 (1965b).
— — Phys. Rev. Letters **14**, 502 (1965c).
Pignotti, A.: Phys. Rev. Letters **10**, 416 (1963).
Polkinghorne, J. C.: J. Math. Phys. **4**, 503 (1963a).
— J. Math. Phys. **4**, 1393 (1963b).
— J. Math. Phys. **4**, 1396 (1963c).
— J. Math. Phys. **5**, 431 (1964).
— Nuovo Cimento **36**, 857 (1965).
Pomeranchuk, I. Ya.: Soviet Phys. JETP **34**(7), 499 (1958).
Prosperi, G. M.: Nuovo Cimento **26**, 541 (1962).
Pursey, D. L., and *L. Sertorio:* Phys. Rev. **155**, 1591 (1967).
Rarita, W., and *B. M. Schwarzschild:* Phys. Rev. **162**, 1378 (1967).

Rarita, W., R. J. Riddell Jr., C. B. Chiu, and *R. J. N. Phillips:* Lawrence Radiation Laboratory report, UCRL-17523. (1967). (Phys. Rev., to be published)
Regge, T.: Nuovo Cimento **14**, 951 (1959).
— Nuovo Cimento **18**, 947 (1960).
Rose, M. E.: Elementary theory of angular momentum. New York: Wiley 1957.
Rosenfeld, A. H., et al.: Rev. Mod. Phys. **39**, 1 (1967).
Rodgers, T. W.: Cambridge thesis and private communication (1967).
Roy, D. P.: Nuovo Cimento **40**A, 513 (1965).
— Phys. Rev. **146**, B 1218 (1966).
Rubinstein, H. R., G. Veneziano, and *M. A. Virasoro:* Fixed poles and compositeness. Weizmann Institute preprint 1967. (Phys. Rev., to be published).
Ryder, L. H.: Nuovo Cimento **52**, 879 (1967).
Sakita, B., and *W. C. Wali:* Phys. Rev. Letters **18**, 31 (1967).
Sakmar, J. A.: Phys. Rev. **135**, B 249 (1964).
Salam, A., and *J. Strathdee:* Phys. Rev. Letters **19**, 39 (1967).
Sciarrino, H., and *M. Toller:* J. Math. Phys. **8**, 1252 (1967).
Sertorio, L., and *M. Toller:* Nuovo Cimento **33**, 413 (1964).
Sharp, D. H., and *W. G. Wagner:* Phys. Rev. **131**, 2226 (1963).
Shepherd, H. K.: Phys. Rev. **159**, 1331 (1967).
Singh, V.: Phys. Rev. **129**, 1889 (1963).
— Phys. Rev. Letters **18**, 36 (1967).
Smithies, F.: Integral equations.: Cambridge University Press 1962.
Sommerfeld, A.: Partial differential equations in physics, 1st paperback edition (1964), p. 282. New York: Academic Press 1949.
Sonderegger, P.: Phys. Letters **20**, 75 (1966).
Squires, E. J.: Nuovo Cimento **25**, 242 (1962).
— Complex Angular momentum and particle physics. New York: Benjamin 1963 a.
— Phys. Letters **7**, 363 (1963 b).
— Nuovo Cimento **34**, 1751 (1964).
— Proceedings of the 1966 Scottish Universities Summer School in Physics, *Priest, T. W.,* and *L. L. J. Vick* (Ed.). London: Oliver and Boyd 1967.
—, and *P. J. S. Watson:* Ann Phys. **41**, 409 (1967).
Srivastava, Y.: Phys. Rev. Letters **19**, 47 (1967).
Srivastava, Y. N., and *P. Nath:* Phys. Rev. **142**, 982 (1966).
Stack, J. D.: Phys. Rev. Letters **16**, 286 (1966).
— Phys. Rev. **164**, 1904 (1967).
Stapp, H. P.: Phys. Rev. **125**, 2139 (1962).
— Lawrence Radiation Laboratory report UCRL-10843 (unpublished). (1963).
— Phys. Rev. **160**, 1251 (1967).
Swift, A. R.: Phys. Rev. Letters **18**, 813 (1967).
Sundaram, A., and *R. Sridhar:* Nuovo Cimento **50**A, 969 (1967).
Taylor, J. C.: Regge poles in invariant amplitudes and families of trajectories. Oxford preprint 1967.
Taylor, J. R.: Phys. Rev. **127**, 2257 (1962).
— Ph. D. Thesis, University of California (unpublished). (1963).
— J. Math. Phys. **7**, 181 (1966).
Thews, R. L.: Phys. Rev. **155**, 1624 (1967).
Tiktopoulos, G.: Phys. Rev. **131**, 480 (1963 a).
— Phys. Rev. **131**, 2373 (1963 b).
Titchmarsh, E. C.: Theory of Fourier integrals.: Oxford University Press 1937.
— The theory of functions, 2nd Edition.: Oxford University Press 1939.
Teplitz, V. L.: Phys. Rev. **137**, B 136 (1965).
Teplitz, D. C., and *V. L. Teplitz:* Phys. Rev. **137**, B 142 (1965).
Toller, M.: Nuovo Cimento **37**, 631 (1965).
— Nuovo Cimento **53**A, 671 (1968).
Trueman, T. L., and *G. C. Wick:* Ann. Phys. **26**, 322 (1964).

Udgaonkar, B. M.: Phys. Rev. Letters **8**, 142 (1962).
— In: High energy physics and elementary particles (IAEA, Vienna) (1965).
Uretsky, J. L.: Phys. Rev. **123**, 1459 (1961).
Volkov, D. V., and *V. N. Gribov:* Soviet Phys. (JETP) **17**, 720 (1963).
Wagner, W. G.: Phys. Rev. Letters **10**, 202 (1963).
Wang, L.-L. C.: Phys. Rev. **142**, 1187 (1966a).
— Phys. Rev. Letters **16**, 756 (1966b).
— Phys. Rev. **153**, 1664 (1967).
Warburton, A. E. A.: Nuovo Cimento **32**, 122 (1964a).
— Phys. Rev. **137**, B 993 (1964b).
Warburton, A. E.: Nuovo Cimento **37**, 266 (1965).
Warnock, R. L.: Nuovo Cimento **50A**, 894 (1967).
Watson, G. N.: Proc. Roy. Soc. **95**, 83 (1918).
Wick, G. C.: Phys. Rev. **96**, 1124 (1954).
Wigner, E. P.: Ann. Math. **40**, 159 (1939).
— In: *Gursey, F.* (Ed.): Group theoretical concepts and methods in elementary particle physics. New York: Gordon and Breach 1964.
Wu, T.-Y., and *T. Ohmura:* The quantum theory of scattering, p. 6. New Jersey: Prentice Hall 1962.
Zachariasen, F.: Phys. Rev. Letters **7**, 112; 268(E) (1961).
— Lectures given at the Pacific International Summer School in Physics, Honolulu, Hawaii (1965).
—, and *C. Zemach:* Phys. Rev. **128**, 849 (1962).
Zweig, G.: unpublished CERN report (1964).

References for Postscript

Ademollo, M., H. R. Rubinstein, G. Veneziano, and *M. A. Virasoro:* Phys. Rev. Letters **19**, 1402 (1967).
Allison, W. W. M., et al.: Phys. Letters **25 B**, 619 (1967).
Barger, V., and *D. Cline:* Phys. Rev. Letters **19**, 1504 (1967).
—, and *L. Durand:* Phys. Rev. Letters **19**, 1295 (1967).
Borgese, A., M. Colocci, M. Lusignoli, M. Restignoli, and *G. Violini:* University of Rome preprint n. 130 (1967).
Brozan, J. B., and *C. E. Jones:* Phys. Rev. **160**, 1494 (1967).
Buschhorn, G., et al.: Phys. Letters **25 B**, 622 (1967).
Chiu, C. B., S-Y. Chu, and *L. L. Wang:* Phys. Rev. **161**, 1563 (1967).
—, and *Chung-I. Tan:* Phys. Rev. **162**, 1701 (1967).
Collins, P. D. B., R. C. Johnson, and *E. J. Squires:* Phys. Letters **26B**, 223 (1968).
Contigorous, A. P., and *J. Tran Thanh Van:* Phys. Rev. Letters **19**, 1353 (1967).
Cutkosky, R. E., and *B. B. Deo:* Phys. Rev. Letters **19**, 1345 (1967).
Drechsler, W.: CERN preprint Th. 841.
Dunne, S. A.: Phys. Rev. Letters **19**, 1299 (1967).
Durand, L.: Phys. Rev. Letters **19**, 1563 (1967).
Feldman, G., and *P. T. Matthews:* Imperial College, London preprint ICTP/67/2.
Filthuth, H. (Ed.): Proceedings of the Heidelberg International Conference on Elementary Particles, North-Holland, Amsterdam (1968).
Finkelstein, J., and *Chung-I. Tan:* Phys. Rev. Letters **19**, 1061 (1967).
Goldberg, H.: Phys. Rev. Letters **19**, 1391 (1967).
Gross, J. D.: Phys. Rev. Letters **19**, 1303 (1967).
Hwa, R.: Phys. Rev. **162**, 1708 (1967).
Iverson, G. I.: Nuovo Cimento **51**, 289 (1967).
Jackson, J. D., and *G. E. Hite:* Berkeley preprint UCRL 17959.
Komy, S. R., and *M. Samiullah:* Trieste internal report 26/1967.

Mandelstam, S., Berkeley Preprint. The relation between PCAC, axial-charge commutation relations and conspiracy theory (1967).

Matsuda, S., and *K. Igi:* Phys. Rev. Letters 19, 928 (1967).

Mitter, P. K.: Phys. Rev. 162, 1624 (1967).

Nwachuku, C. O., M. Samiullah, and *K. T. Shah:* Trieste internal report 24/1967; 15/1967.

Olsson, M. G.: Phys. Rev. Letters 19, 550 (1957).

Rashid, M. A., and *M. Samiullah:* Trieste preprint IC/67/65 (1967).

Rothe, H.: Phys. Rev. 159, 1471 (1967).

Sawyer, W.: Phys. Rev. Letters 19, 137 (1967).

Schwarz, J.: Phys. Rev. 162, 1671 (1967).

Squires, E. J: Phys. Letters 26 B, 461 (1968).

Subject Index

Where numbers appear in bold face type a general discussion on the page referred to and the following pages is indicated. Scattering processes are indexed separately on page 292.

Index of Scattering Processes

Dr. *P. D. B. Collins* and Prof. Dr. *E. J. Squires*
Department of Mathematics, Science Laboratories
University of Durham, South Road
Durham/England